Topics in Applied Physics Volume 19

Topics in Applied Physics Founded by Helmut K. V. Lotsch

1 **Dye Lasers** 2nd Ed. Editor: F. P. Schäfer
2 **Laser Spectroscopy** of Atoms and Molecules. Editor: H. Walther
3 **Numerical and Asymptotic Techniques in Electromagnetics** Editor: R. Mittra
4 **Interactions on Metal Surfaces** Editor: R. Gomer
5 **Mössbauer Spectroscopy** Editor: U. Gonser
6 **Picture Processing and Digital Filtering** 2nd Edition. Editor: T. S. Huang
7 **Integrated Optics** 2nd Ed. Editor: T. Tamir
8 **Light Scattering in Solids** Editor: M. Cardona
9 **Laser Speckle** and Related Phenomena Editor: J. C. Dainty
10 **Transient Electromagnetic Fields** Editor: L. B. Felsen
11 **Digital Picture Analysis** Editor: A. Rosenfeld
12 **Turbulence** 2nd Ed. Editor: P. Bradshaw
13 **High-Resolution Laser Spectroscopy** Editor: K. Shimoda
14 **Laser Monitoring of the Atmosphere** Editor: E. D. Hinkley
15 **Radiationless Processes** in Molecules and Condensed Phases. Editor: F. K. Fong
16 **Nonlinear Infrared Generation** Editor: Y.-R. Shen
17 **Electroluminescence** Editor: J. I. Pankove
18 **Ultrashort Light Pulses** Picosecond Techniques and Applications Editor: S. L. Shapiro
19 **Optical and Infrared Detectors** 2nd Ed. Editor: R. J. Keyes
20 **Holographic Recording Materials** Editor: H. M. Smith
21 **Solid Electrolytes** Editor: S. Geller
22 **X-Ray Optics.** Applications to Solids Editor: H.-J. Queisser
23 **Optical Data Processing.** Applications Editor: D. Casasent
24 **Acoustic Surface Waves** Editor: A. A. Oliner
25 **Laser Beam Propagation in the Atmosphere** Editor: J. W. Strohbehn
26 **Photoemission in Solids I.** General Principles Editors: M. Cardona and L. Ley
27 **Photoemission in Solids II.** Case Studies Editors: L. Ley and M. Cardona
28 **Hydrogen in Metals I.** Basic Properties Editors: G. Alefeld and J. Völkl
29 **Hydrogen in Metals II** Application-Oriented Properties Editors: G. Alefeld and J. Völkl
30 **Excimer Lasers** Editor: Ch. K. Rhodes
31 **Solar Energy Conversion.** Solid-State Physics Aspects. Editor: B. O. Seraphin
32 **Image Reconstruction from Projections** Implementation and Applications Editor: G. T. Herman
33 **Electrets** Editor: G. M. Sessler
34 **Nonlinear Methods of Spectral Analysis** Editor: S. Haykin
35 **Uranium Enrichment** Editor: S. Villani
36 **Amorphous Semiconductors** Editor: M. H. Brodsky
37 **Thermally Stimulated Relaxation in Solids** Editor: P. Bräunlich
38 **Charge-Coupled Devices** Editor: D. F. Barbe
39 **Semiconductor Devices** for Optical Communication Editor: H. Kressel
40 **Display Devices** Editor: J. I. Pankove
41 **The Computer in Optical Research** Methods and Applications Editor: B. R. Frieden
42 **Two-Dimensional Digital Signal Processing I.** Linear Filters Editor: T. S. Huang
43 **Two-Dimensional Digital Signal Processing II.** Transforms and Median Filters. Editor: T. S. Huang
44 **Turbulent Reaction Flow** Editors: P. A. Libby and F. Williams
45 **Hydrodynamic Instabilities and the Transition to Turbulence** Editors: H. L. Swinney and J. P. Gollub
46 **Glassy Metals I** Editors: H.-J. Güntherodt and H. Beck
47 **Sputtering by Particle Bombardment I** Editor: R. Behrisch
48 **Optical Information Processing** Fundamentals Editor: S. H. Lee

Optical and Infrared Detectors

Edited by R. J. Keyes

With Contributions by
R. J. Keyes   P. W. Kruse   D. Long   A. F. Milton
E. H. Putley   M. C. Teich   H. R. Zwicker

Second Corrected and Updated Edition

With 115 Figures

Springer-Verlag Berlin Heidelberg New York 1980

Robert J. Keyes

Lincoln Laboratory, Massachusetts Institute of Technology,
Lexington, MA 02173, USA

ISBN 3-540-10176-4 2. Auflage   Springer-Verlag Berlin Heidelberg New York
ISBN 0-387-10176-4 2nd Edition Springer-Verlag New York Heidelberg Berlin

ISBN 3-540-08209-3 1. Auflage Springer-Verlag Berlin Heidelberg New York
ISBN 0-387-08209-3 1st Edition Springer-Verlag Berlin Heidelberg New York

Library of Congress Cataloging in Publication Data. Keyes, Robert J. 1927– Optical and infrared detectors. (Topics in applied physics; v. 19) Includes bibliographies and index. 1. Infra-red detectors. I. Title. TA1570.K48 1980   681'.41   80-17985
ISBN 0-387-10176-4 (U.S.)

This work is subject to copyright. All rights are reserved, whether the whole or part of the material is concerned, specifically those of translation, reprinting, re-use of illustrations, broadcasting, reproduction by photocopying machine or similar means, and storage in data banks. Under § 54 of the German Copyright Law, where copies are made for other than private use, a fee is payable to the publisher, the amount of the fee to be determined by agreement with the publisher.

© by Springer-Verlag Berlin Heidelberg 1977 and 1980
Printed in Germany

The use of registered names, trademarks, etc. in this publication does not imply, even in the absence of a specific statement, that such names are exempt from the relevant protective laws and regulations and therefore free for general use

Monophoto typesetting, offset printing and bookbinding: Brühlsche Universitätsdruckerei, Giessen
2153/3130-543210

Preface to the Second Edition

This edition, as was the first, is written for those who desire a comprehensive analysis of the latest developments in infrared detector technology and a basic insight into the fundamental processes which are important to evolving detection techniques. Each of the most salient infrared detector types is treated in detail by authors who are recognized as leading authorities in the specific areas addressed. In order to concentrate on pertinent aspects of the present state of the detector art and the unique point of view of each author, extensive tutorials of a background nature are avoided in the text but are readily available to the reader through the many references given. The editor highly recommends R. H. Kingston: *Detection of Optical and Infrared Radiation* (Springer, Berlin, Heidelberg, New York 1978) for those who desire a more tutorial presentation of the photon radiation and detection processes.

The volume opens with a broad-brush introduction to the various types of infrared detectors that have evolved since Sir William Herschel's discovery of infrared radiation 175 years ago. The second chapter presents an overall perspective of the infrared detector art and serves as the cohesive cement for the more in-depth presentation of subsequent chapters. Those detector types which, for one reason or other have not attained wide use today, are also discussed in Chapter 2.

The more notable and widely used infrared detectors can be divided into three basic classes which are indicative of the primary effect produced by the photon-detector interaction, i.e., thermal, photoconductive, photovoltaic, and photoemissive. Chapters 3, 4, and 5 offer a detailed treatment of each of these important processes.

The objectives of an infrared detector are to find a signal, measure it, and extract some vital information for subsequent use. Charge-coupled devices (CCD) and nonlinear photon interactions are new techniques for extracting more information from weak infrared signals. These devices, when directly or indirectly coupled to basic detector types, place a tremendous amount of low noise, compact, inexpensive logic at the disposal of infrared sensor designers. The marriage of CCD concepts to detector technology is young, moving fast, and portends great strides in the ability to retrieve small signal information from complex radiation environments. Chapter 5 presents some of the latest developments and concepts in this unique area.

Methods of measuring the phase and frequency of coherent infrared signals have become important in order to extract the maximum information from laser signals. The efficient nonlinear mixing of coherent infrared radiation in standard detector materials has achieved sensitivities approaching the theoretical limit (a few photons per measurement interval) and has stimulated new applications for lasers in the communication, radar, and spectroscopy fields. A volume concerned with infrared detectors would be incomplete without addressing some of the salient aspects of heterodyne or nonlinear detection—hence Chapter 7. Chapter 8 presents those recent developments in the field of optical and infrared detectors which have evolved since the publication of the first edition. It specifically embraces those advances which the authors have deemed necessary in order to provide the reader an up-to-date perspective of detector technology.

I would like to thank the contributors to this volume, P. Kruse, E. Putley, D. Long, H. Zwicker, F. Milton, and M. Teich, for giving of their time, energy, and expertise in preparation of their chapters; I also thank my wife, Gladys, for her assistance.

Lexington, Mass. *Robert J. Keyes*
May, 1980

Contents

1. **Introduction.** By R. J. Keyes 1

2. **The Photon Detection Process.** By P. W. Kruse (With 22 Figures) . . . 5
   2.1 Classification and Phenomenological Descriptions of Selected Photon Detection Mechanisms 7
      2.1.1 Photon Effects 8
      2.1.2 Thermal Effects 26
      2.1.3 Wave Interaction Effects 33
   2.2 Noise in Radiation Detectors 36
      2.2.1 Noise in Semiconductor Detectors 37
      2.2.2 Noise in Photoemissive Devices 41
   2.3 Figures of Merit 42
      2.3.1 Spectral Response 42
      2.3.2 Responsivity 43
      2.3.3 D^* 44
      2.3.4 D^{**} 45
      2.3.5 Noise Equivalent Power 45
      2.3.6 Detectivity 46
      2.3.7 Frequency Response, Response Time, Time Constant, f^* . 46
      2.3.8 Noise Spectrum 47
   2.4 The Signal Fluctuation and Background Fluctuation Limits . . . 47
      2.4.1 Signal Fluctuation Limit 48
      2.4.2 Background Fluctuation Limit 50
      2.4.3 Composite Signal Fluctuation and Background Fluctuation Limits 56
   2.5 State-of-the-Art of Infrared and Optical Detectors 59
   References 65

3. **Thermal Detectors.** By E. H. Putley (With 14 Figures) 71
   3.1 Basic Principles 72
   3.2 The Thermopile 79
   3.3 The Bolometer 82
   3.4 The Golay Cell and Related Detectors 89
   3.5 Pyroelectric Detector 90

3.6 Other Types of Thermal Detectors 95
3.7 The Use of Thermal Detectors in Infrared Imaging Systems . . . 96
References 98

4. Photovoltaic and Photoconductive Infrared Detectors
By D. Long (With 15 Figures) 101
4.1 Basic Theory 102
    4.1.1 Direct Photon Detection 102
    4.1.2 Photocurrent, Gain, and Responsivity 103
    4.1.3 Noise Mechanisms 104
    4.1.4 Detectivity 106
    4.1.5 Other Detector Parameters 107
4.2 Photovoltaic Detectors 109
    4.2.1 Theory 109
    4.2.2 Materials 114
4.3 Intrinsic Photoconductive Detectors 120
    4.3.1 Theory 120
    4.3.2 Materials 124
4.4 Extrinsic Photoconductive Detectors 129
    4.4.1 Theory 129
    4.4.2 Materials 132
4.5 Summary and Conclusions 133
References 145

5. Photoemissive Detectors. By H. R. Zwicker (With 22 Figures) 149
5.1 Introduction 149
    5.1.1 Applications and Advantages 149
    5.1.2 Limitations 150
    5.1.3 Types of Photoemissive Surfaces: Classical and NEA . . . 150
    5.1.4 RM and TM Modes 151
    5.1.5 Outline of Chapter 152
5.2 The Photoemission Process 152
    5.2.1 Fundamentals of Electron Escape Energy 152
    5.2.2 Escape-Energy Parameters for Metals and Semiconductors . 154
    5.2.3 Thresholds of Various Materials and Choice of Photoemissive Substances 155
5.3 Classical Photoemissive Surfaces 159
    5.3.1 The (CsSb) Photoemitter 159
    5.3.2 The S-1 (AgCsO) Photocathode 161
5.4 Negative Electron Affinity (NEA) Devices 163
    5.4.1 Introduction and Advantages 164
    5.4.2 Basic Physics of NEA Operation 165
    5.4.3 NEA GaAs 167

       5.4.4 Modeling the NEA Surface 169
       5.4.5 Fabrication and Optimization of RM and TM NEA GaAs . 172
       5.4.6 Other NEA IR Photocathodes 177
       5.4.7 NEA Silicon 181
   5.5 Photoemissive Devices: The Photomultiplier 182
       5.5.1 Introduction 182
       5.5.2 Photocathode Dark Current 182
       5.5.3 The Electron Multiplier 184
       5.5.4 Spectral Response Data 188
       5.5.5 Specialized NEA Analysis Tools 191
   References 192

6. **Charge Transfer Devices for Infrared Imaging.** By A.F.Milton
   (With 18 Figures) 197
   6.1 Historical 197
   6.2 Charge Coupled Devices 200
       6.2.1 Basic Operating Principles 200
       6.2.2 Limitations 203
   6.3 Time Delay and Integration (TDI) and IR Sensitive CCD . . . 213
   6.4 Direct Injection: Hybrid and Extrinsic Silicon 216
   6.5 Accumulation Mode: Extrinsic Silicon 219
   6.6 IR CID 220
   6.7 Conclusions 227
   References 228

7. **Nonlinear Heterodyne Detection.** By M.C.Teich (With 24 Figures) . . 229
   7.1 Two-Frequency Single-Photon Heterodyne Detection 231
   7.2 Two-Frequency Multiphoton Heterodyne Detection 232
       7.2.1 Multiple-Quantum Direct Detection 232
       7.2.2 Theory of Multiphoton Photomixing 234
       7.2.3 Signal-to-Noise Ratio and Minimum Detectable Number
             of Photons 237
       7.2.4 Experiment 239
       7.2.5 Discussion 242
   7.3 Three-Frequency Single-Photon Heterodyne Detection Using a
       Nonlinear Device 243
       7.3.1 System Configuration 244
       7.3.2 Application to cw Radar with Sinewave Input Signals . . 249
       7.3.3 Application to cw Radar with Gaussian Input Signals
             (Gaussian Spectra) 257
       7.3.4 Application to cw Radar with Gaussian Input Signals
             (Lorentzian Spectra) 263
       7.3.5 Application to an Analog Communications System . . . 266

	7.3.6	Operation at Low Frequencies and in Various Configurations 267
	7.3.7	Numerical Example: A CO_2 Laser Radar 269
	7.3.8	Application to Binary Communications and Pulsed Radar (Vacuum Channel) 270
	7.3.9	Application to Binary Communications and Pulsed Radar (Lognormal Atmospheric Channel) 282
	7.3.10	Discussion 287
7.4	Multifrequency Single-Photon Selective Heterodyne Radiometry for Detection of Remote Species 288	
	7.4.1	Configuration for Two Received Frequencies 289
	7.4.2	n Received Frequencies and the Factor k' 291
	7.4.3	SNR and MDP for Two Gaussian Signals 293
	7.4.4	Numerical Example: Astronomical Radiation from CN . 294
	7.4.5	Discussion 295
References 297		

8. Recent Advances in Optical and Infrared Detector Technology.
By R. J. Keyes 301

8.1 Thermal Detectors 301
    8.1.1 The Thermopile 302
    8.1.2 The Bolometer 302
    8.1.3 The Golay Cell and Related Detectors 303
    8.1.4 Pyroelectric Detector 304
    8.1.5 Other Types of Thermal Detectors 305
    8.1.6 The Use of Thermal Detectors in Infrared Imaging Systems 306

8.2 Photovoltaic, Photoconductive, and Avalanche Diode Detectors 306
    8.2.1 Intrinsic Photovoltaic Detectors 307
    8.2.2 Intrinsic Photoconductive Detectors 307
    8.2.3 Extrinsic Photoconductive Detectors 308
    8.2.4 Avalanche Photodiodes (APD) 308

8.3 Photoemissive Detectors 309
8.4 Charge Transfer Imaging Devices 310
    8.4.1 Near Infrared and Visible CCD Imagers 310
    8.4.2 Thermal Scene CCD Imagers 311
8.5 Heterodyne Detectors 313
8.6 Desirable Optical and Infrared Detector Technology Advances . 314
References 315

Subject Index 321

Contributors

Keyes, Robert J.
   Lincoln Laboratory, Massachusetts Institute of Technology,
   Lexington, MA 02173, USA

Kruse, Paul W.
   Honeywell Corporate Technology Center, Bloomington, MN 55420, USA

Long, Donald
   Honeywell Corporate Technology Center, Bloomington, MN 55420, USA

Milton, Albert F.
   Naval Research Laboratory, Washington, D.C., USA

Putley, Ernest H.
   Royal Signals and Radar Establishment, Malvern, Worcs.,
   WR 14 3PS, U.K.

Teich, Malvin C.
   Department of Electrical Engineering, Columbia University,
   New York, NY 10027, USA

Zwicker, Harry R.
   Lincoln Laboratory, Massachusetts Institute of Technology,
   Lexington, MA 02173, USA

1. Introduction

R. J. Keyes

Optical or visible photons emitted by our sun provide the most important source of radiation on our planet. It has been sensed, felt, and used (although not fully understood) by living things since the onset of evolution. The process of natural evolution of the eye retina has produced an array of optical detectors capable of sensing a few photons. In the past century technology has approached this sensitivity through the development and perfection of photographic emulsions, video intensifier tubes, and charge coupled devices.

Infrared is the second most intense radiation source in our environment; yet it was not until the turn of the nineteenth century (1800) that Sir *William Herschel* (1738–1822) discovered it and gave it a name. Infrared detector development proceeded at a leisurely pace for the next 150 years. In 1947 *William Shockley, John Bardeen,* and *Walter Brattain,* through the discovery of the transistor, sparked large-scale research and development into the band structure and transport processes of semiconductors. This spark, fanned by huge injections of government and industrial funds, affected nearly every aspect of infrared detection technology for the next 30 years (impurity and intrinsic photoconductivity in binary, ternary, and group IV compounds; photoemission of cesiated semiconductors; thermistors; silicon target vidicons; charge-coupled devices; and heterodyne detection, to name a few). The present emphasis of infrared detector research and development is shifting from the perfection of new classes of materials to techniques for information retrieval through charge-coupled device logic and heterodyne signal processing. The shift in emphasis is due in part to the fact that detector materials are asymptotically approaching the theoretical limit of sensitivity (leaving little room for spectacular improvements) while sophisticated signal processing promises unlimited opportunities to extract vital information about objects in complex scenes through their emitted and reflected infrared radiations. Although these methods of signal processing are relatively new, they utilize as the first step in the processing chain many of the basic materials and concepts presented here in Chapters 4 and 5.

There is no unique set of characteristics which can be used to describe the "ideal" infrared detector. The ideal detector has nearly as many definitions as there are applications. For "hot box" detection on railroad-caraxles, reliability and simple maintenance are measures of perfection. For satellite operations compactness, light weight, low power and high sensitivity may be the important yardsticks. In the discussion which follows, an ideal detector is defined as one whose minimum detectible power is determined by the statistical fluctuation in the photon stream impinging on its surface. All infrared sensors (except those operated in a closed cryogenic laboratory environment) are exposed to relatively intense infrared background radiation levels and hence the ability to detect a small signal is always ultimately determined by the photon noise associated with these background radiations. In the spectral region of interest, the background radiators have a Poisson distribution so that the *rms* fluctuation in the number of photons emitted (photon noise) is equal to the square root of the total number of photons emitted, i.e.,

$$\Delta N_B = (N_B)^{1/2}.$$

In terms of power the noise can be expressed as

$$\Delta P_B = (P_B h\nu/t)^{1/2}$$

where $h\nu$ is the average photon energy and t is the period of time over which the power is integrated. Even for an infrared sensor designed to minimize the effect of background radiation, the noise induced by terrestrial and celestial background radiations in the 8–12 μm spectral are significant. As a case in point consider a sensor in which the smallest efficient detector (equal in size to the Airy disc) is placed at the focal point of a perfect $f/1$ telescope. When viewing a terrestrial scene, the detector receives 2×10^{-8} watts of background radiation and approximately 6×10^{-11} watts when exposed to the diffused emission of the celestial sphere. The photon noise produced by these two background extremes is equivalent to 2×10^{-14} watts and 3.5×10^{-18} watts, respectively.

There are numerous other sources of noise associated with the detection process, which can exceed photon noise. These are treated in detail throughout the text. Invariably the perfection of detector materials is such that internally generated noise can be reduced to negligible levels by operating at cryogenic temperatures. For the infrared spectral region beyond the sensitivity of photoemitters (>2 μm) the thermal noise of preamplifiers usually establishes the minimum detectable power that can be realized under low background radiation conditions. It can be shown that a detector must have a noiseless gain

$$G \gtreqqless \frac{h\nu(kTC)^{1/2}}{eP_s t}$$

if the amplifier noise current is to be equal or less than the current produced by a signal of power P_s. For the ultimate detection limit where the signal is *one photon per measurement interval* (quantum efficiency assumed to be unity) the required gain is

$$G \gtreqqless \frac{2(kTC)^{1/2}}{e}$$

where k is the Stefan-Boltzmann constant, e is the charge of an electron, and C is effective capacitance of the preamplifier input. It is interesting to note that an amplifier operating at 300 K with an input capacitance of 10^{-12} farads and an integration time of 1 s, a detector gain of only 2.4 is sufficient to reach the photon noise limit associated with stellar backgrounds. A gain of 800 is needed in order to achieve the *ultimate one photon per measurement interval* limit.

Part of the problem of measuring weak infrared signals, in addition to those mentioned above, is knowing where to look. The charge-coupled device (CCD) formed on binary, ternary or group IV photoconductors has many of the characteristics favorable to small signal detection. The multi-element nature of the CCD facilitates the acquisition problem by providing the large field of view characteristic of an array while retaining the small field of view of each detector element which is necessary for low background photon noise. All the photon-induced charge integrated for a frame time on each element can be sequentially read out by a single, low-noise preamplifier formed directly on the substrate. At present a CCD formed on silicon can provide arrays containing 3 000 000 detector elements with spectral sensitivity out to 1 μm and a noise equivalent to approximately 50 photons per integration interval. There is no apriori reason not to believe that similar performance will be forthcoming at longer wavelengths from CCD detectors fabricated on the binary and ternary photoconductors.

The capability of CCD to store, add, subtract and rearrange signals is perhaps more significant to the future growth of the infrared field than its use as a prime sensor. These assets place at the disposal of the individual researcher and designer a logic capability that previously was only available at large organizations with costly and cumbersome computers. Another quite different method of signal processing of coherent radiation signals also promises to expand the horizons of infrared radiation.

The extremely narrow spectral linewidths of stable lasers coupled to heterodyne detection techniques allow one to extract information that cannot be obtained from the thermal emission of an object. By mixing the reflected laser signal with stable local oscillator radiation in a detector it is possible to measure an object's translational velocity and angular motions. In fact, through an analysis of the heterodyne frequency spectrum, it is possible to determine many of the geometric features of an object even when the object is smaller than the resolution limit of the optical system.

Surprisingly, heterodyne detector systems which approach the theoretical detector limit are less sensitive to imperfections of the detector material than those used for nonheterodyne operations. An ordinary infrared detector which may be dominated by thermal generation or amplifier noise can serve as a nearly theoretically perfect receiver when used in a heterodyne mode. This is a consequence of the gain inherent in the nonlinear mixing of the coherent laser signal and local oscillator radiations in the detector. As long as the photon noise due to local oscillator radiation exceeds all other sources of noise, the detector will be capable of sensing a few photons per measurement interval. It is possible to convert a poor passive detector into a near-ideal heterodyne sensor by increasing the local oscillator power. In principle this can be attained even at room temperature as long as sufficient local oscillator power is available and the heat it generates can be removed from the detector fast enough to prevent thermal runaway.

An attempt to write a timely and comprehensive book on the present state of infrared detector technology is complicated by the rapid advances being made in the field. In order to present the latest developments in a wide range of detector disciplines, this volume is divided into six chapters each of which is authored by a recognized authority in the particular subject addressed. Each author in this volume provides the history, technical analysis, and the tutorial background germane to the subject matter of his contribution. As a general rule long tutorial explanations, which have appeared in the open literature, are referenced rather than presented in the text. However some situations arise where the author feels compelled to use detailed mathematical language to reinforce a point; the treatment given nonlinear detection in Chapter 7 is a case in point.

Although not specifically delineated, the volume is also divided into three general sections. The first addresses the full spectrum of infrared detectors and contains a limited coverage of all the material presented in subsequent chapters. It serves as an introduction to the volume and presents to the reader an overall view of the present state of the infrared technology art. It also serves as the mortar between the more in-depth discussions which follow. The midsection, Chapters 3, 4, and 5, is a detailed analysis of those detector types which are most widely used today: thermal, photoconductive/photovoltaic and photoemissive.

The final section, Chapters 6 and 7, addresses new techniques for the extraction of information from detector elements. These new techniques primarily embrace charge-coupled devices and nonlinear mixing of coherent radiations.

Each scientist and engineer perceives his field of expertise from a different vantage point. The editor suspects that his background in solid-state physics has tinted his perception of the salient infrared detector developments of the past 25 years as well as his judgement of the future course of things to come.

2. The Photon Detection Process

P. W. Kruse

With 22 Figures

As with other types of sensory devices, optical[1] and infrared detectors are capable of being classified as to their operating mode. Their performance can be described in terms of unique figures of merit which enable the system designer to predict, then evaluate, the performance of the system in which they are employed. And, like all sensors, their performance cannot exceed that of an ideal device of their class; fundamental limits exist which constrain their performance.

With the exception of Section 2.4, this chapter is largely descriptive, rather than mathematical. It is the purpose of the author to set before the reader those broad aspects of optical and infrared detector technology which are common to the subject as a whole. Section 2.1 introduces the basic methods of classifying infrared detectors and presents a phenomenological discussion of the important detection mechanisms. Although a very large number of photoeffects are described there, the emphasis is on the few which have been widely exploited. A simplified analysis of these effects is presented.

To make a practical photodetector, it is not sufficient to study and evaluate the interaction of radiation with materials giving rise to a photoeffect. As with all types of sensors, internal noise limits the ability to detect a very small signal in the detector output. Thus accompanying the study of photoeffects in materials is one of noise in materials. Since the effects of greatest utility are those in which the signal is manifested as a change in the electrical properties of the material, Section 2.2 presents a description of electrical noise in photosensitive materials.

In order to describe quantitatively the performance of photodetectors it is necessary to define figures of merit. The development of figures of merit for infrared detectors, which took place during the 1950s and 1960s, is largely complete. Those pertaining to optical detectors have been developed in parallel with those for infrared detectors. There has been some effort to incorporate the figures of merit for infrared detectors into those defined for optical detectors, but this has not been entirely accomplished. Section 2.3 presents figures of merit appropriate to infrared and optical detectors.

Infrared and optical detector technology is a deterministic area of knowledge in which it is possible to analyze the performance of detectors in terms of the properties of the materials employed in the detectors. It is also possible to determine analytically the ultimate limits to the detector performance set by the statistical nature of the radiation to which it responds. There are two such limits,

---

[1] Optical detectors include those operating in the ultraviolet and visible regions of the spectrum.

set by fluctuations in the signal radiation and in the background radiation falling on the detector. Many detectors in widespread use today operate at one or the other limit. Section 2.4 presents a discussion of these fundamental limits. Section 2.5 is a brief review of the performance of state-of-the-art detectors, including a comparison of measured performance with the fundamental limits.

There is a certain imbalance in the level of presentation of the several sections which is due to the nature of the following chapters. The discussion of the photon, thermal, and wave interaction effects is at an introductory level, the purpose of which is to present an overview of the many effects without a detailed analysis of them. Subsequent chapters will treat in depth the important ones. For a similar reason the discussion of noise mechanisms is also at an introductory level. On the other hand, the derivation of the fundamental limits has been reserved to this chapter, since it is common to the later chapters. Thus Section 2.4 is of a much more mathematical nature than the other sections.

There are a number of excellent references which will enable the reader to obtain an overview of infrared and optical detector research and development. Among the books published in the 1960s which contain much basic information still of value are those by *Smith* et al. [2.1], *Holter* et al. [2.2], *Kruse* et al. [2.3], *Jamieson* et al. [2.4], *Wolfe* [2.5], *Conn* and *Avery* [2.6], and *Hudson* [2.7]. The theoretical basis for photoconductivity and the other photon effects can be found in *Ryvkin* [2.8], *Bube* [2.9], *Pell* [2.10], and an issue of *Applied Optics* devoted to photon effects [2.11]. Imaging devices are well covered in the two volumes by *Biberman* and *Nudelman* [2.12, 13] and in six volumes in the series *Advances in Electronics and Electron Physics* [2.14]. Of more recent interest is *Moss* et al. [2.15], *Willardson* and *Beer* [2.16], *Materials for Radiation Detection* [2.17], *Dimmock* [2.18], *Seib* and *Aukerman* [2.19], *Emmons* et al. [2.20], *Anderson* and *McMurtry* [2.21], *Melchior* et al. [2.22], an issue of the *Proceedings of the IEEE* devoted to remote sensing [2.23], and a symposium on unconventional infrared detectors [2.24].

It is assumed that the reader is broadly familiar with the properties of semiconductors, from which most detectors of importance are prepared, and with semiconductor technology which allows the preparation of interfaces such as $p-n$ junctions and Schottky barriers. For those who are not familiar with these concepts, many good books are available, including those by *Sze* [2.25], *Grove* [2.26], *Kittel* [2.27], and *Moll* [2.28]. It is also assumed that the reader is aware of the nature of electromagnetic radiation, including the properties of monochromatic and black body sources. For further information on this topic see the appropriate chapters in [2.1–7].

2.1 Classification and Phenomenological Descriptions of Selected Photon Detection Mechanisms

Electromagnetic radiation can interact with materials in many ways. In general, these can be categorized as photon effects, thermal effects, and wave interaction effects. In the first category, photons interact directly with the electrons in a material. Since the electrons can be bound to lattice atoms or to impurity atoms or they can be free, a variety of interactions is possible. The second category, thermal effects, is characterized by changes in certain properties of a material due to a change in temperature arising from absorption of radiation. The third category of effects is based upon the interaction of the electromagnetic field with the material, resulting in changes in certain internal properties of the material.

Although all three types of interaction are observed, the choice of materials and the experimental arrangement usually cause one of the effects to predominate. The development of infrared and optical detectors has followed the dual paths of phenomena and materials, with early emphasis on observation of new phenomena and later emphasis on preparation of new materials. Most of the important photodetection phenomena have been established for decades; the emphasis during the past ten to twenty years has been on exploiting these well-established phenomena in new materials. In practice, the most important phenomenon is photoconductivity, a photon effect manifested by an increase in the free electron (and/or hole) concentration which causes an increase in the electrical conductivity. Beginning with the discovery early in the 1950s that the III–V compounds were semiconductors, there has been an extremely large effort worldwide to synthesize new materials from this and other classes [e.g., alloy semiconductors such as (Hg, Cd)Te and (Pb, Sn)Te] and to study photoconductivity in them. Although new effects have been discovered and demonstrated experimentally (e.g., photon drag, Josephson junctions, negative electron affinity photocathodes), by and large the emphasis has been on materials.

In addition to the manner in which they respond to radiation, detectors can also be classified as elemental or imaging, depending upon whether the output signal is an average over the entire sensitive area or whether it represents the spatial variation over the area. Imaging and elemental detectors generally make use of the same phenomena. For example, vidicons employ photoconductivity as the detection phenomenon; the spatial variation of it is read out by an electron beam. Charge coupled imaging devices employ either photoconductivity or the photovoltaic effect to produce free carriers which are then injected into the CCD structure for signal processing, such as by time delay integration or parallel-to-serial scan conversion. Other types of scanning systems, such as those real time thermal imaging systems known as FLIRs (Forward Looking InfraRed) employ moving mirrors to scan radiation from the scene across an array of elemental detectors, the temporal output of which is a two-dimensional representation of the thermal emission from the scene being viewed. Because imaging devices in general employ the same photoeffects exhibited by elemental detectors, they will not be discussed here. Among the many references to imaging devices are *Photoelectronic Imaging Devices* [2.12, 13] and selected volumes in the series *Advances in Electronics and Electron Physics* [2.14]. The interested reader is referred to them for a broad overview of the field.

2.1.1 Photon Effects

Those effects which have received the greatest emphasis in infrared and optical detector development are the photon ones, having the many forms listed in Table 2.1. Photon effects include all interactions of incident photons with electrons within the material, whether bound to lattice atoms or free. In almost all cases the material employed is a semiconductor. The effects can be subdivided into two classes, internal and external. The internal ones are those in which the photoexcited carrier (electron or hole) remains within the sample. Internal photon effects can be subdivided in three ways: a) those in which the incident photon interacts with a bound electron, either at a lattice atom or an impurity atom, producing a free electron-hole pair (intrinsic photoeffect) or a free electron and a bound hole (extrinsic photoeffect; the converse is also included); b) those in which the incident photons interact with carriers which are already free; c) those in which the incident photon produces a localized excitation of an electron into a higher energy state of an atom, without the electron ever leaving the atom. The external photoeffect, also known as the photoemissive effect, is one in which an incident photon causes the emission of an electron from the surface of the absorbing material, known as the photocathode. Detectors employing the external photoeffect make use of several methods for providing gain following photoemission.

Certain generalities can be made about these photoeffects. They are rules of thumb which are true in many instances, but are not universally true. They include the following: Photon detectors are characterized by high speed of response when compared to thermal detectors. They show a wavelength dependence of the photosignal per unit incident radiant power which increases with wavelength to a long wavelength limit, beyond which the photosignal falls to zero. Detectors having long wavelength limits in the ultraviolet, visible, or near infrared (say, up to 2–3 µm) operate uncooled at room temperature (295 K). Detectors with somewhat longer long wavelength limits (say, up to 4–5 µm) require cooling to dry ice temperature (195 K). Many types of infrared detectors require cooling to liquid nitrogen temperature (77 K); it is in this category that intrinsic detectors operating in the 8 to 14 µm atmospheric window are found. Some detectors having even longer long wavelength limits, say 30 µm, must operate at even lower temperatures; the lowest practical one is that of liquid helium (4.2 K). In general, extrinsic detectors require a lower operating temperature than intrinsic ones for reasons set forth by *Long* [2.29].

Although the list of photon effects in Table 2.1 is extensive, only the photoconductive, photovoltaic, and photoemissive ones have been widely

Table 2.1. Classification of photon effects

1. Internal	1.2 Free carrier interactions
1.1 Excitation of additional carriers	Photon drag
Photoconductivity	Hot electron bolometer
Electrically biased	Putley detector
Intrinsic	1.3 Localized interactions
Extrinsic	Infrared quantum counter
Microwave biased	Phosphor
Photovoltaic effect	Photographic film
$p-n$ junction	**2. External (Photoemissive)**
Avalanche	2.1 Photocathodes
$p-i-n$	Conventional
Schottky barrier	Negative electron affinity
Heterojunction	2.2 Gain mechanisms
Bulk	Gas avalanche
Photoelectromagnetic effect	Dynode multiplication (photomultipliers)
Dember effect	Channel electron multiplication
Phototransistor	

exploited. Thus, the subsections which follow will concentrate on these, presenting enough detail to provide the reader an insight into the phenomena. The other photoeffects listed in Table 2.1 will be described briefly, including references to the literature in which more detailed treatment can be found.

Photoconductivity

The most widely employed effect is photoconductivity, in which the radiation changes the electrical conductivity of the material upon which it is incident. In nearly all cases the change in conductivity is measured by means of electrodes attached to the sample. It is possible, however, to observe the effect by inserting the sample in a microwave cavity, such as *Sommers* [2.30] has done. Because the microwave method of detecting photoconductivity is very seldom used, it will not be discussed here.

The sample geometry and circuit employed for detecting the photoconductive effect is shown in Fig. 2.1. In almost all cases the transverse geometry is used, in which the direction of the incident radiation is perpendicular to the direction in which the change in current is measured. The photosignal is detected either as a change in voltage developed across a load resistor in series with the detector as illustrated, or as a change in current through the sample. Frequently the signal is detected as a change in voltage across a load resistor matched to the dark resistance of the detector. *Penchina* [2.31] points out that this may not optimize the signal-to-noise ratio under all conditions.

Intrinsic and Extrinsic. Photoconductivity can be observed in virtually all semiconductors. Intrinsic photoconductivity requires the excitation of a free

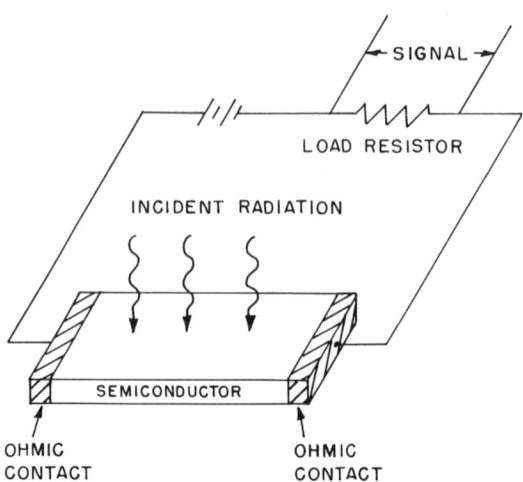

Fig. 2.1. Transverse photoconductivity geometry and circuit

hole-electron pair by a photon whose energy is at least as great as the energy gap E_g, see Fig. 2.2a. Thus, the basic requirement is

$$hv \geq E_g, \tag{2.1a}$$

or

$$\frac{hc}{\lambda} \geq E_g, \tag{2.1b}$$

where h is Planck's constant, v is the radiation frequency, c is the speed of light, and λ is the radiation wavelength. The long wavelength limit λ_0 of an intrinsic photoconductor is therefore

$$\lambda_0 = \frac{hc}{E_g}. \tag{2.2}$$

No intrinsic photoconductivity occurs for radiation of wavelength greater than λ_0. A convenient expression to determine the long wavelength limit in micrometers of an intrinsic photoeffect for a semiconductor whose energy gap is expressed in electron volts is

$$\lambda_0 (\mu m) = \frac{1.24}{E_g (eV)}. \tag{2.3}$$

Table 2.2 lists energy gaps of selected semiconductors which have been employed as photodetectors at the temperatures indicated. The corresponding long wavelength limits are shown.

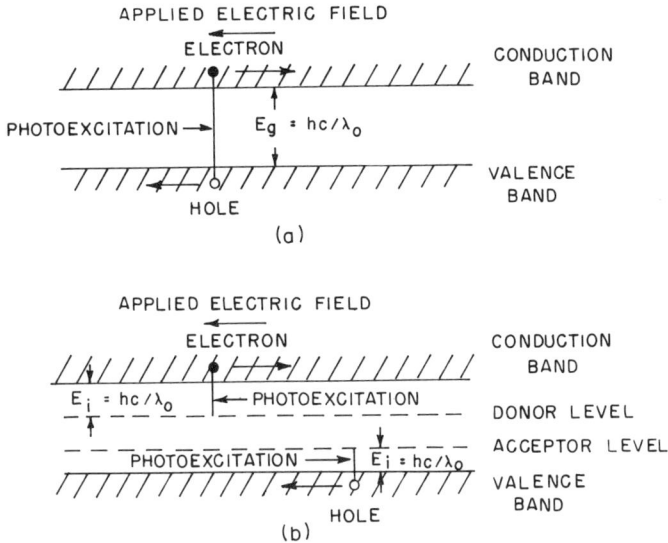

Fig. 2.2a and b. Photoconductive processes: (a) Intrinsic. (b) Extrinsic

Table 2.2. Forbidden energy gaps E_g and long wavelength limits λ_0 for the intrinsic photoeffect of selected semiconductors at the indicated temperatures

Semiconductor	T [K]	E_g [eV]	λ_0 [μm]
CdS	295	2.4	0.52
CdSe	295	1.8	0.69
CdTe	295	1.50	0.83
GaP	295	2.24	0.56
GaAs	295	1.35	0.92
Si	295	1.12	1.1
Ge	295	0.67	1.8
PbS	295	0.42	2.9
PbSe	195	0.23	5.4
InAs	195	0.39	3.2
InSb	77	0.23	5.4
$Pb_{0.2}Sn_{0.8}Te$	77	0.1	12
$Hg_{0.8}Cd_{0.2}Te$	77	0.1	12

Extrinsic photoconductivity occurs when an incident photon, lacking sufficient energy to produce a free hole-electron pair, can produce excitation at an impurity center in the form of either a free electron-bound hole or a free hole-bound electron, see Fig. 2.2b. The long wavelength limit of an extrinsic photoeffect is thus given by

$$\lambda_0 = \frac{hc}{E_i}, \tag{2.4}$$

Table 2.3. Ionization energies E_i and long wavelength limits λ_0 of impurities in germanium and silicon

Semiconductor: Impurity	E_i [eV]	λ_0 [µm]
Ge:Au	0.15	8.3
Ge:Hg	0.09	14
Ge:Cd	0.06	21
Ge:Cu	0.041	30
Ge:Zn	0.033	38
Ge:B	0.0104	120
Si:In	0.155	8
Si:Ga	0.0723	17
Si:Bi	0.0706	18
Si:Al	0.0685	18
Si:As	0.0537	23
Si:P	0.045	28
Si:B	0.0439	28
Si:Sb	0.043	29

where E_i is the impurity ionization energy. In practical units this becomes

$$\lambda_0[\mu m] = \frac{1.24}{E_i[eV]}. \tag{2.5}$$

Table 2.3 lists ionization energies of selected impurities in Ge and Si, and the corresponding long wavelength limits.

Analysis. The basic expression describing either intrinsic or extrinsic photoconductivity in semiconductors under equilibrium excitation (i.e., steady state) is

$$i_{so} = \eta q N_\lambda G, \tag{2.6}$$

where i_{so} is the short circuit photocurrent at zero frequency (dc), i.e., the increase in current above the dark current accompanying irradiation; η is the quantum efficiency, i.e., the number of excess carriers produced per absorbed photon; q is the electronic charge; N_λ is the number of photons of wavelength λ absorbed in the sample per unit time; and G is the internal gain, i.e., the number of electrons which flow through the external circuit for each absorbed photon. In general, photoconductivity is a majority carrier phenomenon, i.e., it is the increase in the number of majority carriers (electrons in n-type material, holes in p-type) accompanying irradiation which makes up the photocurrent. Photoexcited minority carriers will also contribute, but because minority carrier lifetimes are usually shorter than majority carrier ones, their contribution is proportionately less.

The photoconductive gain is given by the ratio of free carrier lifetime to transit time.

$$G = \frac{\tau}{T_r}. \tag{2.7}$$

In this simplified model τ is the majority carrier lifetime and T_r is the transit time for majority carriers between the sample electrodes. The latter is given by

$$T_r = \frac{l^2}{\mu V_A}. \tag{2.8}$$

Here l is the distance between electrodes, μ is the majority carrier mobility, and V_A is the bias voltage applied to the sample. Combining (2.6), (2.7) and (2.8) results in

$$i_{so} = \frac{qN_\lambda \mu \tau V_A}{l^2}. \tag{2.9}$$

The photon absorption rate N_λ is related to the absorbed monochromatic power P_λ by

$$N_\lambda = \frac{P_\lambda \lambda}{hc}, \tag{2.10}$$

so that (2.9) becomes

$$i_{so} = \frac{\eta q P_\lambda \mu \tau \lambda V_A}{hcl^2}. \tag{2.11}$$

Thus the dc short circuit photocurrent is proportional to the absorbed radiant power and depends linearly on the wavelength for values less than the long wavelength limit λ_0 given by (2.3) or (2.5).

When the detector is employed with a load resistor as shown in Fig. 2.1, the photocurrent is reduced below i_{so} by an amount equal to the ratio of the detector resistance to the sum of the detector and load resistance. Thus the photovoltage appearing across the load resistor, which equals in magnitude the photovoltage appearing across the detector, is

$$\Delta V_L = \frac{i_{so} R_d R_L}{R_d + R_L}, \tag{2.12}$$

where R_L is the load resistance and R_d is the detector resistance. The detector resistance is related to the conductivity σ, majority carrier concentration n, length l, width w, and thickness d, by

$$R_d = \frac{l}{\sigma w d} = \frac{l}{nq\mu w d}, \tag{2.13}$$

assuming that the change in conductivity upon irradiation is small compared to the dark conductivity. When R_L greatly exceeds R_d the photovoltage observed across either the detector or the load resistor is essentially the open circuit photovoltage v_{SO}. When this approximation is introduced into (2.12) and combined with (2.11) and (2.13), v_{SO} is found to be

$$v_{SO} = i_{SO} R_d = \frac{\eta P_\lambda \lambda \tau V_A}{hcnlwd}.\tag{2.14}$$

Equations (2.11) and (2.14) describe the magnitude of the photosignal in terms of the properties of the detector and the photon arrival rate. Another parameter of interest is the speed of response, which determines the extent to which the photosignal can follow modulation of the incident photon stream. The speed of response is determined in many instances by the majority carrier lifetime; in other cases, by carrier trapping times. If the lifetime or trapping time is very small, say, a microsecond, then the detector can respond to radiation modulated at frequencies up to high values, say, about a megahertz. In general, the frequency response of photoconductors is given by

$$i_S = \frac{i_{SO}}{(1+\omega^2\tau^2)^{1/2}},\tag{2.15}$$

and

$$v_S = \frac{v_{SO}}{(1+\omega^2\tau^2)^{1/2}},\tag{2.16}$$

where i_S and v_S are the short circuit current and the open circuit voltage at frequency f, given in terms of the angular frequency ω by

$$\omega = 2\pi f,\tag{2.17}$$

and τ is the appropriate lifetime or trapping time.

Thus the signal at low frequencies is independent of frequency; at high frequencies it is inversely proportional to frequency. The transition between the two occurs near where $\omega\tau$ is unity.

Photovoltaic Effect

The second photon effect of general utility is the photovoltaic effect. Unlike the photoconductive effect, it requires an internal potential barrier with a built-in electric field to separate a photoexcited hole-electron pair. Although it is possible to have an extrinsic photovoltaic effect, see *Ryvkin* [2.32], almost all practical photovoltaic detectors employ the intrinsic photoeffect. Usually this occurs at a simple $p-n$ junction. However, other structures employed include those of an avalanche, $p-i-n$, Schottky barrier and heterojunction photodiode. There is also a photovoltaic effect occuring in the bulk. Each will be discussed, with emphasis on the $p-n$ junction photoeffect.

p−n Junction. The most common form of photovoltaic effect occurs at a $p-n$ junction prepared in the semiconductor by standard techniques, such as bulk doping during growth, impurity diffusion, or growth of an epitaxial layer of one type upon a substrate of the opposite type. Figure 2.3 illustrates one geometry employed in a $p-n$ junction photodiode whereas Fig. 2.4 shows the energy band diagram, illustrating the field-induced charge separation accompanying photoexcitation.

Whereas the current-voltage characteristic of a photoconductor is linear, that of a $p-n$ junction exhibits rectification. As shown in Fig. 2.5, the photocurrent is an additive contribution to the dark current. This has given rise to a certain confusion in terminology. By definition, the photovoltaic effect is obtained without bias, i.e., it is the open circuit voltage obtained upon

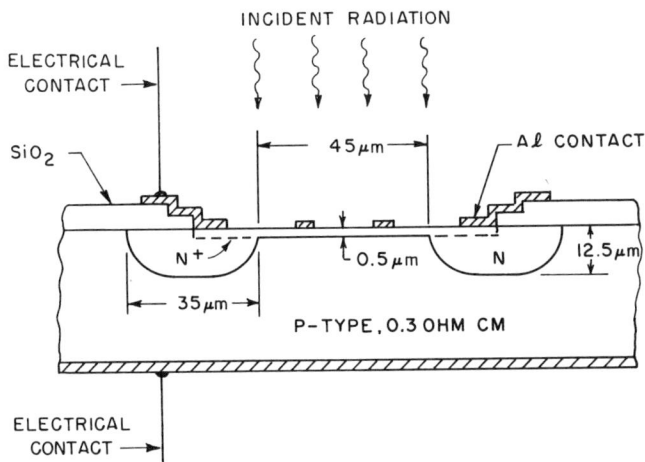

Fig. 2.3. Silicon planar photodiode with diffused guard ring structure (after *Anderson* et al. [2.33])

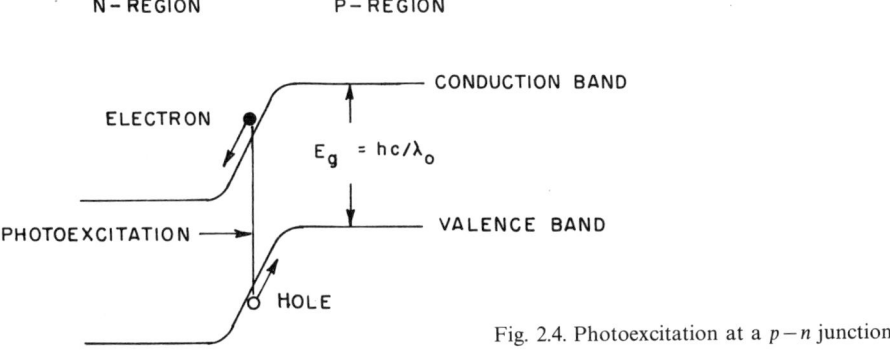

Fig. 2.4. Photoexcitation at a $p-n$ junction

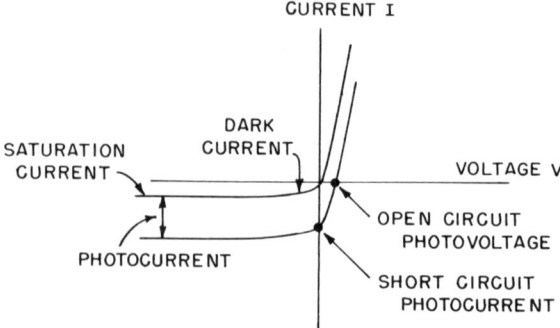

Fig. 2.5. Current-voltage characteristic of a photodiode

irradiation. However, junction detectors are frequently operated under reverse bias ("back bias"), so that the observed photosignal is a photocurrent rather than photovoltage. In this case they are said to be operated in the photoconductive mode, even though the configuration is not that of the photoconductor already discussed. The electrical circuits appropriate to a photovoltaic detector and a back-biased junction photodetector are illustrated in Fig. 2.6.

The basic equation of photoconductivity, (2.6), also applies to a photovoltaic detector, but the gain G is unity. Thus

$$i_{so} = \eta q N_\lambda \qquad (2.18)$$

is the expression for the dc photocurrent of a photovoltaic detector. Introducing the relationship between photon rate and power given by (2.10) results in

$$i_{so} = \frac{\eta q P_\lambda \lambda}{hc}. \qquad (2.19)$$

Thus when a $p-n$ junction or Schottky diode is operated in the photoconductive mode the short circuit current is given by (2.19). The open circuit voltage, i.e., the photovoltaic signal, is obtained by multiplying the short circuit current by the dynamic resistance at zero bias, that is, the slope dV/dI at the origin. In general, the current voltage characteristic of a diode is given by

$$I_d = I_s [\exp(qV/\beta kT) - 1], \qquad (2.20)$$

where I_d is the dark current, I_s is the dark reverse-bias saturation current, V is the applied voltage (positive in the forward direction), k is Boltzmann's constant, β is a constant of the order of unity, and T is the absolute temperature. The slope at the origin is

$$\frac{1}{R} = \frac{dI}{dV}\bigg|_{V=0} = \frac{qI_s}{\beta kT}, \qquad (2.21)$$

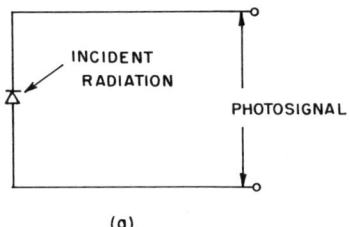

(a)

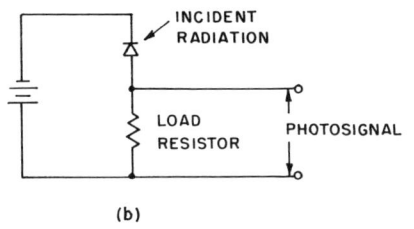

(b)

Fig. 2.6a and b. Photodiode circuits: (a) Open circuit photovoltage (photovoltaic effect). (b) Reverse-bias photoconductivity

where R is the dynamic resistance at zero bias. Thus the open circuit photovoltage, i.e., the photovoltaic signal v_{SO}, is given by

$$v_{SO} = \frac{\eta P_\lambda \lambda \beta k T}{hc I_s}. \qquad (2.22)$$

As was true for the photoconductive signal, the photovoltaic signal per unit radiant power is proportional to wavelength. Note, however, that the carrier lifetime does not appear in the expressions for the short circuit photocurrent or open circuit photovoltage.

In contrast to photoconductivity, the photovoltaic effect depends largely upon the minority carrier lifetime. This is because the presence of both the photoexcited electron and photoexcited hole is required for the intrinsic effect to be observed. Because the minority carrier lifetime is usually shorter than the majority carrier one, the photovoltaic signal terminates when the minority carrier recombines. The time dependent photosignal is given by (2.15) and (2.16), but the appropriate lifetime is the minority carrier one. For this reason, photovoltaic detectors are usually faster than photoconductive ones made from the same material.

The above discussion has been confined to the photovoltaic effect arising at a $p-n$ junction in a semiconductor due to photoexcitation of carriers on either side of the junction. Among the related areas of interest are a) avalanche photodiodes, b) $p-i-n$ photodiodes, c) Schottky barrier photodiodes, d) heterojunction photodiodes, and e) bulk photovoltaic effect. Each of these will be discussed briefly.

Avalanche Photodiode. Whereas a $p-n$ junction photodiode has no internal gain mechanism, an avalanche photodiode does. Thus the photosignal of an avalanche photodiode is larger than that of a $p-n$ junction photodiode made from the same material and of the same area in response to the same incident radiation power. Although the internal gain cannot increase the detector signal-to-noise ratio, and may even reduce it (see [2.34]), internal gain is useful in that it reduces the requirements on the high gain low noise amplifier employed with detectors. Furthermore, under certain conditions avalanche gain can be achieved without compromising greatly the response time of the photodiode. In a limited sense, an avalanche photodiode can be considered to be the solid state analogue of a photomultiplier.

Avalanching, (avalanche breakdown) occurs in a $p-n$ junction of moderate doping levels under reverse bias. (At high doping levels, Zener tunneling prevents buildup of a sufficent field to cause avalanche breakdown to occur.) In the absence of radiation, the thermally excited carriers normally present in the semiconductor are accelerated within the high field region of the junction to velocities so high that their collision with lattice atoms transfer energy sufficient to free additional electrons. These are then accelerated, undergo additional collisions, and free more electrons. Thus an avalanche of electrons occurs within the high field region of the junction. Because avalanching can be initiated by photoexcited electrons or holes as well as by thermally excited ones, the effect produces an increase in the number of photoexcited carriers.

Much effort has been spent in studying avalanche breakdown in many semiconductors and in designing avalanche photodiodes. Because it is difficult to cause uniform avalanching over a broad area, avalanche photodiodes usually have small sensitive areas. Both $p-n$ and $p-i-n$ configurations (see below) have been employed. Many papers are available describing their preparation and performance (see, for example *McIntyre* [2.35], *Melchior* and *Lynch* [2.36], *Ruegg* [2.37], *Lindley* et al. [2.38], and *Lynch* [2.39]).

p−i−n Photodiode. Instead of employing an abrupt $p-n$ junction, a $p-i-n$ photodiode incorporates an intrinsic region between the p and n sides of the junction. By making the surface region (say, the p-region) thin compared to the optical absorption depth, the incident radiation will penetrate into the intrinsic region, where absorption there will produce hole-electron pairs. Because the high electric field present there will cause the carriers to drift quickly through that region, the frequency response of $p-i-n$ photodiodes can be made to be higher than that of $p-n$ photodiodes made from the same material. See, for example, *Mather* et al. [2.40] for further details.

Schottky Barrier Photodiode. A photoeffect similar to that obtained in a $p-n$ junction can be found at a Schottky barrier, formed at a metal-semiconductor interface. As with a $p-n$ junction, a metal-semiconductor interface when properly made provides a potential barrier which causes separation of photoexcited holes and electrons, thereby giving rise to a short circuit photocurrent and an open circuit photovoltage. In most instances the metal is in the form of a very thin film which is semitransparent to the incident radiation. Photoexcitation can occur within the semiconductor or over the potential barrier at the metal-semiconductor interface. In some instances the photocell is illuminated from the semiconductor side, which must be thin enough so that that photoexcited minority carriers can diffuse to the interface region.

Not all semiconductors can be prepared in both n- and p-types. Schottky barrier photodiodes are of special interest in those materials in which $p-n$ junctions cannot be formed. They also find application as UV and visible radiation detectors, especially for laser receivers where their high frequency response (in the gigahertz range in many cases) is of particular usefulness. See *Ahlstrom* and *Gartner* [2.41], *Schneider* [2.42] and *Sharpless* [2.43] for more detailed descriptions.

Heterojunction Photodiode. In the early 1960s there was considerable interest in semiconductor heterojunctions for use as wide gap emitters in transistors. The heterojunctions were formed by epitaxial growth of one semiconductor, e.g. GaAs, on another, e.g. Ge, having a similar lattice structure and periodicity. Rectification properties could be observed not only in the $p-n$ junction configuration but also in $n-n$ and $p-p$ configurations. The reasons for this were detailed by *Anderson* [2.44].

It was natural that photoeffects were explored in heterojunctions and new photodevices considered, see for example *Kruse* et al. [2.45, 46] and *Perfetti* et al. [2.47]. However, the development of Si technology removed the need for a wide gap emitter transistor, and the problem of preparing heterojunctions with electrical and optical properties conforming to the ideal caused a drastic decrease in the interest in this area. There remains some interest in heterojunctions prepared from alloy semiconductors, e.g., GaAs-Ga$_{1-x}$Al$_x$As and PbTe-Pb$_{1-x}$Sn$_x$Te, where lattice mismatch can be minimized.

Bulk Photovoltaic Effect. Although the potential gradient giving rise to the photovoltaic effect is usually an abrupt one, e.g., at a $p-n$ junction or a Schottky barrier, it is also possible to observe a photovoltaic effect arising at a diffuse gradient in nonhomogeneous semiconductors. *Tauc* [2.48] showed that an impurity gradient can effectively separate photoexcited electrons and holes; he termed this the "bulk photovoltaic effect". *Kruse* [2.49] showed that a spatial variation in energy gap in an alloy semiconductor such as (Hg, Cd)Te can also give rise to a bulk photovoltaic effect. *Marfaing* and *Chevallier* [2.50] analyzed the bulk photovoltaic effect arising from an effective mass gradient in inhomogeneous alloy semiconductors. In any form, the bulk photovoltaic effect has found little practical application.

Photoemissive Effect

The third of the principal photon effects is the photoemissive one, also known as the external photoeffect. As the name implies, the action of the incident radiation is to cause the emission of an electron from the surface of the photocathode into the surrounding space, there to be collected by an anode. The spectral responses of selected photocathodes are illustrated in Fig. 2.7.

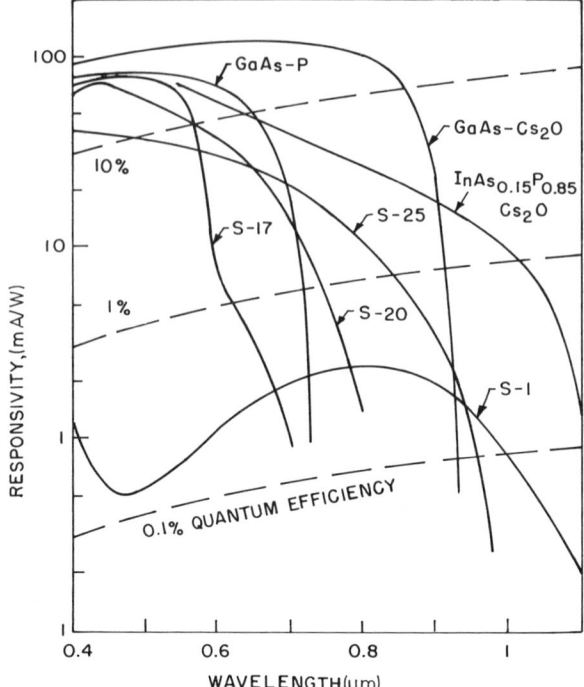

Fig. 2.7. Sensitivity vs wavelength for several representative photocathodes

Applications of the photoemissive effect are numerous. Vacuum phototubes, consisting only of a photocathode and anode, are employed for very fast response. Gas filled phototubes, which rely upon an avalanche effect arising from impact ionization of the gas, are used for high gain without external amplification. Image orthicon tubes employing the photoemissive effect are found in television cameras. That most widely used is the photomultiplier, in which the photoemitted electrons impinge upon a ladderlike structure of electrically biased dynodes having secondary emitting properties. Each electron incident upon a dynode causes the emission of more than one additional electron, so that an electron gain or signal amplification mechanism exists. A variation of this is the channel electron multiplier, a tube having an applied electrical bias along its length whose internal surface possesses secondary emitting properties. An incident electron makes repeated collisions with the internal walls while traversing the tube, giving rise to an avalanche of electrons. Methods have been devised to prepare of the order of a million channels in a cross-sectional area of the order of an inch. Such devices, termed microchannel plates, are used with photoemissive surfaces and phosphors in image intensification devices to see under the low level illumination available at night.

Conventional Photocathodes. The photocathode controls the spectral properties of the phototube. Photocathodes can be divided into two categories, conventional ones and negative electron affinity ones. The energy band diagrams of

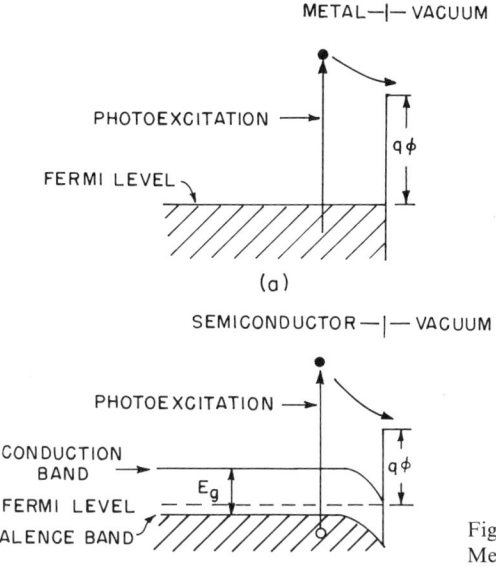

Fig. 2.8a and b. Photoemissive processes: (a) Metal photocathode. (b) Semiconductor photocathode with positive electron affinity

the two types of conventional photocathodes are illustrated in Fig. 2.8. Metal photocathodes, see Fig. 2.8a, are characterized by a photoelectric work function $q\phi$, representing the energy needed to raise an electron from the Fermi level within the interior of the metal to a level above the potential barrier at the surface of the metal, thereby allowing the electron to escape from the metal. Thus the minimum energy which a photon must possess to cause photoemission equals the work function of the metal. Work functions of metals are several electron volts, so that metallic photocathodes are useful only for visible or ultraviolet radiation detection.

The other type of conventional photocathode, the energy band diagram of which is illustrated in Fig. 2.8b, employs a semiconductor material with a positive electron affinity. Again, the minimum energy required for a photon to cause photoemission is that which will raise the electron to an energy level higher than the potential barrier at the surface. In general, semiconductor photo-cathodes require less energy than metallic photocathodes for photoexcitation, and therefore have spectral responses extending to longer wavelengths than metallic photocathodes. However, even these photocathodes extend at best only into the very near infrared.

Negative Electron Affinity Photocathodes. The other category of photocathodes, those having a negative electron affinity, resulted from the discovery by *Scheer* and *Van Laar* [2.51] that it was possible to overcoat the surface of selected p-type semiconductors with an evaporated layer of a low work function material such that the resulting structure has a negative electron affinity as shown in Fig. 2.9. In this case the incident photon energy must only equal or exceed the energy gap of

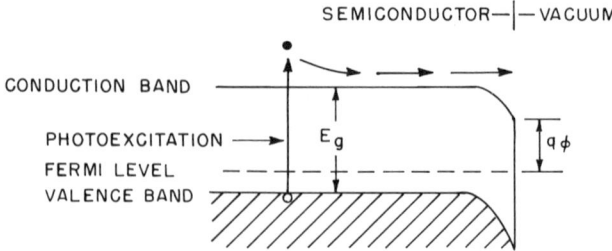

Fig. 2.9. Photoemissive process: semiconductor photocathode having negative electron affinity

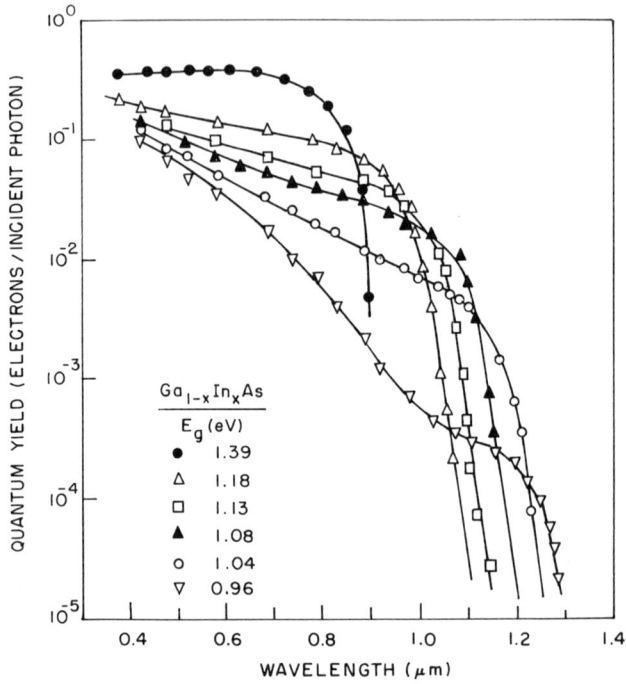

Fig. 2.10. Quantum yields for selected $Ga_{1-x}In_xAs$ alloys with decreasing bandgaps (after *Fisher* [2.52])

the semiconductor for photoemission to occur. By selecting appropriate semiconductors and coatings it is possible to prepare photocathodes in which the spectral responses extend into the near infrared. In practice it appears that the quantum efficiency decreases with increasing long wavelength limit, see Fig. 2.10. Thus no photocathodes are presently available which are efficient detectors of infrared radiation at wavelengths exceeding about 1 µm.

Other Photon Effects

In addition to the three principal photon effects, there exist many others of lesser importance. These are discussed below.

Photoelectromagnetic Effect. Photons of sufficiently short wavelength incident upon a semiconductor sample immersed in a transverse magnetic field produce hole-electron pairs by intrinsic excitation. Because the radiation is absorbed exponentially with depth into the material, a carrier concentration gradient is produced which is directed normal to the surface. Thus the photoexcited carriers diffuse from the surface into the bulk, crossing lines of magnetic flux as they do. Because these oppositely charged carriers are moving in the same direction, they are deflected by the magnetic field toward opposite ends of the sample, thereby establishing a longitudinal electric field. Electrodes attached to the sample ends can detect this photovoltage [2.53].

The photoelectromagnetic (PEM) effect has been exploited in InSb as an uncooled detector of radiation of wavelength shorter than about 7 µm [2.54]. Its low sensitivity limits its use to industrial applications.

Dember Effect. Radiation whose wavelength is sufficiently short which is incident upon the surface of a semiconductor produces hole-electron pairs by intrinsic excitation. If the sample thickness and carrier diffusion lengths are large compared to the absorption depth of the radiation, the carriers will diffuse away from the front surface toward the back. If the carrier mobilities and diffusion lengths are unequal, a charge separation will occur giving rise to an electric field between the front and back surfaces. Thus a photocurrent will flow in the external circuit connecting these surfaces. *Moss* [2.55] has derived the expression for the Dember photocurrent in terms of the parameters of the semiconductor. The effect has not been found to be useful in practical applications.

Hot Electron Bolometer, Putley Detector. Whether it is more appropriate to include the hot electron bolometer and Putley detector in a list of detectors employing photon effects, or to instead list them with thermal effects, is somewhat arbitrary. Both employ photon effects in that incident photons interact with free electrons in a semiconductor. However, they are thermal (as the name bolometer implies) in that the effects are explainable in terms of a change in the effective temperature of the free electrons. Because the interpretation is mainly from the viewpoint of a photon-electron interaction, they are included here in the list of photon effects.

Although the most widely exploited photon effects depend upon increasing the number of free carriers in a semiconductor by intrinsic or extrinsic absorption, it is also possible to detect photons by the change in the mobility of electrons or holes accompanying irradiation of a semiconductor with photons whose wavelength is such that free carrier absorption is the dominant absorption mechanism. This requires that the wavelength of the incident radiation is longer than that of the absorption edge of the semiconductor. Since the free carrier absorption coefficient depends upon the square of the wavelength in the spectral interval past the absorption edge out to the submillimeter region for semiconductors of interest (after which it becomes independent of wavelength), detectors exploiting this effect operate in the far infrared and submillimeter regions. *Rollin*

[2.56] proposed, and *Rollin* and *Kinch* [2.57] developed a sensitive detector based on this effect, termed a hot electron bolometer, the operation of which requires temperatures of the order of 4 K. Basically, the absorption of far infrared radiation by free electrons in a semiconductor such as InSb causes excitation of the electrons into higher energy states in the conduction band. This causes a change in the electron mobility, therefore a change in conductivity of the sample, which can be detected as a short circuit current or open circuit voltage. The response time of the effect is the lattice relaxation time, which can be extremely short, much less than nanoseconds.

Putley [2.58] showed that it was possible to tune this effect in wavelength through application of an external magnetic field. This quantizes the continuum of energy levels within the conduction and valence band into the so-called Landau levels, which are separated by an energy known as the cyclotron resonance energy, the magnitude of which is proportional to the magnetic field strength. Far infrared photons whose energy equals that needed for cyclotron resonance absorption at the magnetic field in which a sample is immersed will cause excitation of electrons from one Landau level into the next higher one; the resultant mobility change can be detected as a change in conductivity. Whether the effects observed in practice are those predicted theoretically is not clear, see the discussion by *Shenker* [2.59].

Photon Drag Effect. Because photons carry momentum, they transfer momentum to free carriers in semiconductors during free carrier absorption. *Gibson* et al. [2.60] have exploited this effect in a very fast but very insensitive detector for use in detecting pulsed laser radiation. In heavily doped semiconductors, free carrier absorption is the dominant absorption mechanism for wavelengths longer than the absorption edge. An incident stream of photons having wavelengths consistent with free carrier absorption will transfer momentum to the free carriers, effectively pushing them in the direction of the Poynting vector. Thus a longitudinal electric field is established within the semiconductor which can be detected through electrodes attached to the sample. Detectors exploiting this photon drag effect are useful as laser detectors because of their rapid response and ability to absorb large amounts of power (such as from a CO_2 laser) without damage because the radiation is absorbed over a large volume of material. Their lack of sensitivity makes them of little use as infrared detectors for most applications.

Phototransistor. A phototransistor consists of a $p-n-p$ or $n-p-n$ transistor structure so configured that photoexcited excess majority carriers can be produced within the base region, either by excitation therein or by diffusion from the adjacent emitter or collector regions. Their presence within the base causes a forward bias of the emitter much as the base-emitter junction can be forward biased by means of voltage applied to external electrodes. This in turn produces gain as in normal transistor action. As *Bube* [2.61] points out, the gain so achieved can be expressed in terms of the lifetime divided by the transit time [see (2.7)] as is employed in photoconductivity.

The only semiconductors which have been employed in phototransistors in any large quantities are Ge and Si. The former has a limited infrared response. The latter responds in the visible and very near infrared. Neither has found application where high performance is essential.

Infrared Quantum Counter. Bloembergen [2.62] has proposed a concept for detecting infrared radiation which he has termed an infrared quantum counter. The basic idea requires a solid having three or four appropriately spaced energy levels. Signal radiation whose energy is resonant between the lower and middle level of a three-level system causes excitation of the ground state electrons to the middle state. A second source of radiation, the pump, which is resonant with the energy difference between the middle and upper level, then excites those electrons from the middle state to the upper. They then relax to the lower level, emitting a photon of appropriate energy, or to a fourth level between the middle and upper ones again emitting a photon. Detection of these emitted photons occurs by means of a separate photodetector.

The advantage of the infrared quantum counter scheme lies in its inherently noiseless detection. A disadvantage, derived by *Gelinas* [2.63] (see also *Kruse* et al. [2.64]), is that under normal operating conditions its background limited performance is poor. *Brown* and *Shand* [2.65] have reviewed potential systems in which infrared quantum counter action could take place, including dielectrics with various dopants such as rare earth ions, semiconductors, liquids and gases. Quantum counter action has been realized experimentally by *Porter* [2.66] and *Grandrud* and *Moos* [2.67, 68] in rare earth doped $LaCl_3$, by *Brown* and *Shand* [2.69] in $SrF_2:Er^{3+}$, and by *Varsanyi* [2.70] in $PrCl_3:Nd^{3+}$. None of the results is indicative of a high performance detector.

Phosphors. Infrared radiation can stimulate the emission of visible radiation from certain types of phosphors which have been previously excited by ultraviolet radiation. *Becker* and *Risgin* [2.71] have reported that self-activated ZnS crystals have exhibited both visible luminescence and photoconductivity when stimulated by radiation in the 1–3 μm interval. *Geusic* et al. [2.72] have shown that rare earth phosphors can convert very near infrared (0.93 μm) radiation into the visible with an efficiency as high as 1%. Although phosphors can be employed to convert images directly, their lack of sensitivity has limited their usefulness.

Photographic Film. The most widely employed photon detector is, of course, photographic film, the spectral response of which is largely confined to the ultraviolet and visible. Certain dyes, however, have been found to extend its response to about 1.2 μm. Numerous references are available concerning the photographic mechanism, see for example *Mees* [2.73], *Fujisawa* [2.74] and *Mott* and *Gurney* [2.75]. *Jones* [2.76, 77] has studied the factors which establish the sensitivity limits.

2.1.2 Thermal Effects

The second principal category of effects, known as thermal effects, includes those which depend upon the changes in the properties of materials resulting from temperature changes caused by the heating effect of the incident radiation. The effects do not depend upon the photon nature of the incident radiation. Thus thermal effects are generally wavelength independent; the photosignal depends upon the radiant power (or its rate of change) but not upon its spectral content. This assumes, of course, that the mechanism responsible for the absorption of the radiation is itself independent of wavelength, which is not strictly true in most instances.

Because heating and cooling of a macroscopic sample is a relatively slow process, thermal detectors as a class are slower in their rate of response than photon ones, although some thermal detectors are faster than selected photon ones. It is convenient to think of thermal detectors as having millisecond response times and photon ones as having microsecond ones, but this is clearly only a rough rule to which there are many exceptions.

In the discussions which follow, attention will be directed toward two approaches which have found the greatest utility in infrared systems, namely, bolometers and the pyroelectric effect. Others will be discussed briefly. A list of thermal effects is included in Table 2.4.

Bolometer Effect

A bolometer is the thermal analogue of a photoconductor. The effect is that of change in resistivity of a material in response to the heating effect of incident radiation. In contrast to photoconductors, bolometers can be made of any material which exhibits a temperature dependent change of resistance. The temperature dependence is specified in terms of the temperature coefficient of resistance α defined as

$$\alpha = \frac{1}{R}\frac{dR}{dT}, \qquad (2.23)$$

where R is the resistance and T is the absolute temperature of the sample.

The geometry of a bolometer is similar to that of the photoconductor shown in Fig. 2.1. The electrical circuit requires a voltage source to measure the change in resistance due to the heating effect of the radiation. Five types of bolometers have been devised, namely, thermistor, metallic, superconducting, superinducting, and cryogenic. Of these, only the thermistor and cryogenic semiconductor ones are of importance. Each of the five is described below.

Thermistor Bolometer. Thermistor (thermally sensitive resistor) material is an oxide of manganese, cobalt, or nickel which exhibits a negative temperature coefficient of resistance; the resistance decreases as the temperature increases. Because it is employed under electrical bias, the self-heating effects due to

Table 2.4. Classification of thermal effects

1. Bolometers	4. Golay cell
1.1 Thermistor	5. Gas filled condenser mircrophone
1.2 Metal	6. Pyromagnetic
1.3 Superconducting	7. Nernst
1.4 Superinducting	8. Liquid crystals
1.5 Cryogenic semiconductor	9. Absorption edge image converter
2. Pyroelectric	10. Evaporagraph
3. Thermoelectric	

internal power dissipation from the bias current must be considered. With increasing bias the temperature rises rapidly and the dynamic resistance (slope of the current-voltage characteristic) decreases. Above a critical bias the dynamic resistance becomes negative. Unless a ballast resistor is employed, thermal runaway will occur in this region, and the thermistor will burn up.

Thermistor bolometers are operated below this critical bias, where they have a positive dynamic resistance. They are usually operated uncooled, so are relatively inexpensive. Because the ambient temperature change will also affect the resistance, a bridge circuit is employed in which one of the two identical elements is shielded from the incident radiation. Thus radiant energy will imbalance the bridge but changes in the ambient temperature will not. Thermistor bolometer technology is well established; relevant papers derive from the 1950s. See *Wormser* [2.78] and *Dewaard* and *Wormser* [2.79] for a detailed description. The speed of response is dictated by the time required for the bolometer to decrease in temperature to its ambient value upon sudden removal of the incident radiation. Heat is lost through radiation to the surroundings (the ideal which may be difficult to achieve) or conduction to the substrate upon which it is mounted, or by conduction through the electrical leads. Most frequently conduction to the substrate is the dominant mechanism. In this case the response time is controlled by the product of heat capacity of the sensitive element and the thermal conductance between the sensitive element and substrate. *Wormser* [2.78] shows that the time dependent signal voltage v_S for a thermistor bolometer in a matched bridge circuit in response to a pulse of radiation of energy E_i absorbed by the thermistor is given approximately by

$$v_S = \frac{E_i \alpha V_B [1 - \exp(-t/\tau)]}{2C}, \qquad (2.24)$$

where V_B is the bias voltage, C is the rate of heat conduction from the thermistor to the substrate, and τ is the thermal response time, it being assumed that the duration of the radiant energy pulse is of the same order as τ. A detailed analysis of the operation of the thermistor bolometer is provided by *Kruse* et al. [2.80].

Metal Bolometer. In contrast to thermistor materials, metals have a positive temperature coefficient of resistance, i.e., the resistance increases as the temperature increases. Usually the absolute value of the temperature coefficient of resistance of metals is much less than that of thermistor materials. Furthermore, metals do not exhibit the negative resistance effects which thermistor materials do. Thus, in general metallic bolometers are less sensitive detectors of radiation than are thermistor bolometers. They do not exhibit thermal runaway.

On the other hand, it is easy to prepare metals in the form of evaporated thin films on a large variety of substrates. Their geometries are readily dictated by the evaporation masks employed. By making the evaporated films very thin, heat capacities can be made to be very low, and the speed of response can be high. Because of the high reflectivity of metals it is necessary to overcoat the films with an evaporated film of a material such as platinum or gold black which will absorb the incident radiation. Like thermistor bolometers, most metal film bolometers operate at room temperature. See *Dewaard* and *Wormser* [2.79] for additional details.

Superconducting Bolometer, Superinductor. At one time there was considerable interest in superconducting bolometers as infrared radiation detectors because of the large temperature dependence of resistance at the superconducting transition temperature [2.81, 82]. By maintaining the ambient temperature of a superconducting sample just below the transition temperature, a small temperature increment arising from the absorption of incident radiation will cause a dramatic change in the sample resistance, which can be used to generate a large output signal. Problems with system complexity, including maintaining the sample at the transition temperature, together with advances in other types of thermal detectors, have caused interest largely to disappear.

Naugle and *Porter* [2.83] have proposed using the increase in inductance of a superconducting thin film as the transition temperature is approached from below to detect radiation. Their analysis reveals high sensitivity for a granular aluminum thin film superinducting bolometer operated at 2 K.

Cryogenic Bolometers. Although the superconducting bolometer has not proved to be a useful device, other cryogenic bolometers have been developed which are of great utility, especially as detectors of far infrared radiation. Among the earliest was the carbon bolometer developed by *Boyle* and *Rodgers* [2.84], consisting of a flat slab cut from the core of a carbon resistor which was mounted on a copper heat sink maintained at liquid helium temperature (4 K). Because the temperature coefficient of resistance of a carbon composition resistor has a large (and negative) value at cryogenic temperatures and the specific heat is low, the carbon cryogenic bolometer is a highly sensitive detector. It is also relatively easy to prepare. More recent information on carbon resistance bolometers is reported by *Corsi* et al. [2.85].

Low [2.86, 87] has devised a cryogenic bolometer which employs Ga doped single crystal Ge as the sensitive element operated at 4 K. *Oka* et al. [2.88] have devised one employing In doped Ge, whereas *Bachman* et al. [2.89] have reported on a Si one, *Nayar* [2.90] on a TlSe one and *Shivanandan* and *McNutt* [2.91] on an InSb one. The advantages of these semiconductor bolometers, all of which operate near 4 K, are high sensitivity, low noise, and reproducibility. All employ single crystals of semiconductors. The theory is well established, dating from the initial work of *Low* [2.86], so that it is possible to design the detectors to have properties which are experimentally realized. *Dall'Oglio* et al. [2.92] have compared the experimental performance of the Ge, Si, and carbon bolometers with the Golay detector, discussed below. All are usefull as far infrared detectors; the cryogenic semiconductor bolometers are especially useful for astronomical applications.

Pyroelectric Effect

One of the most useful detectors of recent origin is the pyroelectric one, proposed by *Chynoweth* [2.93] and reduced to practice by *Cooper* [2.94, 95] and others. Many investigators have contributed to the development of pyroelectric detector technology in the past decade, among whom are *Astheimer* and *Schwarz* [2.96], *Beerman* [2.97, 98], *Glass* [2.99], *Liu* et al. [2.100–103], *Lock* [2.104], and *Phelan* et al. [2.105]. *Putley* [2.106] and *Liu* [2.107] have reviewed the theory of operation and state-of-the-art of pyroelectric detectors.

The pyroelectric effect is exhibited by temperature sensitive pyroelectric crystals, among which are certain ferroelectric crystals including TGS (triglicine sulfate), SBN ($Sr_{1-x}Ba_xNb_2O_6$), PLZT (lanthanum doped lead zirconate titanate) and $LiNbO_3$. Such crystals exhibit spontaneous electric polarization which can be measured as a voltage at electrodes attached to the sample. However, at a constant temperature the internal charge distribution will be neutralized by free electrons and surface charges, so no voltage is detectable. If the temperature is rapidly changed the internal dipole moment will change, producing a transient voltage. This pyroelectric effect can be exploited as a sensitive detector of modulated radiation, operating at ambient temperature.

Pyroelectric detectors are capacitors, having metallic electrodes applied to opposite surfaces of the temperature-sensitive ferroelectric crystal. Modulated radiation incident on the detector gives rise to an alternating temperature change ΔT. Accompanying the temperature change is an alternating charge ΔQ on the external electrodes given by

$$\Delta Q = pA\Delta T, \tag{2.25}$$

where p is known as the pyroelectric coefficient of the material, and A is the area over which the incident radiation is absorbed. Thus the photocurrent is proportional to the rate of change of temperature:

$$i_S = pA\frac{d(\Delta T)}{dt}. \tag{2.26}$$

It can readily be shown (e.g., see [2.80]) that the rate of change of temperature of a material having heat capacity C_H and thermal conductance to its surroundings K when modulated radiation falls upon it is described by

$$C_H\frac{d(\Delta T)}{dt} + K\Delta T = \eta P(t), \tag{2.27}$$

where η is the fraction of the incident power converted to heat. Here $P(t)$ is the incident radiant power modulated at angular frequency ω:

$$P(t) = P_1 \exp(j\omega t), \tag{2.28}$$

where P_1 is a constant and $j = \sqrt{-1}$.
The solution to (2.27) is

$$\Delta T = \frac{\eta P_1 \exp(j\omega t)}{K + j\omega C_H}. \tag{2.29}$$

Thus

$$\frac{d(\Delta T)}{dt} = \frac{j\omega \eta P_1 \exp(j\omega t)}{K + j\omega C_H}, \tag{2.30}$$

so that, from (2.26)

$$i_s = \frac{j\omega P A \eta P_1 \exp(j\omega t)}{K + j\omega C_H}. \tag{2.31}$$

The rms amplitude of the signal current is therefore

$$i_{s,\text{rms}} = \frac{\eta p A P_1 \omega \tau}{\sqrt{2} C_H (1 + \omega^2 \tau^2)^{1/2}}, \tag{2.32}$$

where the thermal time constant τ is

$$\tau = \frac{C_H}{K} = \frac{c \varrho_d l A}{K}. \tag{2.33}$$

Here c is the specific heat, ϱ_d is the density, and l is the sample thickness. Strictly speaking, the concept of thermal time constant is valid only when the thermal loss mechanism is by radiation to the surroundings or conduction to the substrate. This covers the cases of interest here. The rms signal voltage is given by

$$v_{s,\text{rms}} = \frac{i_{s,\text{rms}} R}{(1 + \omega^2 R^2 C^2)^{1/2}}, \tag{2.34}$$

where R and C are the parallel equivalent resistance and capacitance of the sensitive element. Thus

$$v_{s,\text{rms}} = \frac{\eta p A P_1 \omega \tau R}{\sqrt{2} C_H (1 + \omega^2 \tau^2)^{1/2} (1 + \omega^2 R^2 C^2)^{1/2}}. \tag{2.35}$$

Under many operating conditions, $(\omega\tau)^2 \gg 1$ and $(\omega RC)^2 \gg 1$. Since

$$C = \frac{\kappa\varepsilon_0 A}{l}, \tag{2.36}$$

where κ is the dielectric constant of the material and ε_0 is the permittivity of free space, then

$$v_{\text{s,rms}} = \frac{\eta p P_1}{\sqrt{2c\varrho_d A \omega \kappa \varepsilon_0}}. \tag{2.37}$$

Equation (2.37) shows that large signals are obtained from pyroelectric materials which have a large ratio of pyroelectric coefficient to dielectric constant.

Other Thermal Effects

In addition to the various types of bolometers and the pyroelectric effect, several other thermal effects have been exploited as radiation detectors. They are described below.

Thermoelectric Effect. In a circuit consisting of two different conductors, usually metals, and the junctions between them, preferential heating of one junction will generate a voltage which is a measure of the temperature difference. This thermoelectric effect is employed in laboratories in the form of thermocouples as a precise measurement of temperature. When the temperature difference arises from the absorption of radiation at one junction, the device is known as a radiation thermocouple. In order to increase the signal voltage, radiation thermocouples are connected in series to form a radiation thermopile.

Radiation thermopiles are usually made by evaporating a film of one metal upon a substrate, then evaporating a second metal such that it overlaps part of the first, thereby forming the junctions. Although their photosignal is low relative to many photon detectors, radiation thermopiles are rugged and require no electrical bias. They have been found to be useful in spaceborne applications. The preparation and properties of radiation thermopiles have been reviewed by *Astheimer* and *Weiner* [2.108] and by *Stevens* [2.109].

Golay Cell, Gas-Filled Condenser Microphone. One of the earliest types of detectors employed for far infrared sensing is the Golay cell, also known as the pneumatic infrared detector [2.110, 111]. It consists of a gas-filled enclosure in which one end is a thin membrane blackened so as to absorb radiation focused upon it. The energy absorbed by the membrane heats the gas within the enclosure, thereby distending a small flexible membrane mirror at the opposite end of the chamber. An optical system is employed to detect the very small distortions of the mirror accompanying absorption of energy by the gas. Although the Golay cell has seen widespread use as a detector of far infrared radiation in laboratory spectrometers, it lacks sensitivity and ruggedness for most field applications.

A second type of pneumatic infrared detector is the gas-filled condenser microphone, differing principally from the Golay cell in the method of absorbing incident radiation and in the signal readout means. Instead of the radiation being absorbed by a blackened membrane, it is absorbed directly by the gas filling the chamber. Rather than employing an optical method for detecting the deflection of a flexible membrane, an electrical method is employed. Heating of the gas due to absorption of the incident radiation distorts the membrane, which forms one plate of a capacitor. The capacitance change is determined by an electrical oscillator/amplifier system. Because the absorption spectrum is unique to the fill gas, the detector is employed in nondispersive infrared gas analyzers which might use, for example, a CO_2 fill gas to determine the concentration of CO_2 in an unknown gas. See *Hill* and *Powell* [2.112] for additional information.

Pyromagnetic Effect. An analogue to the pyroelectric effect is the pyromagnetic one, in which certain materials exhibit changes in their magnetic properties as a function of temperature. *Bené* et al. [2.113, 114] have employed this effect to detect infrared radiation. Amplitude modulated radiation falling on a sample of such materials as single crystal gadolinium, polycrystalline Mn_5Ge_3, FeRh, and others causes a time-modulated magnetization within the material, thereby inducing a current in a coil surrounding the sample. Like the pyroelectric effect, the pyromagnetic one occurs only in response to changes in the infrared radiation absorbed by the material. The phenomenon has not been widely exploited.

Nernst Effect. The Nernst effect appears in materials containing a temperature gradient which are immersed in a magnetic field. The geometry employed to detect the effect is the same as that of the photoelectromagnetic effect; incident radiation falling on the front surface of a sample in which the magnetic field is normal to the direction of the radiation generates a voltage in the direction normal to the plane of the magnetic field and Poynting vectors. For this reason it is possible sometimes to confuse the PEM and Nernst effects when detecting a photosignal, as *Kruse* et al. [2.115] have pointed out.

Washwell et al. [2.116] have developed the pertinent equations for the Nernst photosignal in terms of the material parameters. They have exploited the effect in Bi and $Bi_{97}Sb_3$ as an uncooled infrared detector. Like other uncooled thermal detectors, the speed of response is somewhat slow. The authors believe it can be improved to the microsecond range.

Liquid Crystals. Of the three types of liquid crystals, namely, smetic, nematic, and cholesteric, it is the cholesteric which has found application in temperature sensing, and therefore in infrared detection. These materials have rheological properties similar to fluids but optical behavior similar to crystals, arising from their planar rod-like molecular arrangement. Cholesteric liquid crystals exhibit irridescence when illuminated due to internal scattering of circularly polarized light. Their reflectivity changes markedly with small changes in their internal temperature, thereby making them useful for direct conversion of thermal imagery to visible imagery. However, their performance is inferior to conventional thermal imaging systems employing mechanical scanning of semiconductor photon detectors. They find laboratory application to detect infrared laser radiation. See *McColl* [2.117], *Ennulat* and *Fergason* [2.118], *Woodmansee* [2.119], and *Assouline* et al. [2.120] for additional information.

Absorption Edge Image Converter, Evaporagraph. Two devices of interest at one time were the absorption edge image converter, which employed the temperature dependence of the absorption edge of a semiconductor as the sensing mechanism, and the Evaporagraph, which depended on the temperature dependent thickness of a thin film of oil. See *Kruse* et al. [2.121] for further details.

2.1.3 Wave Interaction Effects

The third principal category of effects is that of wave interaction, which arises from the interaction of the electromagnetic field of the incident radiation with the sensing material. The principal wave interaction effects are optical heterodyne detection and optical parametric effects. Among the others are effects at Josephson junctions and metal-metal oxide-metal contacts. These are listed in Table 2.5 and discussed below.

Optical Heterodyne Detection

Whether optical heterodyne detection should be classed as a wave interaction effect or a photon effect is not obvious. Because it depends upon the interaction of the electric field vector of the signal radiation with that from a reference source, it is listed here as a wave interaction effect.

The excitation rate of free carriers in a semiconductor depends upon the rate of absorption of photons, which is a measure of the intensity of the absorbed radiation. Because the intensity is proportional to the square of the electric field vector, a photoconductor or a photovoltaic detector is a square law detector. Therefore, an alternative to the conventional way of viewing photoexcitation is that the semiconductor acts as a mixer element, beating the electric field vector against itself in a homodyne manner. Thus if two coherent sources of radiation having different frequencies (wavelengths) are superimposed upon a semiconductor, mixing action will occur. The resultant intensity will contain the four terms shown below:

$$E_t^2 = E_1^2 \cos^2(\omega_1 t) + E_2^2 \cos^2(\omega_2 t) + E_1 E_2 \cos[(\omega_1 - \omega_2)t]$$
$$+ E_1 E_2 \cos[(\omega_1 + \omega_2)t]. \qquad (2.38)$$

Here E_1 and E_2 are the amplitudes and ω_1 and ω_2 are the angular frequencies of the two waves. Electron lifetimes in semiconductors are orders of magnitude too large to permit photoexcitation at the sum frequency (of the order, say, of 10^{14} Hz). However, if the difference frequency is less than, say, 1 GHz, then

Table 2.5. Classification of wave interaction effects

1. Optical heterodyne detection
2. Optical parametric
2.1 Upconversion
2.2 Downconversion
3. Josephson junction photoeffect
4. Metal-metal oxide-metal photodiode

photoexcitation at the difference frequency can occur if the response time is sufficiently short, e.g., in the nanosecond range. This implies that the wavelengths of the two coherent sources are nearly equal.

An optical heterodyne detector consists of a conventional photoconductor or photovoltaic detector employed with a laser, termed the local oscillator, which irradiates the detector at a relatively intense level. The frequency (wavelength) of the local oscillator is so chosen that the difference frequency between it and the coherent signal source meets the requirement stated above. Both frequencies must, of course, produce photoexcitation. Thus the weaker coherent signal source, which is collinear with the local oscillator and absorbed within the same area of the semiconductor, will be mixed with the local oscillator, producing the difference frequency, see Fig. 2.11. It can be shown that the signal-to-noise ratio of an ideal optical heterodyne detector is independent of all material parameters except for quantum efficiency, i.e., no internal semiconductor noise is evident in the output. The limit to the minimum detectable signal power arises from the fluctuations in the stream of signal photons (signal fluctuation limit, see Sect. 2.4). This is several orders of magnitude better than that which can be achieved by conventional methods of incoherent detection employing photoconductivity or the photovoltaic effect, the ultimate limit to which is set (in the infrared) by fluctuations in the stream of background photons (background fluctuation limit, see Sect. 2.4).

Optical heterodyne detection has become of practical use for systems in which the signal source is a laser, for example, in optical communication systems [2.123] and laser radar [2.124]. *Teich* [2.125] and *Arams* et al. [2.126] have reviewed the theoretical basis and experimental results. *Keyes* and *Quist* [2.127] include a discussion of optical heterodyne detection in their review of coherent detection.

Parametric Effects

A purely optical analogue to optical heterodyne detection is the optical parametric effect, in which mixing of two coherent radiation beams in an optical crystal produces sum and difference frequencies. Parametric mixing is employed to increase the intensity of a weak signal and to change the optical frequency to a spectral interval where it is more easily detectable by a photocell having high sensitivity. Although both upconversion (sum frequency generation) and downconversion (difference frequency generation) can be achieved, parametric up-

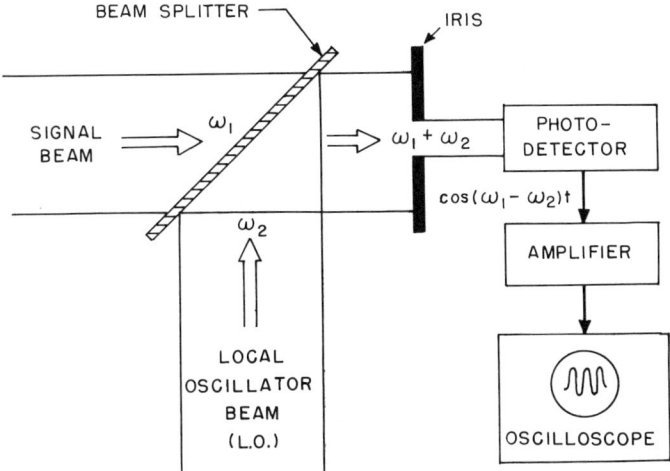

Fig. 2.11. The generalized optical heterodyne receiver (after *Teich* [2.122])

conversion is of greater interest. *Milton* [2.128] has prepared a detailed review of the field. Birefringent phase matching is employed in upconversion to meet the conditions conservation of energy and momentum by the two waves moving through the crystal. These conditions require that the sum frequency equals the sum of the reference (pump) and signal frequencies, and the sum wave vector equals the sum of the pump and signal wave vectors. It is also necessary that the interacting waves remain in phase for an appreciable distance throughout the crystal to achieve a reasonable increase in the signal. In birefringent materials the optical dispersion (dependence of refractive index upon wavelength) of the ordinary and extraordinary ray differs with crystal direction and polarization. By proper choice of material and crystal direction for given signal and pump frequencies, the phase matching condition can be achieved over an appreciable crystal volume, thereby providing a large increase in the intensity of the signal radiation.

Crystals useful for parametric upconversion should exhibit nonlinear optical properties, be birefringent, and be optically transparent at the frequencies of interest. Examples of crystals employed in the visible and near infrared are KDP (potassium dihydrogen phosphate) and $LiNbO_3$. Ag_3AsS_3, HgS, Ag_3SbS_3, and $ZnGeP_2$ have been used to upconvert 10.6 μm radiation into the visible, using a 1.06 μm pump [2.129–131].

An alternative to upconversion by birefringent crystals involves the use of phase matching in optical waveguides made from selected materials of which GaAs is an example [2.132]. The optical dispersion within the waveguide is employed to obtain broadband operation and long interaction lengths. Down-conversion has also been employed in GaAs to generate 100 μm radiation [2.133].

Other Wave Interaction Effects

In addition to optical heterodyne detection and parametric effects, other wave interaction effects include those occurring at Josephson junctions and metal-metal oxide-metal contacts.

Josephson Junction Photoeffect. The Josephson effect, which occurs in metal-insulator-metal thin films, metal-metal point contacts, and geometrical constrictions, in superconductors below the transition temperature, involves the flow of strongly correlated electron Cooper pairs across the barrier between the superconductors [2.134]. Both an ac and a dc Josephson current exist. Electromagnetic radiation whose energy is equal to the single particle energy gap of the Josephson junction can reduce the dc tunneling current by exciting Cooper pairs to the normal state. The radiation also affects the ac current by changing both its amplitude and frequency and changes the phase relationship of Cooper pairs on opposite sides of the junction [2.135].

In principle, Josephson junction detectors are useful only for energies corresponding to the single particle energy gap, i.e., wavelengths in the far infrared beyond 100 μm. It is theoretically possible to detect single photons by this means [2.136].

Metal-Metal Oxide-Metal Photodiode. *Javan* and coworkers have developed an uncooled infrared-sensitive point contact diode useful as a mixing element having a frequency response of 10^{14} Hz [2.137, 138]. It consists of a metal-metal oxide-metal structure in which electrons tunnel from one metal through the oxide to the other metal under the influence of the electric field vector of the incident radiation. Using a thin wire antenna and an oxide thickness the order of angstroms, alternating voltages at infrared frequencies have been developed when the diode is irradiated by an infrared laser. Thus the M–O–M structure serves as a mixer when irradiated by two lasers of differing wavelengths, generating sum and difference frequencies. Its utility as a sensitive infrared detector suffers from the problem of coupling the radiation to the diode, but its extremely high speed of response should find application as an uncooled heterodyne detector [2.139].

2.2 Noise in Radiation Detectors

All detectors are limited in the minimum radiant power which they can detect by some form of noise which may arise in the detector itself, in the radiant energy to which the detector responds, or in the electronic system following the detector. Careful electronic design including that of low noise amplification can reduce system noise below that in the output of the detector. That topic will not be treated here; see *Motchenbacher* and *Fitchen* [2.140] for details.

The other sources are radiation noise and noise internal to the detector. It is the objective of optimum detector design to reduce the internal noise of the detector to a level at which only the noise arising from the radiant energy to which the detector responds can be detected. This cannot be done in all cases. The topic of the radiant power limits, including background fluctuation and signal fluctuation noise, is treated in Section 2.4. This section will discuss noise mechanisms internal to the detector. Since most detectors of importance are made from semiconductors, the emphasis is upon noise in semiconductors, but shot noise in photoemissive devices will also be discussed.

As was true for that of photoeffects, the objective of this discussion of noise mechanisms is to acquaint the reader with the broad concepts of noise in detectors without deriving in great detail the appropriate equations. See *Van Vliet* [2.141] for a detailed treatment. Nevertheless, it will be necessary to present certain equations which describe the dependence of noise upon internal material parameters and external system parameters. The discussion will consider initially noise in semiconductor detectors, followed by noise in photoemissive devices.

2.2.1 Noise in Semiconductor Detectors

In the absence of electrical bias, the absolute minimum internal noise exists, termed Johnson noise, Nyquist noise or thermal noise. This form of noise arises from the random motion of the current carriers within any resistive material and is always associated with a dissipative mechanism. The Johnson noise power is dependent only upon the temperature of the material and the measurement bandwidth, although the noise voltage and current depend upon the value of the resistance.

Any other form of internally generated noise must depend upon bias. Since they add (quadratically) to Johnson noise, all other types of noise are referred to as excess noise. Three principal forms of excess noise exist. One amenable to analysis which is found in photoconductors is generation-recombination or $g-r$ noise. A second, also amenable to analysis, which is found in photodiodes, i.e., $p-n$ junctions and Schottky barrier diodes, is referred to as shot noise of diffusing carriers, or simply as shot noise. The third form of excess noise, not amenable to exact analysis, is called $1/f$ (one over f) noise because it exhibits a $1/f$ power law spectrum to a close approximation. It has also been called flicker noise, a term carried over from a similar power law form of noise in vacuum tubes.

The topics included here are limited to the usual types of noise in the common types of infrared photon detectors. Noise in thermal detectors, such as temperature noise in bolometers, is not included. Noise associated with the avalanche process is omitted. The detailed noise theory of phototransistors, an extension of shot noise in photodiodes, is not included. Modulation noise, an example of which arises from conductivity modulation by means of carrier trapping in slow surface states, is not included. Pattern noise, due to the nonuniform nature of the impurity distribution in either photoconductors or photovoltaic detectors, is omitted. The reader is referred to standard references on noise [2.141–146] for information on these processes.

Johnson Noise

Johnson noise is found in all resistive materials, including semiconductors. It occurs in the absence of electrical bias as a fluctuating voltage or current depending upon the method of measurement. There are several ways of analyzing the power spectrum, beyond the scope of this review. Among the more interesting is that which views Johnson noise as a one-dimensional form of black body radiation (see the summary by *Jones* [2.147]). The mean square Johnson noise power P_N in a measurement bandwidth B is given by

$$P_N = kTB, \tag{2.39}$$

where k is Boltzmann's constant and T is the absolute temperature of the sample. The open circuit noise voltage v_N and the short circuit noise current i_N are given by

$$v_N = (4kTRB)^{1/2}, \tag{2.40}$$

and

$$i_N = \left(\frac{4kTB}{R}\right)^{1/2}, \tag{2.41}$$

where R is the resistance of the sample being measured.

Generation-Recombination (G−R) Noise

G−R noise is caused by fluctuations in the rates of thermal generation and recombination of free carriers in a semiconductor, thereby giving rise to a fluctuation in the average carrier concentration. Thus the electrical resistance fluctuates; this is observed as a fluctuating voltage across the sample when a bias current flows through it. Some consider the term g−r noise to include background fluctuation noise also, since the latter is manifested as a fluctuation in carrier concentration arising from a fluctuation in the photoexcitation rate. In the discussion here, the term g−r noise will be the conventional usage in which the fluctuations are due to thermal processes only.

The analysis of g−r noise has been the subject of study for nearly 40 years. The most rapid development of the theory came during the period 1950–1960, principally by *Van der Ziel* [2.142, 143, 148], *Van Vliet* [2.149, 150], and *Burgess* [2.151–153]. *Long* [2.154] has clarified the application of the theory to infrared detectors. Many forms of the g−r noise expression exist, depending upon the internal properties of the semiconductor. Two of the most useful are those applicable to a simple one-level extrinsic semiconductor and to a near-intrinsic semiconductor.

Consider first the simple extrinsic photoconductor. Here the sample is a semiconductor containing a single impurity level, the source of the free electrons (or holes) present in the sample. Thus the fluctuation in the number of the free carriers arises from the fluctuation in the generation and recombination rates through that level. If it is assumed that the temperature is so low that very few of the extrinsic centers are thermally ionized (which is valid for most extrinsic cooled photoconductive infrared detectors), then the short circuit g−r noise current i_N and the open circuit g−r noise voltage v_N, which appear only in the presence of a bias current I_B, are given by

$$i_N = 2I_B \left[\frac{\tau B}{N_0(1+\omega^2\tau^2)} \right]^{1/2}, \tag{2.42}$$

and

$$v_N = i_N R = 2I_B R \left[\frac{\tau B}{N_0(1+\omega^2\tau^2)} \right]^{1/2}. \tag{2.43}$$

Here R is the sample resistance, τ is the lifetime of the free carriers, N_0 is the total number of free carriers in the sample, ω is the angular frequency, and B is the measurement bandwidth.

The second example of interest is that of the near intrinsic semiconductor, applicable to the intrinsic infrared photoconductor. Here the extrinsic levels are so shallow as to be completely ionized at the operating temperature. Thus the fluctuations in the free carrier concentration arise from intrinsic generation, i.e., production of free electron-hole pairs, by the thermal energy of the lattice. In the usual case the semiconductor is slightly n- or p-type, so that the numbers of free electrons and holes are not equal. The photoexcitation process is across the forbidden energy gap, i.e., the sample is an intrinsic photoconductor even though it is a slightly extrinsic semiconductor. In this case, the g−r short circuit noise current i_N and open circuit noise voltage v_N appearing in the presence of a bias current I_B are

$$i_N = 2I_B \left[\frac{(b+1)}{(bN+P)} \right] \left[\left(\frac{NP}{N+P} \right) \frac{\tau B}{(1+\omega^2\tau^2)} \right]^{1/2}, \tag{2.44}$$

and

$$v_N = i_N R = 2I_B R \left[\frac{(b+1)}{(bN+P)} \right] \left[\left(\frac{NP}{N+P} \right) \frac{\tau B}{(1+\omega^2\tau^2)} \right]^{1/2}. \tag{2.45}$$

Here b is the ratio of electron to hole mobility, N is the total number of free electrons in the sample, P is the total number of free holes in the sample, τ is the free carrier lifetime (assumed to be the same for electrons and holes), and B is the measurement bandwidth.

In some intrinsic photoconductors one type of thermally excited carrier is present in far greater numbers than the other. If electrons dominate, and if their mobility is greater than that of the holes (the usual case), then (2.44) and (2.45) reduce to

$$i_N = \frac{2I_B}{N} \left(\frac{P\tau B}{1+\omega^2\tau^2} \right)^{1/2}, \tag{2.46}$$

and

$$v_N = \frac{2I_B R}{N} \left(\frac{P\tau B}{1+\omega^2\tau^2} \right)^{1/2}. \tag{2.47}$$

Shot Noise

The third form of noise in semiconductors is shot noise, found in photodiodes under bias. The usual form of the diode equation is

$$I = I_0[\exp(qV/kT) - 1], \tag{2.48}$$

where I is the diode current, I_0 is the reverse bias saturation current, V is the applied voltage, q is the electronic charge, k is Boltzmann's constant, and T is the absolute temperature.

In general the shot noise current is given by

$$I_N = [(2qI + 4qI_0)B]^{1/2}, \tag{2.49}$$

where B is the measurement bandwidth.

The zero bias resistance, i.e., the slope of the $I-V$ characteristic at the origin, is

$$R = \frac{kT}{qI_0}. \tag{2.50}$$

Thus at zero bias, $I=0$ and (2.49) becomes

$$i_N = \left(\frac{4kTB}{R} \right)^{1/2}, \tag{2.50a}$$

which is recognized as the expression for the Johnson noise current. Under sufficient reverse bias, $I = -I_0$ and (2.49) becomes

$$i_N = (2qI_0 B)^{1/2}. \tag{2.51}$$

which is the form of the shot noise current most frequently employed.

1/f Power Law Noise

The other form of noise found in semiconductors is $1/f$ power law noise, characterized by a spectrum in which the noise power depends approximately inversely upon frequency. The general expression for the noise current is

$$i_N = \left(\frac{K_1 I_B^\alpha B}{f^\beta}\right)^{1/2}, \tag{2.52}$$

where K_1 is a proportionality factor, I_B is the bias current, B is the measurement bandwidth, f is the frequency, α is a constant whose value is about two, and β is a constant whose value is about unity.

Infrared detectors usually exhibit $1/f$ noise at low frequencies. At higher frequencies the amplitude drops below that of one of the types of white (frequency independent) noise, including Johnson, g–r, and shot noise. In general, $1/f$ noise appears to be associated with the presence of potential barriers at the contacts, interior, or surface of the semiconductor. In single crystal semiconductor samples having ohmic contacts the noise arises largely from the surface. Reduction of $1/f$ noise to an acceptable level is an art which depends greatly on the processes employed in preparing the contacts and surfaces.

Composite of Noise Sources in a Photoconductor

Figure 2.12 illustrates an idealized noise spectrum in a photoconductor in the absence of irradiation. At low frequencies $1/f$ noise dominates, at intermediate frequencies g–r noise dominates, and at high frequencies Johnson noise dominates. The transition points vary with semiconductor material, doping, and processing technology, but for infrared detectors they are very roughly near 1 kHz and 1 MHz. Thus the usual case for photoconductive infrared detectors in the frequency range of most interest is to be limited in their detection sensitivity (if they are not background fluctuation limited) by g–r noise.

2.2.2 Noise in Photoemissive Devices

Photoemissive devices exhibit shot noise in the emitted stream of electrons. The expression for the noise current in the absence of irradiation is similar to that for shot noise in a semiconductor:

$$i_N = (2qI_0 B)^{1/2}, \tag{2.53}$$

where q is the electronic charge and B is the measurement bandwidth. In a photoemissive device, I_0 represents the dark current from the photocathode. This can arise from thermionic emission, field emission, or anode-cathode leakage currents. Phototubes can also exhibit $1/f$ noise, termed flicker noise, having a spectrum similar to that of $1/f$ noise in semiconductors. See *Van der Ziel* [2.155] for additional details.

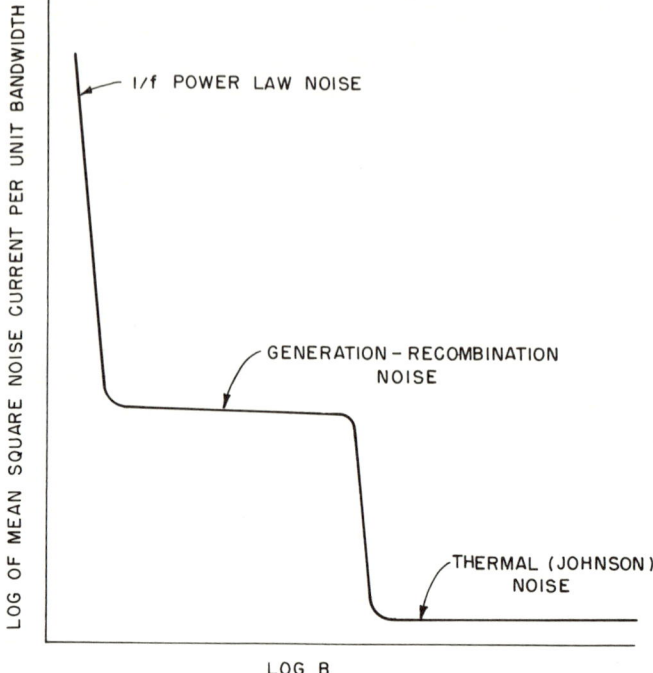

Fig. 2.12. Relative contributions of noise sources in a photoconductor

2.3. Figures of Merit

In order to properly compare the performance of infrared detectors it is necessary to define certain figures of merit which describe the signal and signal-to-noise ratio from the detector in terms of the wavelength and power of the incident radiation. The development of the appropriate figures of merit required much effort with rapid strides being made during the 1950s, especially through the efforts of *Jones* [2.147]. This section contains a review of these figures of merit. See *Wolfe* [2.156] for more details.

2.3.1 Spectral Response

Probably the most basic of all figures of merit is the spectral response, also known as the relative response, which describes in a relative sense the manner by which a detector changes its output signal in response to changes in wavelength of the input signal. It is important to note whether the spectral response is in reference to unit rate of arrival of photons per unit wavelength interval or unit incident radiant power per unit wavelength interval. The latter is the more common convention for infrared detectors, whereas the former seems to be more widely employed for optical detectors.

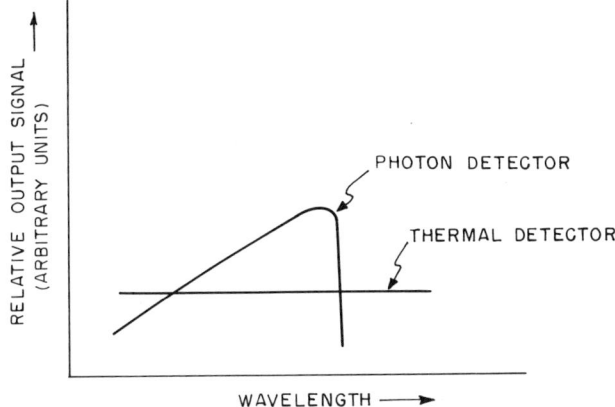

Fig. 2.13. Ideal spectral responses of photon and thermal detectors for unit radiant power per unit wavelength interval

Photon detectors ideally exhibit a relative response per unit radiant power per unit wavelength interval which rises linearly with increasing wavelength, peaks, and then drops off sharply to zero. This is because their response is generally proportional to photon arrival rate for energies greater than (wavelengths less than) the semiconductor absorption edge (see Sect. 2.1.1). For equal radiant power per unit wavelength interval the photon rate is proportional to wavelength. On the other hand, thermal detectors, which respond to the radiant power rather than the photon rate, exhibit ideally a spectral response to unit radiant power per unit wavelength interval which is wavelength independent. Idealized representations of these two types of response are illustrated in Fig. 2.13.

2.3.2 Responsivity

The responsivity is a measure of the dependence of the signal output of a detector upon the input radiant power. It is important to specify the nature of the source. The convention for infrared detectors is to speak of a black body responsivity and a spectral responsivity. The black body reference temperature is usually 500 K. The output may be either a voltage or current signal. Thus the usual units of responsivity for infrared detectors are amps/watt or volts/watt. In order to identify the reference source, the usual notation for the responsivity of infrared detectors is of the form $R(T,f)$ or $R(\lambda,f)$. Here $R(T,f)$ represents the output signal voltage (or current) in response to input radiation from a black body at absolute temperature T, with the signal measured at frequency f in response to modulation of the reference source at frequency f. Similarly, $R(\lambda,f)$ represents the output signal measured at frequency f in response to monochromatic radiation of wavelength λ modulated at frequency f.

Thus, if v_S is the rms signal voltage at the output of a detector of area A measured at frequency f in response to incident radiation of rms power P and irradiance H modulated at frequency f from a black body of temperature T, the black body responsivity is

$$R(T, f) = \frac{v_S}{P} = \frac{v_S}{HA}. \tag{2.54}$$

Although responsivity is also a valid figure of merit for detectors operating in the visible spectrum, the units are sometimes different. Since the lumen is a standard unit of visible radiant power, it is common practice to measure the responsivity of detectors such as photomultipliers in units of amps/lumen. Again, either a spectral or a black body reference can be employed. A useful black body reference temperature for visible detectors is 2870 K, the temperature at which the peak emission is at 1 μm.

2.3.3 D*

Because the performance of infrared detectors is limited by noise, it is important to be able to specify a signal-to-noise ratio in response to incident radiant power. An area-independent figure of merit is D^* ("dee-star") defined as the rms signal-to-noise ratio in a 1 Hz bandwidth per unit rms incident radiant power per square root of detector area. D^* can be defined in response to a monochromatic radiation source or in response to a black body source. In the former case it is known as the spectral D^*, symbolized by $D_\lambda^*(\lambda, f, 1)$ where λ is the source wavelength, f is the modulation frequency, and 1 represents the 1 Hz bandwidth. Similarly, the black body D^* is symbolized by $D^*(T, f, 1)$, where T is the temperature of the reference black body, usually 500 K. Unless otherwise stated, it is assumed that the detector field of view is hemispherical (2π ster). The units of D^* are cm Hz$^{1/2}$/watt. The relationship between D_λ^* measured at the wavelength of peak response and $D^*(500 \text{ K})$ for an ideal photon detector is illustrated in Fig. 2.14. For an ideal thermal detector, $D_\lambda^* = D^*(T)$ at all wavelengths and temperatures.

In terms of measurement parameters, D^* is given by

$$D^* = \frac{(A_D B)^{1/2}}{P} \left(\frac{v_S}{v_N}\right), \tag{2.55}$$

where A_D is the detector area in cm^2, B is the electrical bandwidth in Hz, (v_S/v_N) is the rms signal-to-noise voltage ratio (the noise measured in the bandwidth B) and P is the incident radiant power. [When measuring current instead of voltage, (v_S/v_N) is replaced by (i_S/i_N), the rms signal-to-noise current ratio.] Whether D^* is monochromatic or black body depends on whether the source is monochromatic or a black body.

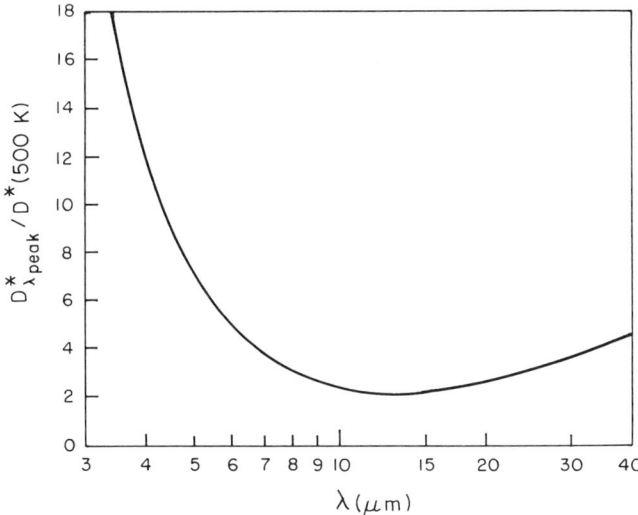

Fig. 2.14. Ratio of D_λ^* at wavelength of peak response (assumed to be long wavelength limit) to D^* (500 K) as a function of long wavelength limit for an ideal photon detector

2.3.4 D**

Jones [2.157] has defined D^{**} ("dee-double-star"), a figure of merit appropriate to background limited detectors, which removes the need to specify the field of view when listing D^*. As discussed in Section 2.4, the idealized dependence of D^* upon angular subtense of the emitting background is given by

$$D^*(\theta) = \frac{D^*(2\pi)}{\sin \theta}, \tag{2.56}$$

where $D^*(\theta)$ is the D^* value obtained when the aperture through which the radiation reaches the detector has a half angle of θ, and $D^*(2\pi)$ is the D^* value obtained when the radiation reaches the detector from the entire forward hemisphere. D^{**} is related to D^* through

$$D^{**} = D^* \sin \theta \tag{2.57}$$

for an ideal detector. Thus for a hemispherical field of view, $D^{**} = D^*$. The units of D^{**} are $\text{cm} \, \text{Hz}^{1/2} \, \text{ster}^{1/2}/\text{watt}$.

2.3.5 Noise Equivalent Power

A figure of merit which has been largely superseded by D^* for infrared detectors, but still finds application in optical detectors, is the noise equivalent power, or NEP, frequently symbolized as P_N. The NEP is the rms incident radiant power which gives rise to an rms signal voltage (or current) equal to the rms noise voltage (or current). The reference bandwidth must be specified; so too must the detector area and the field of view. Thus the NEP is related to D^* by

$$\text{NEP} = \frac{(A_D B)^{1/2}}{D^*}. \tag{2.58}$$

As is true for responsivity and detectivity, the NEP can be either monochromatic (spectral) or black body, depending upon the reference source. The units of NEP are watts.

2.3.6 Detectivity

Detectivity is defined as the signal-to-noise ratio per unit radiant power. Thus it is the reciprocal of the noise equivalent power, having units of (watts)$^{-1}$. Confusion has arisen over the term detectivity, which has also been applied to D^*. The above definition will be employed herein.

2.3.7 Frequency Response, Response Time, Time Constant, f^*

These figures of merit relate to the manner by which a detector responds to modulated radiation. As the name implies, the frequency response is a plot of signal output (voltage or current) as a function of modulation frequency of the input radiation. Many detectors exhibit an idealized "flat" response from zero frequency to a value in, say, the MHz range, above which the signal power "rolls off" at 6 db per octave, i.e., the signal voltage is inversely proportional to frequency. The equation governing this response is

$$v_s = \frac{v_{so}}{[1+(2\pi f\tau)^2]^{1/2}}, \tag{2.59}$$

where v_{so} is the signal voltage at zero frequency, f is the frequency, and τ is the response time, also known as the time constant. It can be seen that τ is given by

$$\tau = \frac{1}{2\pi f_{3\,db}}, \tag{2.60}$$

where $f_{3\,db}$ is the frequency at which the signal power is 3 db below the value at zero frequency, i.e., the voltage is 0.707 that of v_{so}.

An alternative definition of τ is sometimes employed to describe the response to pulsed radiation. If the rise and decay times of the radiation pulse are so short that the pulse appears as a square wave to the detector, then the response time τ is the time required for the photosignal to rise to $(1-e^{-1})$, i.e. 63% of the peak value, or to decay to e^{-1}, i.e. 37% of the peak value.

For many photon detectors, D^* is independent of modulation frequency at low frequencies, but above a critical frequency it becomes inversely proportional to frequency. *Borrello* [2.158] has termed this corner frequency f^*. He shows that the D^*f^* product is a constant whose value depends upon the photon capture cross section and minimum energy required to absorb a photon in a given material. The D^*f^* product has been found to be useful in evaluating infrared photon detectors.

2.3.8 Noise Spectrum

In many cases it is useful to determine how the detector noise depends upon frequency. A plot of noise voltage (or current) as a function of frequency is known as the noise spectrum. Because the various types of detector noise exhibit different dependencies upon frequency, measurement of the noise spectrum is useful in determining the mechanism giving rise to the noise. Furthermore, the noise spectrum and frequency response (i.e., signal spectrum) can be used to calculate the dependence of signal-to-noise ratio on frequency, thereby determining the dependence of D^* on frequency.

2.4 The Signal Fluctuation and Background Fluctuation Limits

The ultimate performance of infrared and optical detectors is reached when there is no amplifier noise, no noise generated within the detector itself, and there is no radiating background against which the signal must be detected. Under these conditions, the only events produced within the detector are due to signal photons. This limit is known as the signal fluctuation limit, since the random nature of the rate of arrival of signal photons in a given observation time determines the minimum power which can be detected.

The practical operating limit for most infrared detectors is not the signal fluctuation limit but the background fluctuation limit, also known as the photon noise limit and the BLIP (Background Limited Infrared Photodetector) limit. The source to be detected is surrounded by a radiating background. The radiation falling on the detector is a composite of that from the target and that from the background. If the background radiation were not fluctuating, the signal due to it could be subtracted electrically from the sum of the background and target radiation. However, the background level fluctuates about the average value due to the random nature of the photon emission. Thus the signal radiation must be compared to the fluctuations, or noise, in the background radiation.

The radiating background which is most commonly encountered is that arising from objects at an ambient temperature of 295 K (also listed as 290 K or 300 K). This characterizes the earth itself, structures, most tactical targets, and even some of the internal surfaces of the detector housing. The thermal emission from a 295 K body peaks at about 10 µm; one-fourth of the energy is below that wavelength and three-fourths above. Thus under most operating conditions the background fluctuation limit is operative for infrared detectors, whereas the signal fluctuation limit is operative for ultraviolet and visible detectors. It will be seen that by employing optical heterodyne detection it is possible to achieve the signal fluctuation limit with infrared detectors even in the presence of ambient temperature backgrounds.

The analysis of the signal and background limits which follows, while detailed, is not exhaustive. Since thermal detectors respond to radiant power rather than photon arrival rate, the signal and background fluctuation limits derived below do not apply to them. Rather, it is necessary to repeat the derivations in terms of power rather than photon rate. An analysis of the background fluctuation limit for thermal detectors is found elsewhere [2.159].

2.4.1 Signal Fluctuation Limit

Consider a photodiode viewing a monochromatic source of wavelength λ. The short circuit signal current i_{SO} is given by [see (2.18)]

$$i_{SO} = \eta q N_\lambda, \tag{2.61}$$

where η is the quantum efficiency, q is the electronic charge, and N_λ is the number of absorbed photons per second of wavelength λ which give rise to signal events. If all the internal noise is negligible, and the background radiation falling on the detector is negligible, then the detector output short circuit noise current i_N will be simply the shot noise in the signal [see (2.51)], i.e.,

$$i_N = [2q i_{SO} B]^{1/2}, \tag{2.62}$$

where B is the measurement bandwidth. Therefore, the signal-to-noise current ratio is

$$\frac{i_{SO}}{i_N} = \left(\frac{\eta N_\lambda}{2B}\right)^{1/2}. \tag{2.63}$$

But N_λ is given by [see (2.10)]

$$N_\lambda = \frac{P_\lambda \lambda}{hc}, \tag{2.64}$$

where P_λ is the incident power of wavelength λ, h is Planck's constant, and c is the speed of light. Thus

$$P_\lambda = \frac{2hcB}{\lambda \eta} \left(\frac{i_{SO}}{i_N}\right)^2. \tag{2.65}$$

Equation (2.65) implies that if the signal-to-noise power ratio is unity, then the minimum detectable power is just the energy of a photon multiplied by twice the measurement bandwidth and divided by the quantum efficiency. For any real situation, however, the observer can wait only a finite length of time for a photon to be detected. There must be a reasonably high probability that for a given average rate of arrival, at least one photon will arrive during the observation period. Since the threshold value involves low numbers of photons per observation interval, the Poisson density function must be employed to describe the detection probability, i.e.,

$$p(N) = \frac{(\bar{N})^N \exp(-\bar{N})}{N!}, \tag{2.66}$$

where $p(N)$ is the probability of N photons being detected in a given observation time in which the average number of photons detected is $\bar{N}$.

In a practical sense, a value must be set for the probability of detecting a photon in a given observation interval. Let it be that if a source is emitting $\bar{N}$ photons per observation interval which reach the detector, the probability of one or more being detected in that interval must be at least 0.99. This is equivalent to saying that there is a one percent probability that no photons will be detected. Thus

$$p(0) = \exp(-\bar{N}) = 0.01, \tag{2.67}$$

or $\bar{N} = 4.61$. In other words, the average number of photons detected per observation interval must be at least 4.61 if the probability of detecting no photons in any given observation interval is to be less than one percent.

This result can be expressed in terms of a minimum detectable power. Since the energy per photon is hc/λ, the minimum detectable power $P_{\lambda,\min}$ required in order that there be at least a 99% probability of detecting a photon in an observation time τ_0 is

$$P_{\lambda,\min} = \frac{4.61 hc}{\tau_0 \lambda \eta}, \tag{2.68}$$

where η is the quantum efficiency for detection. To illustrate the magnitude of this, if $\lambda = 1\,\mu\text{m}$, $\eta = 1$, and the observation time is 1 s then $P_{\lambda,\min} = 9.15 \times 10^{-19}$ watts.

Electrical measurements are expressed in terms of modulation bandwidth rather than observation times. It can be shown that the bandwidth B corresponding to an observation time τ_0 is the reciprocal of $2\tau_0$. Thus,

$$P_{\lambda,\min} = 9.22 \frac{hcB}{\eta \lambda} \tag{2.69}$$

is the minimum power required to achieve a 99% probability that a photon will be detected. Note that the detector area does not enter into the expression for

Table 2.6. Signal fluctuation limit for 500° K black body source, 1 Hz bandwidth, and unit quantum efficiency in terms of long wavelength limit λ_0

λ_0 [μm]	$P_{T,\min}$ [watts]
1.0	1.0×10^{-9}
1.5	2.0×10^{-13}
2.0	2.8×10^{-15}
3.0	5.0×10^{-17}
4.0	7.8×10^{-18}
10.0	4.4×10^{-19}
∞	1.7×10^{-19}

$P_{\lambda,\min}$ and that $P_{\lambda,\min}$ depends linearly upon the bandwidth, which differs from the case in which the detection limit is set by internal or background noise.[2]

When the emitting source is a black body rather than monochromatic, the minimum detectable power $P_{T,\min}$ is given by (see [2.159])

$$P_{T,\min} = \frac{9.22B}{\eta} \frac{\int_{v_0}^{\infty} M(v, T_S) dv}{\int_{v_0}^{\infty} \frac{M(v, T_S) dv}{hv}}, \quad (2.70)$$

where it has been assumed that η is independent of wavelength between v_0 and infinity. Here $M(v, T_S)$ is the Planck distribution function expressed in terms of the optical frequency $v = c/\lambda$ and the source temperature T_S. The frequency v_0 corresponds to the detector long wavelength limit λ_0. $M(v, T_S)$ is given by

$$M(v, T_S) = \frac{2\pi h v^3}{c^2 [\exp(hv/kT_S) - 1]}. \quad (2.71)$$

To indicate the magnitude of the power, Table 2.6 illustrates $P_{T,\min}$ as a function of λ_0 for $T_S = 500$ K, $\eta = 1$, and $B = 1$ Hz.

2.4.2 Background Fluctuation Limit

Dependence of D^* Upon Wavelength and Background Temperature

The background limit for a photon detector depends upon the spectral distribution of the target, the spectral distribution of the background, the spectral response of the detector, the temperature of the detector, the mode of

---

[2] By assuming the signal to be sinusoidally modulated, and ignoring the argument concerning the probability of detecting at least one photon per counting time interval, *Seib* and *Aukerman* [2.19] derived an expression for the signal fluctuation limit identical to (2.69) except that the multiplicative constant is not 9.22 but 2.83 for an ideal photoemissive or photovoltaic detector and 5.66 for a photoconductor. A similar expression can be derived for the ultimate performance of an optical heterodyne detector, see for example *Teich* [2.125], but the multiplicative constant is unity for photoemissive and reverse-biased photodiode detectors, and two for photoconductors and photovoltaic detectors. This difference in the constant arises from the differing assumptions as to the manner in which the detector is employed and the minimum detectable signal-to-noise ratio.

operation of the detector, and the field of view within which the detector receives background radiation. Consider first that the source is monochromatic. To determine D_λ^* the rms fluctuations in the rate of arrival of detectable background photons must be equated to the average rate of arrival of signal photons. The average rate of arrival of background photons of frequency v at a detector of sensitive area A is given by $\bar{N}_B A$, where

$$\bar{N}_B A = \frac{M(v, T_B)}{hv} A dv. \qquad (2.72)$$

Here T_B is the temperature of the background and $M(v, T_B)$ is the Planck distribution function in terms of T_B instead of T_S.

The variance σ^2 in $\bar{N}_B$ is

$$\sigma^2 = \bar{N}_B \frac{\exp(hv/kT_B)}{\exp(hv/kT_B) - 1}. \qquad (2.73)$$

The modulation frequency dependence of the mean square fluctuations in the rate of generation of current carriers due to the arrival of detectable photons is

$$P_N(f) = A \int_{v_0}^\infty \eta(v) \sigma^2 dv = A \int_{v_0}^\infty \eta(v) \frac{M(v, T_B) \exp(hv/kT_B)}{hv[\exp(hv/kT_B) - 1]} dv, \qquad (2.74)$$

where $v_0 = hc/\lambda_0$ is the optical frequency corresponding to the long wavelength limit λ_0 of the detector and $\eta(v)$ is the quantum efficiency for photons of frequency v.

The rate of generation of carriers excited by photons from a monochromatic source of power $P_{\lambda,S}$ incident on the detector is given by

$$N_S = \frac{\eta(v_S) P_{\lambda,S}}{hv_S}, \qquad (2.75)$$

where $\eta(v_S)$ is the quantum efficiency for photons of frequency v_S.

The rms fluctuations in the bandwidth B are given by the square root of the quantity $2B$ times the frequency dependence of the mean square fluctuations. Thus, for signal equal to noise,

$$N_S = [2B P_N(f)]^{1/2}. \qquad (2.76)$$

Combining (2.72), (2.73), (2.74), and (2.75) gives

$$P_{S,\min} = \frac{hv_S}{\eta(v_S)} \left\{ 2AB \int_{v_0}^\infty \frac{\eta(v) 2\pi v^2 \exp(hv/kT_B) dv}{c^2 [\exp(hv/kT_B) - 1]^2} \right\}^{1/2}. \qquad (2.77)$$

The value of D^* is found by dividing $P_{S,\min}$ into $(AB)^{1/2}$.

$$D^*_\lambda = \frac{\eta(v_S)}{hv_S \left\{ \int_{v_0}^\infty \frac{4\pi\eta(v)v^2 \exp(hv/kT_B)dv}{c^2[\exp(hv/kT_B)-1]^2} \right\}^{1/2}}. \tag{2.78}$$

If $hv_S/kT \gg 1$, then (2.77) and (2.78) simplify to

$$P_{S,\min} = \frac{hv_S(2ABN_B)^{1/2}}{\eta(v_S)} = \frac{hc(2ABN_B)^{1/2}}{\lambda_S \eta(v_S)} \tag{2.79}$$

and

$$D^*_\lambda = \frac{\eta(v_S)}{hv_S(2N_B)^{1/2}}. \tag{2.80}$$

The value of D^*_λ as a function of detector long wavelength limit, evaluated from (2.78), is illustrated in Fig. 2.15, from *Jacobs* and *Sargent* [2.160].

It can be shown (e.g., see [2.159]) that when the signal source is a black body at temperature T_S, and the radiating background is a black body at temperature T_B, then the background noise limited black body D^* is

$$D^*(T_S) = \frac{\int_{v_0}^\infty \frac{\eta(v)M(v,T_S)dv}{hv}}{\sigma T_S^4 \left\{ \int_{v_0}^\infty \frac{4\pi\eta(v)v^2 \exp(hv/kT_B)dv}{c^2[\exp(hv/kT_B)-1]^2} \right\}^{1/2}}. \tag{2.81}$$

Again, if the approximation $hv_0/kT \gg 1$ is valid, where v_0 is the frequency corresponding to the detector long wavelength limit λ_0, then

$$D^*(T_S) = \frac{G\eta(v_S)}{hv_S(2N_B)^{1/2}}, \tag{2.82}$$

where G is defined as

$$G = \frac{hv_S}{\sigma T_S^4 \eta(v_S)} \int_{v_0}^\infty \frac{\eta(v)M(v,T_S)dv}{hv}. \tag{2.83}$$

The value of G as a function of detector long wavelength limit for three representative background temperatures is illustrated in Fig. 2.16.

Dependence Upon Other Parameters

The calculations in the previous subsection assumed that the noise in the detector arose from fluctuations in the generation rate of carriers corresponding to fluctuations in the background radiation arriving at the detector from a

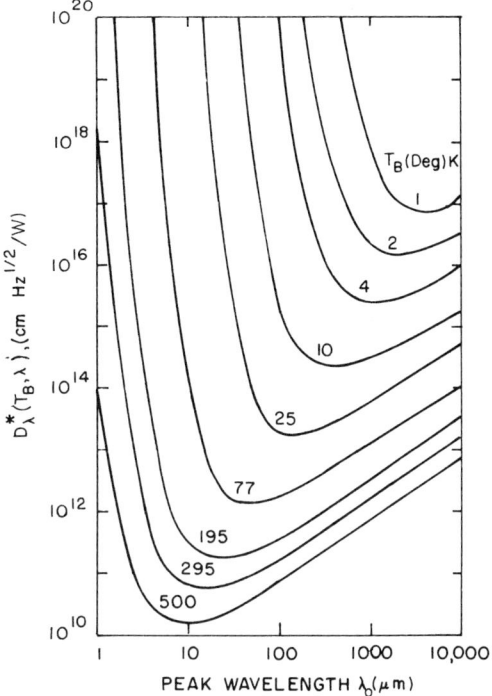

Fig. 2.15. Photon noise limited D_λ^* at peak wavelength (assumed to be cutoff wavelength), for a photovoltaic detector for selected background temperatures T_B. Values for a cooled photoconductive detector are 0.71 times those shown. (Assumes 2π steradian field of view and $\eta = 1$) (after *Jacobs* and *Sargent* [2.160])

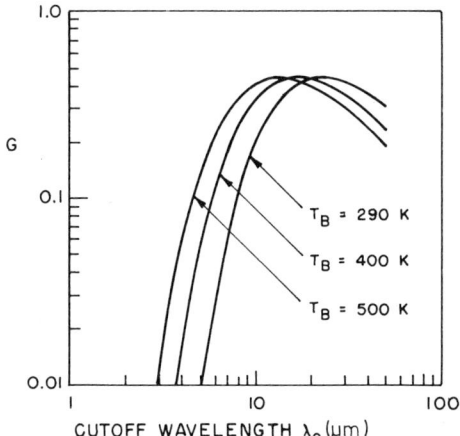

Fig. 2.16. Dependence of the function G upon detector cutoff wavelength λ_0 for black body source temperatures of 290 K, 400 K, and 500 K and field of view of 2π steradians (after *Kruse* et al. [2.3, p. 363])

hemispherical field of view. Neither the operating temperature of the detector nor the mode of operation influenced the calculation. This subsection is concerned with the dependence of the background limited D^* upon operating temperature, operating mode, and field of view. It shall be seen that it can be lower by a factor no greater than two, depending upon mode of operation and temperature, and can be higher by orders of magnitude at reduced fields of view.

Mode of Operation. The calculations in the previous section assumed that only fluctuations in the carrier generation rate due to fluctuations in the rate of arrival of background photons caused fluctuations in the signal from a detector. This is true for photoemissive and photovoltaic detectors, where the signal is developed as the excited carriers move across a boundary at the photocathode or the $p-n$ junction. In the photoconductive effect, however, the signal arises from measuring the change in conductance, which is proportional to the change in carrier concentration. Fluctuations in both the generation and the recombination rates cause fluctuations in the instantaneous value of the carrier concentration. It can be shown that the noise power due to recombination fluctuations must equal that due to generation fluctuations; i.e., $P_N(f)$ of (2.74) must be doubled. Since $P_N(f)$ enters as the square root into the denominator of the D^* expression, the background limited D_λ^* and $D^*(T_S)$ are reduced by the square root of two for photoconductors.

Operating Temperature. The calculations in the previous sections assumed that only fluctuations in the rate of arrival of photons from the forward hemisphere were important. This is evidenced by the employment of $M(v, T_B)$, which applies to a hemisphere. If the sensitive element of the detector is at the same temperature as the background, it will receive radiation not only from the forward hemisphere but from the reverse as well. Even though the back side of the element is mounted on a substrate, radiation will enter either through or from the substrate. Whether or not this is important is determined by whether the detector responds only to radiation incident on the front surface. In most photovoltaic detectors the back surface is much farther from the junction than the sum of the optical absorption depth and the carrier diffusion length. Thus most photovoltaic detectors have a preferred surface, and the background limit does not depend upon the mode of operation.

On the other hand, photoconductive devices in principle respond only to the number of photoexcited carriers, regardless of where they are generated within the material. Thus they will receive equal contributions of background noise from both hemispheres if both are at the same temperature. This will cause another reduction of D_λ^* and $D^*(T_S)$ by the square root of two. Photoemissive detectors having translucent photocathodes will be sensitive also to radiation from both hemispheres. Those with opaque photocathodes will not.

By cooling the detector element substantially below the background temperature, the radiation from the back hemisphere can be ignored. Thus the value of $D^*(T_S)$ and D_λ^* need not be reduced for the photoconductive and translucent photoemissive detectors when the operating temperature is well below ambient. Cooling to 77 K is sufficient for a 295 K background limit.

Combined Effects of Mode of Operation and Operating Temperature—The expressions for D_λ^* and $D^*(T_S)$ given in (2.78) and (2.81) apply directly to photovoltaic detectors, and to photoemissive detectors having opaque photocathodes.

The Photon Detection Process 55

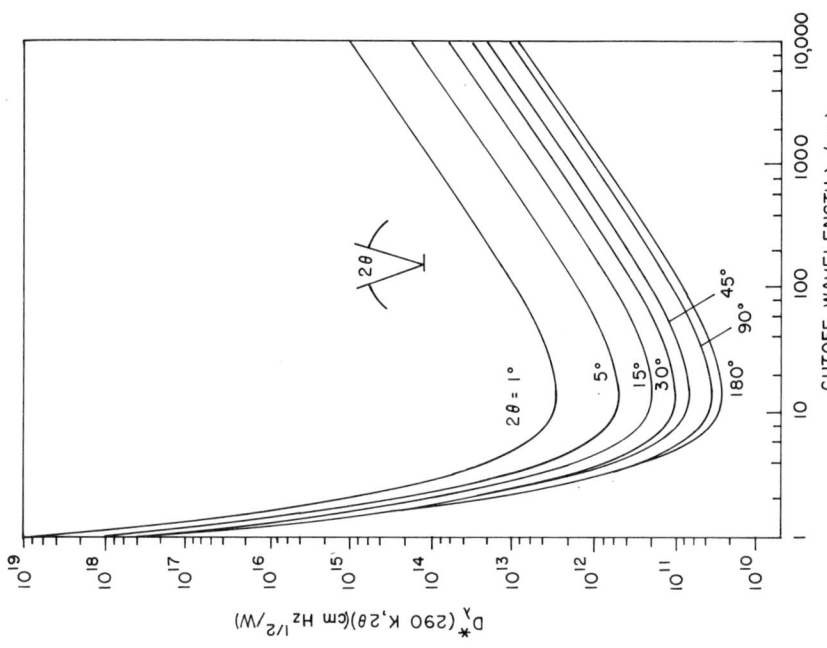

Fig. 2.18. Dependence of D_λ^* upon λ and field of view for a background limited photoconductor. Background temperature is 290 K

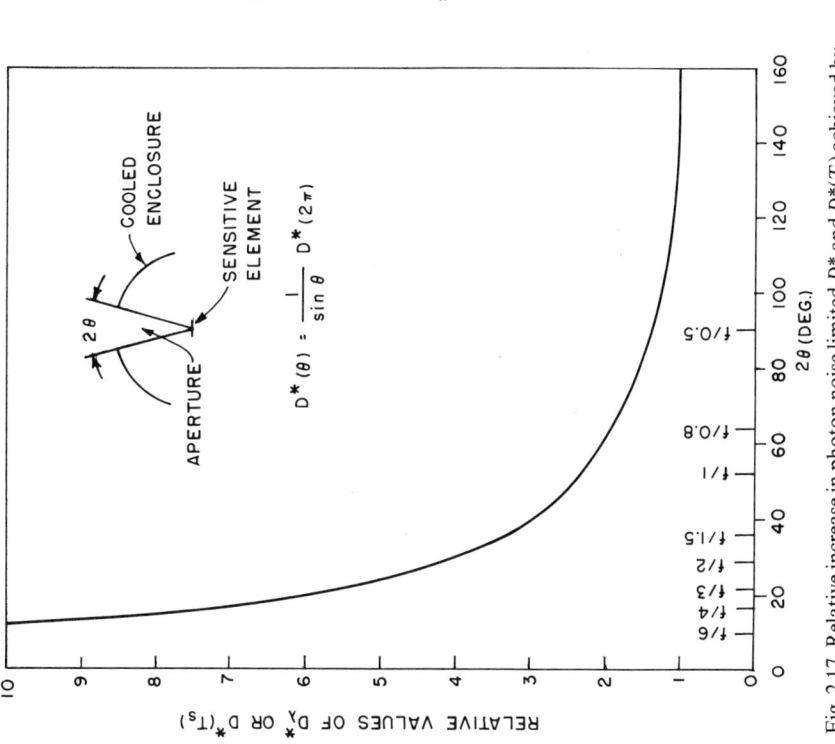

Fig. 2.17. Relative increase in photon noise limited D_λ^* and $D^*(T)$ achieved by using cooled aperture in front of detector (after Kruse et al. [2.3, p. 36])

— The expressions must be reduced by a factor of two for photoconductors operating at the background temperature, and by the square root of two for photoconductors which are cooled significantly below the background temperature.

— The expressions apply to a cooled photoemissive detector employing a translucent photocathode, but must be reduced by the square root of two if the operating temperature equals the background temperature.

Field of View. The background limited values of D^* and $D^*(T_s)$ expressed in (2.78) and (2.81) were based upon an optical acceptance angle of 2π steradians, otherwise known as a hemispherical field of view. The background noise was established by the fluctuations in the rate of arrival of photons from all angles within this field of view. If the field of view is reduced by means of a cooled aperture, the background noise will be reduced. Thus the values of D^* and $D^*(T_s)$ depend upon the field of view. It can be shown [2.159] that this dependence is given by

$$D_\lambda^*(\theta) = \frac{D_\lambda^*(2\pi)}{\sin\theta},\qquad(2.84)$$

and

$$D^*(T_s, \theta) = \frac{D^*(T_s, 2\pi)}{\sin\theta},\qquad(2.85)$$

where θ is the half-angle describing the field of view. Note that the definition of D^{**} (see Subsect. 2.3.4) removes the need to specify the field of view for a detector which is background limited at that field of view.

Figure 2.17 illustrates the improvement in D_λ^* and $D_\lambda^*(T_s)$ obtained by reducing the field of view. Figure 2.18 shows D_λ^* as a function of λ_0 and field of view for a background limited cooled photoconductor. The background temperature is 290 K.

2.4.3 Composite Signal Fluctuation and Background Fluctuation Limits

It is instructive to determine the composite signal fluctuation and background fluctuation limits. Because of their differing dependencies upon area and bandwidth, it is necessary to specify these values and calculate a noise equivalent power for each limit. Thus Fig. 2.19 illustrates the spectral noise equivalent power over the wavelength range from 0.1 µm to 20 µm assuming a background temperature of 290 K, detector areas of 1 cm² and 1 mm² (applicable only to the background fluctuation limit), a 2π steradian field of view (applicable only to the background fluctuation limit), and bandwidths of 1 Hz and 10^4 Hz.

The intersections of three pairs of curves for which the bandwidths of the signal and background fluctuation limits are equal are emphasized in Fig. 2.19. Note that all lie between 1.0 µm and 1.5 µm. To illustrate the composite, that for an area of 1 cm² (applicable to the background fluctuation limit) and a

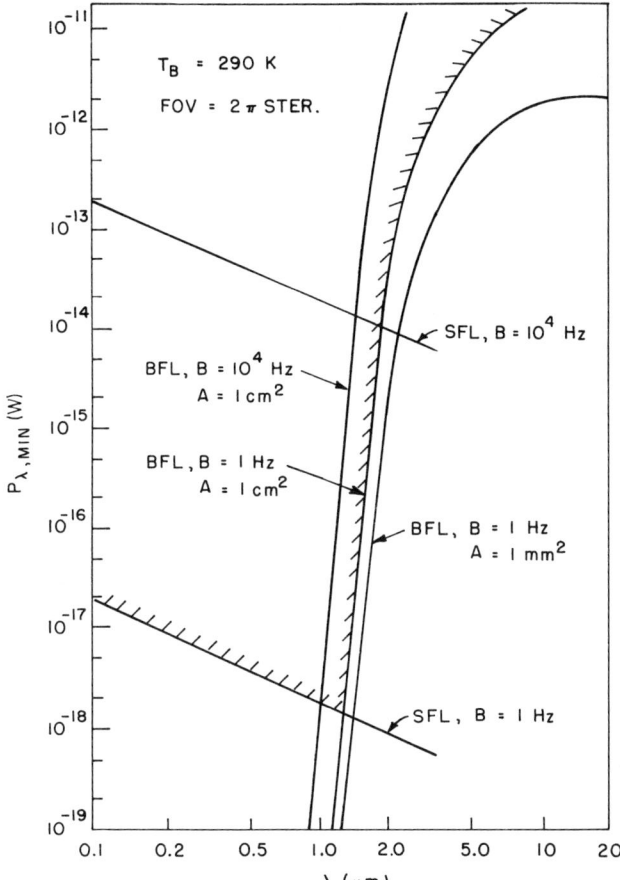

Fig. 2.19. Minimum detectable monochromatic power as a function of wavelength for composite of signal fluctuation limit (SFL) and background fluctuation limit (BFL) for two detector areas and electrical bandwidths. Background temperature is 290 K and field of view is 2π steradians. Detector long wavelength limit is assumed equal to source wavelength

bandwidth of 1 Hz has been scored. For this pair, at wavelengths below 1.2 μm the signal fluctuation limit dominates; the converse is true above 1.2 μm. The minimum detectable monochromatic radiant power at 1.2 μm is 1.5×10^{-18} watts in a 1 Hz bandwidth. Below 1.2 μm the wavelength dependence is small. Above 1.2 μm it is very large, due to the steep dependence upon wavelength of the short wavelength end of the 290 K background spectral distribution.

Figure 2.20 illustrates the minimum detectable power from a 500 K black body as a function of wavelength, illustrating the composite of the signal fluctuation and background fluctuation limits. The same values of the param-

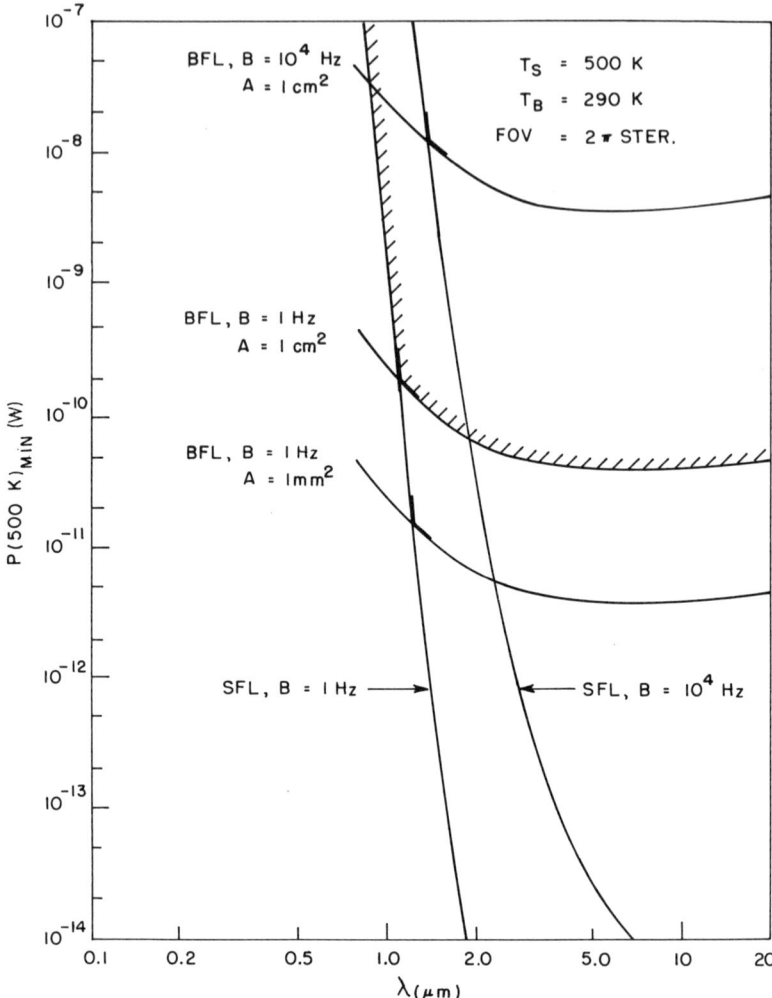

Fig. 2.20. Minimum detectable power from a 500 K black body as a function of detector long wavelength limit set by signal fluctuation limit (SFL) and background fluctuation limit (BFL) for two detector areas and electrical bandwidths. Background temperature is 290 K and field of view is 2π steradians

eters have been employed as in Fig. 2.19. Again the intersections have been emphasized, and the composite for an area of 1 cm² and a bandwidth of 1 Hz has been scored. In this case also the intersections lie between 1.0 µm and 1.5 µm. For the scored composite, signal fluctuations limit the detectability of a 500 K body against a 290 K background for wavelengths less than 1.2 µm; for longer wavelengths, background fluctuations impose the limit.

2.5 State-of-the-Art of Infrared and Optical Detectors

To complete the review of the photon detection process, this section presents the state-of-the-art of optical and infrared detectors. Figure 2.21 after *Seib* and *Aukerman* [2.19] shows the spectral detectivity of optical detectors responding in the 0.1 to 1.2 μm region. Note that this is not D^*, but rather the reciprocal of

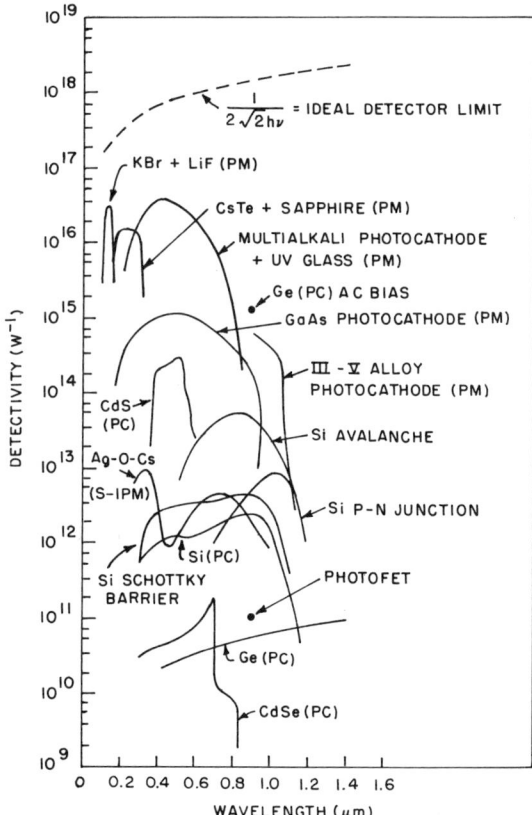

Fig. 2.21. Representative detectivity vs wavelength values of $0.1-1.0\,\mu m$ photodetectors. PC indicates a photoconductive detector and PM indicates a photomultiplier. Detector areas are given in Table 2.7. Bandwidth is 1 Hz (after *Seib* and *Aukerman* [2.19])

Table 2.7. Areas of detectors illustrates in Fig. 2.21

Detector	Area [cm^2]
CdS photoconductor (PC)	1
CdSe photoconductor (PC)	1
Si Schottky barrier photodiode	0.03
Si $p-n$ junction photodiode	0.25
Si photoconductor	0.25
Si avalanche photodiode	0.07
Ge photoconductor (PC)	0.20
Ge ac bias photoconductor (PC)	2.4×10^{-5}
Photomultipliers (PM)	1.0

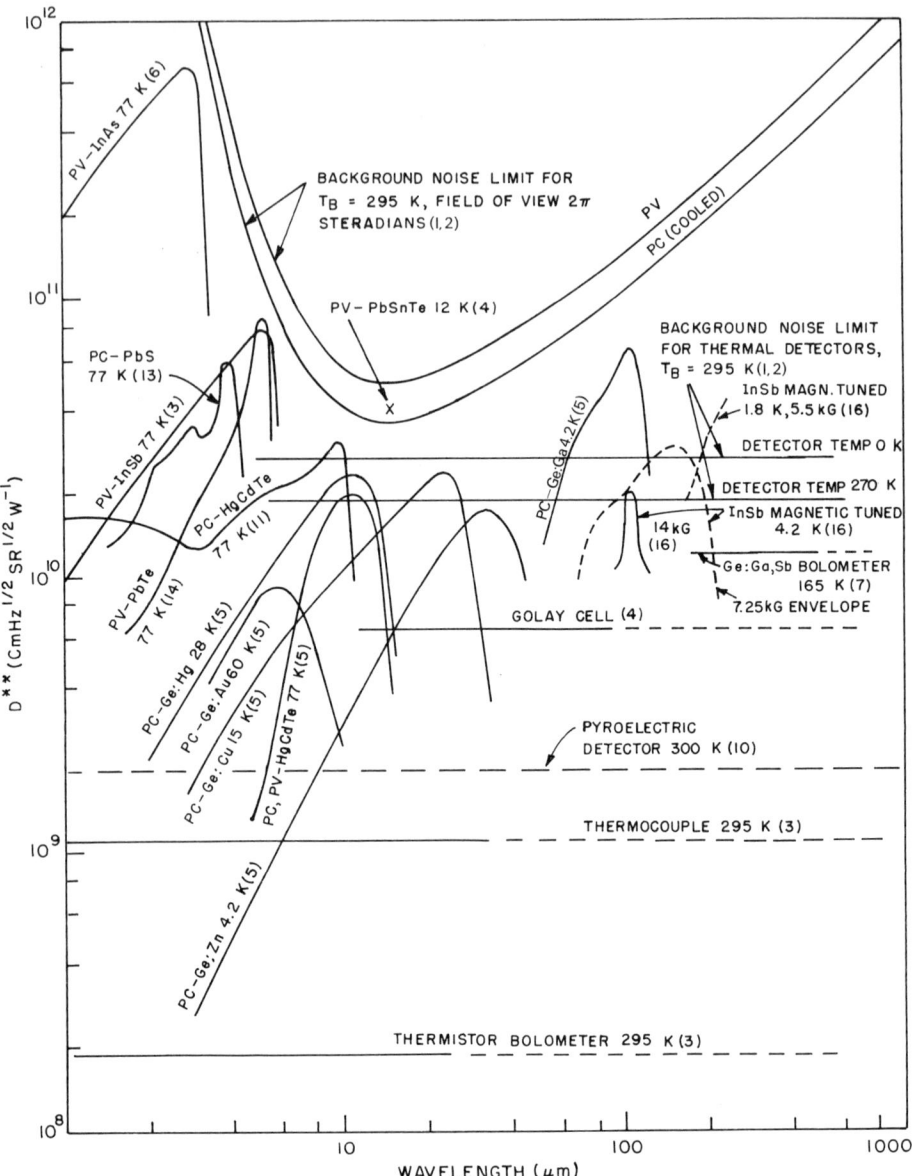

Fig. 2.22. Spectral D^{**} of detectors in the 1 to 1000 μm interval (after [2.17])

Table 2.8. State of the art of infrared detectors (after [2.17])

Material	Operating mode[a]	Maximum temperature for BLIP [K]	Operating temperature [K]	λ_0 [µm]	$D^{**}_{\lambda_0}$ [cm Hz$^{1/2}$Sr$^{1/2}$/watt]	Response time [s]	References	Additional notes
ATGS	Pyroelectric	—	300	—	2×10^9	$\gtrsim 1 \times 10^{-1}$	(10)	
Ge:Ga(Sb)	Bolometer	—	1.65	—	1.22×10^{10}	—	(7)	Optimum modulation frequency 200–500 Hz
Thermistor	Bolometer	—	295	—	1.98×10^8	1×10^{-3} – 5×10^{-3}	(3)	
Thermocouple		—	295	—	1.1×10^9	1.3×10^{-5} – 3×10^{-5}	(3)(4)	
InSb	Hot electron bolometer	—	4.2	Tunable 60 to 200 µm	3×10^{10}	2×10^{-7} – 5×10^{-7}	(16)	60° field of view assumed for calculation of D^{**}
Ge:B	PC	—	4.2	108	4.6×10^{10}	1×10^{-8}	(15)	Field of view assumed same (0.75 sr) as Ref. (8) for calculation of D^*
Ge:Ga	PC	—	4.2	104	6.8×10^{10}	4×10^{-8}	(8)	

Table 2.8 (continued)

Material	Operating mode[a]	Maximum temperature for BLIP [K]	Operating temperature [K]	λ_0 [μm]	$D^{**}_{\lambda_0}$ [cm Hz$^{1/2}$ Sr$^{1/2}$/watt]	Response time [s]	References	Additional notes
Ge:Cu	PC	17	4.2	23	2×10^{10} – 4×10^{10}	3×10^{-9} – 1×10^{-8}	(5)(6) (17)(21)	
Ge:Cu(Sb)	PC	17	4.2	23	2×10^{10}	$<2.2 \times 10^{-9}$	(6)(24)	
Pb$_{1-x}$Sn$_x$Te	PV	—	12	16	4×10^{10}	1×10^{-8} – 1×10^{-7}	(4)	$x = 0.17$–0.20
	PC, PV	—	4.2	14	1.7×10^{10}	1.2×10^{-6}	(4)(5) (29)	
			77	10	3×10^{8}	1.5×10^{-8}		
Hg$_{1-x}$Cd$_x$Te	PC, PV	—	77	12	1×10^{10} – 6×10^{10}	$<2 \times 10^{-7}$ – 4×10^{-6}	(4)(6)(9) (25)(26) (27)(28)	$x = 0.2$
	PC	—	77	9.6	3.1×10^{10}	8×10^{-7}	(11)	
Ge:Hg(Sb)	PC	35	4.2	11	1.8×10^{10}	3×10^{-10} – 2×10^{-9}	(5)(6)	
Ge:Hg	PC	35	4.2	11	7×10^{9} – 4×10^{10}	1×10^{-9} – 3×10^{-8}	(5)(6)(17) (9)(20)(22)	
	PC	—	27	10.5	4×10^{10}	—	(23)	

Table 2.8 (continued)

Material	Operating mode[a]	Maximum temperature for BLIP [K]	Operating temperature [K]	λ_0 [μm]	$D^{**}_{\lambda_0}$ [cm Hz$^{1/2}$ Sr$^{1/2}$/watt]	Response time [s]	References	Additional notes
Ge:Au	PC	60	77	6	$3 \times 10^9 - 1 \times 10^{10}$	3×10^{-8}	(3)(6)(17)(19)(20)	
Ge:Au(Sb)	PC	60	77	6	6×10^9	1.6×10^{-9}	(5)(6)	
InSb	PC, PV	110	77	5.3	$6 \times 10^{10} - 1 \times 10^{11}$	5×10^{-6}	(4)(6)(17)(18)	
	PV	—	77	4.9	1×10^{11}	$<2 \times 10^{-8}$	(12)	
PbTe	PV	—	77	5	8.7×10^{10}	2.5×10^{-8}	(14)	
PbS	PC	—	77	3.8	6×10^{10}	3.2×10^{-5}	(13)	
InAs	PV	—	77	2.8	7×10^{11}	5×10^{-7}	(5)(6)	

[a] PC—Photoconductive, PV—Photovoltaic.

References for the data in Table 2.8 and Fig. 2.22

1. S.F.Jacobs, M.Sargent,III: Infra. Phys. **10**, 233 (1970)
2. P.W.Kruse, L.D.McGlauchlin, R.B.McQuistan: *Elements of Infrared Technology* (John Wiley and Sons, New York 1969)
3. R.Hudson: *Infrared Engineering* (John Wiley and Sons, New York 1969)
4. A.C.Beer, R.K.Willardson (eds.): *Semiconductors and Semimetals 5, Infrared Detectors* (Academic Press, New York 1970)
5. H.Melchior, M.B.Fisher, F.R.Arams: Proc. IEEE **58**, 1466 (1970)
6. Manufacturer's Data from Honeywell, Philco-Ford, Santa Barbara Research Center, Texas Instruments, and Raytheon
7. S.Zwerdling, R.A.Smith, J.P.Theriault: Infra. Phys. **8**, 271 (1968)
8. W.J.Moore, H.Shenker: Infra. Phys. **5**, 96 (1965)
9. A.Cohen-Solal, Y.Riant: Appl. Phys. Lett. **19**, 436 (1971)
10. P.J.Lock: Appl. Phys. Lett. **19**, 390 (1971)
11. W.L.Eisenman: *Properties of Photodetectors, 78th Report* (Naval Weapons Center, Corona Laboratories, Corona, California, Dec. 1968)
12. A.G.Foyt, W.T.Lindley, J.P.Donelly: Appl. Phys. Lett. **16**, 335 (1970)
13. R.B.Schoolar: Appl. Phys. Lett. **16**, 446 (1970)
14. J.P.Donnelly, T.C.Harman, A.G.Foyt: Appl. Phys. Lett. **18**, 259 (1971)
15. H.Shenker, W.J.Moore, E.M.Swiggard: J. Appl. Phys. **35**, 2965 (1964)
16. E.H.Putley: Appl. Optics **4**, 649 (1965)
17. H.Levinstein: Appl. Optics **4**, 639 (1965)
18. F.D.Morton, R.E.J.King: Appl. Optics **4**, 659 (1965)
19. L.Johnson, H.Levinstein: Phys. Rev. **117**, 1191 (1960)
20. W.Beyen, P.Bratt, H.Davis, L.Johnson, H.Levinstein, A.Macrae: J. Opt. Soc. Amer. **49**, 686 (1959)
21. D.Bode, H.Graham: Infra. Phys. **3**, 129 (1963)
22. S.R.Borrello, H.Levinstein: J. Appl. Phys. **33**, 2947 (1962)
23. Y.Darviot, A.Sorrentino, B.Joly, B.Pajot: Infra. Phys. **7**, 1 (1967)
24. J.T.Yardley, C.B.Moore: Appl. Phys. Lett. **7**, 311 (1965)
25. J.Schlickman: In *Proceedings of the Electro-Optical Systems Conference, September 1969* (Industrial Scientific Conference Management, Chicago 1969)
26. P.W.Kruse: Appl. Optics **4**, 687 (1965)
27. B.E.Bartlett, D.E.Charlton, W.E.Dunn, P.C.Ellen, M.D.Jenner, M.H.Jervis: Infra. Phys. **9**, 35 (1969)
28. H.Mocker: Appl. Optics **8**, 677 (1969)
29. I.Melngailis, T.C.Harman: Appl. Phys. Lett. **13**, 180 (1968)

the noise equivalent power for a 1 Hz bandwidth. The authors employed this figure of merit because they wished to include photomultipliers whose noise does not in all cases depend upon the square root of the photocathode area. Table 2.7 lists the areas which *Seib* and *Aukerman* state are proper to the various detectors illustrated. The reader can convert to the D^* values appropriate to the photoconductive and photovoltaic detectors by multiplying the detectivity value illustrated by the square root of the detector area. The signal fluctuation limit shown in the figure (see Footnote 2 in Subsect. 2.4.1) is independent of area.

Figure 2.22 illustrates the performance of infrared detectors operating in the 1 to 1000 µm interval and the photon noise limits for cooled and uncooled detectors. The data are from a report of the National Materials Advisory Board [2.17]. Table 2.8 presents additional information concerning the detectors of Fig. 2.22, including the references from which the data were obtained.

Acknowledgements. The author is indebted to his colleagues Dr. *Paul E. Petersen* and Dr. *S. T. Liu* for their careful review of the manuscript. He also appreciates the efforts of *Diane Ellringer* in typing the several drafts, and encouragement and support from Honeywell.

References

2.1 R. A. Smith, F. E. Jones, R. P. Chasmar: *The Detection and Measurement of Infra-Red Radiation* (Oxford, London 1957)
2.2 M. R. Holter, S. Nudelman, G. H. Suits, W. L. Wolfe, G. J. Zissis: *Fundamentals of Infrared Technology* (The Macmillan Company, New York 1962)
2.3 P. W. Kruse, L. D. McGlauchlin, R. B. McQuistan: *Elements of Infrared Technology* (John Wiley and Sons, New York 1962)
2.4 J. A. Jamieson, R. H. McFee, G. N. Plass, R. H. Grube, R. G. Richards: *Infrared Physics and Engineering* (McGraw-Hill Book Co., New York 1963)
2.5 W. L. Wolfe (ed.): *Handbook of Military Infrared Technology* (Office of Naval Research, Department of the Navy, Washington, D.C. 1965)
2.6 G. K. T. Conn, D. G. Avery: *Infrared Methods, Principles and Applications* (Academic Press, New York 1960)
2.7 R. D. Hudson, Jr.: *Infrared System Engineering* (John Wiley and Sons, New York 1969)
2.8 S. M. Ryvkin: *Photoelectric Effects in Semiconductors* (Consultants Bureau, New York 1964)
2.9 R. H. Bube: *Photoconductivity of Solids* (John Wiley and Sons, New York 1960)
2.10 E. M. Pell (ed.): *Proceedings of the Third International Conference on Photoconductivity, Stanford, August 12–15, 1969* (Pergamon Press, Oxford 1971)
2.11 Appl. Opt. **4**, No. 6 (June 1965)
2.12 L. M. Biberman, S. Nudelman (ed.): *Photoelectronic Imaging Devices*, Vol. 1: *Physical Processes and Methods of Analysis* (Plenum Press, New York 1971)
2.13 L. M. Biberman, S. Nudelman (eds.): *Photoelectronic Imaging Devices*, Vol. 2: *Devices and Their Evaluation* (Plenum Press, New York 1971)
2.14 J. D. McGee, W. L. Wilcock (eds.): *Photo-Electronic Image Devices: Proceedings of a Symposium Held at London September 3–5, 1958*, Vol. 12 of *Advances in Electronics and Electron Physics*, ed. by L. Marton (Academic Press, New York 1958)
J. D. McGee, W. L. Wilcock, L. Mandel (eds.): *Photo-Electronic Image Devices: Proceedings of the Second Symposium Held at Imperial College, London, September 5–8, 1961*, Vol. 16 of *Advances in Electronics and Electron Physics*, ed. by L. Marton (Academic Press, New York 1962)
J. D. McGee, D. McMullan, E. Kahan (eds.): *Photo-Electronic Image Devices: Proceedings of the Third Symposium Held at Imperial College, London, September 20–24, 1965*, Vol. 22A, B of *Advances in Electronics and Electron Physics*, ed. by L. Marton (Academic Press, New York 1966)
J. D. McGee, D. McMullan, E. Kahan, B. L. Morgan (eds.): *Photo-Electronic Image Devices: Proceedings of the Fourth Symposium Held at Imperial College, London, September 16–20, 1968*, Vol. 28A, B of *Advances in Electronics and Electron Physics*, ed. by L. Marton (Academic Press, New York 1969)
J. D. McGee, D. McMullan, E. Kahan (eds.): *Photo-Electronic Image Devices: Proceedings of the Fifth Symposium Held at Imperial College, London, September 13–17, 1971*, Vol. 33A, B of *Advances in Electronics and Electron Physics*, ed. by L. Marton (Academic Press, New York 1972)
B. L. Morgan, R. W. Airey, D. McMullan (eds.): *Photo-Electronic Image Devices: Proceedings of the Sixth Symposium Held at Imperial College, London, September 9–13, 1974*, Vol. 40A, B of *Advances in Electronics and Electron Physics*, ed. by L. Marton (Academic Press, New York 1976)

2.15 T.S.Moss, G.J.Burrell, B.Ellis: *Semiconductor Opto-Electronics* (John Wiley and Sons, New York 1973)
2.16 R.K.Willardson, A.C.Beer (eds.): *Semiconductors and Semi-Metals 5, Infrared Detectors* (Academic Press, New York 1970)
2.17 *Materials for Radiation Detection*, Publication NMAB 287 (National Materials Advisory Board, National Research Council, National Academy of Sciences-National Academy of Engineering, Washington, D.C. 1974)
2.18 J.O.Dimmock: J. Electron. Mater. **1**, 255 (1972)
2.19 D.H.Seib, L.W.Aukerman: In *Advances in Electronics and Electron Physics 34*, ed. by L.Marton (Academic Press, New York 1973)
2.20 R.B.Emmons, S.R.Hawkins, K.F.Cuff: Opt. Engr. **14**, 21 (1975)
2.21 L.K.Anderson, B.J.McMurtry: Appl. Opt. **5**, 1573 (1966)
2.22 H.Melchior, M.B.Fisher, F.R.Arams: Proc. IEEE **58**, 1466 (1970)
2.23 Proc. IEEE **63**, No. 1 (January 1975)
2.24 *Proceedings of the Special Meeting on Infrared Detectors, 16 March 1971* (Infrared Information and Analysis Center, Environmental Research Institute of Michigan, Ann. Arbor, Michigan 1971)
2.25 S.M.Sze: *Physics of Semiconductor Devices* (Wiley-Interscience, New York 1969)
2.26 A.S.Grove: *Physics and Technology of Semiconductor Devices* (John Wiley and Sons, New York 1967)
2.27 C.Kittel: *Introduction to Solid State Physics*, 4th ed. (John Wiley and Sons, New York 1971)
2.28 J.L.Moll: *Physics of Semiconductors* (McGraw-Hill Book Co., New York 1964)
2.29 D.Long: Infrared Phys. **7**, 121 (1967)
2.30 H.S.Sommers, Jr.: In *Semiconductors and Semimetals 5, Infrared Detectors*, ed. by R.K.Willardson and A.C.Beer (Academic Press, New York 1970)
2.31 C.M.Penchina: Infrared Phys. **15**, 9 (1975)
2.32 S.M.Ryvkin: op. cit., p. 369ff.
2.33 L.K.Anderson, P.G.McMullin, L.A.D'Asaro, A.Goetzberger: Appl. Phys. Lett. **6**, 62 (1965)
2.34 R.J.McIntyre: IEEE Trans. Electron Devices **ED-13**, 164 (1966)
2.35 R.J.McIntyre: IEEE Trans. Electron Devices **ED-17**, 347 (1970)
2.36 H.Melchior, W.T.Lynch: IEEE Trans. Electron Devices **ED-13**, 829 (1966)
2.37 H.W.Ruegg: IEEE Trans. Electron Devices **ED-14**, 239 (1967)
2.38 W.T.Lindley, R.J.Phelan, Jr., C.M.Wolfe, A.G.Foyt: Appl. Phys. Lett. **14**, 197 (1969)
2.39 W.T.Lynch: IEEE Trans. Electron Devices **ED-15**, 735 (1968)
2.40 D.P.Mather, R.J.McIntyre, P.P.Webb: Appl. Opt. **9**, 1843 (1970)
2.41 E.Ahlstrom, W.W.Gartner: J. Appl. Phys. **33**, 2602 (1962)
2.42 M.V.Schneider: Bell System Tech. J. **45**, 1611 (1966)
2.43 W.M.Sharpless: Appl. Opt. **9**, 489 (1970)
2.44 R.L.Anderson: Solid-State Electron. **5**, 341 (1962)
2.45 P.W.Kruse, F.C.Pribble, R.G.Schulze: J. Appl. Phys. **38**, 1718 (1967)
2.46 P.W.Kruse, R.G.Schulze: In *Proceedings of the Third International Conference on Photoconductivity, Stanford, August 12–15, 1969*, ed. by E.M.Pell (Pergamon Press, Oxford 1971)
2.47 P.Perfetti, M.Antichi, F.Capasso, G.Margaritondo: Infrared Phys. **14**, 255 (1974)
2.48 J.Tauc: Rev. Mod. Phys. **29**, 308 (1957)
2.49 P.W.Kruse: Appl. Opt. **4**, 687 (1965)
2.50 Y.Marfaing, J.Chevallier: In *Proceedings of the Third International Conference on Photoconductivity, Stanford, August 12–15, 1969*, ed. by E.M.Pell (Pergamon Press, Oxford 1971)
2.51 J.J.Scheer, J.Van Laar: Solid State Commun. **3**, 189 (1965)
2.52 D.G.Fisher, R.E.Enstrom, B.F.Williams: Appl. Phys. Lett. **18**, 371 (1971)
2.53 S.W.Kurnick, R.N.Zitter: J. Appl. Phys. **27**, 278 (1956)
2.54 P.W.Kruse: J. Appl. Phys. **30**, 770 (1959)
2.55 T.S.Moss: *Optical Properties of Semiconductors* (Academic Press, New York 1959) p. 61ff.
2.56 B.V.Rollin: Proc. Phys. Soc. (London) **77**, 1102 (1961)

2.57 M.A.Kinch, B.V.Rollin: Brit. J. Appl. Phys. **15**, 672 (1963)
2.58 E.H.Putley: Appl. Opt. **4**, 649 (1965)
2.59 H.Shenker: Appl. Phys. Lett. **5**, 183 (1964)
2.60 A.F.Gibson, M.F.Kimmitt, A.C.Walker: Appl. Phys. Lett. **17**, 75 (1970)
2.61 R.H.Bube: op. cit., p. 80
2.62 N.Bloembergen: Phys. Rev. Lett. **2**, 84 (1959)
2.63 R.W.Gelinas: *RAND Report P-1844* (RAND Corporation, Santa Monica, California, October 24, 1959)
2.64 P.W.Kruse, L.D.McGlauchlin, R.B.McQuistan: op. cit., p. 371 ff.
2.65 M.R.Brown, W.A.Shand: In *Advances in Quantum Electronics*, Vol. 1, ed. by D.W.Goodwin (Academic Press, New York 1970)
2.66 J.F.Porter: IEEE J. Quant. Electron. **QE-1**, 113 (1965)
2.67 W.B.Grandrud, M.W.Moos: J. Chem. Phys. **49**, 2170 (1968)
2.68 W.B.Grandrud, M.W.Moos: IEEE J. Quant. Electron. **4**, 249 (1968)
2.69 M.R.Brown, W.A.Shand: Phys. Lett. **18**, 95 (1965)
2.70 F.Varsanyi: Phys. Rev. Lett. **14**, 786 (1965)
2.71 A.G.Becker, O.Risgin: In *Proceedings of the Third International Conference on Photoconductivity, Stanford, August 12–15, 1969*, ed. by E.M.Pell (Pergamon Press, Oxford 1971)
2.72 J.E.Geusic, F.W.Ostermayer, H.M.Marcos, L.G.Van Uitert, J.P.van der Ziel: J. Appl. Phys. **42**, 1958 (1971)
2.73 C.E.K.Mees: *The Theory of the Photographic Process* (Macmillan, New York 1945)
2.74 S.Fujisawa (ed.): *Photographic Sensitivity: Proceedings of a Symposium Held at Tokyo, Japan in October 1957* (Maruzen, Tokyo 1958)
2.75 N.F.Mott, R.W.Gurney: *Electronic Processes in Ionic Crystals* (Oxford University Press, London 1953)
2.76 R.C.Jones: J. Opt. Soc. Am. **45**, 799 (1955)
2.77 R.C.Jones: J. Opt. Soc. Am. **48**, 874 (1958)
2.78 E.M.Wormser: J. Opt. Soc. Am. **43**, 15 (1953)
2.79 R.DeWaard, E.M.Wormser: Proc. IRE **47**, 1508 (1959)
2.80 P.W.Kruse, L.D.McGlauchlin, R.B.McQuistan: op. cit., p. 345 ff.
2.81 D.H.Andrews, R.M.Milton, W.DeSorbo: J. Opt. Soc. Am. **36**, 518 (1946)
2.82 D.H.Martin, D.Blon: Cryogenics **1**, 159 (1961)
2.83 D.G.Naugle, W.A.Porter: In *Proceedings of the Special Meeting on Unconventional Infrared Detectors, 16 March 1971* (Infrared Information and Analysis Center, Environmental Research Institute of Michigan, Ann Arbor, Michigan 1971)
2.84 W.S.Boyle, K.F.Rodgers, Jr.: J. Opt. Soc. Am. **49**, 66 (1959)
2.85 S.Corsi, G.Dall'Oglio, G.Fantoni, F.Melchiorri: Infrared Phys. **13**, 253 (1973)
2.86 F.J.Low: J. Opt. Soc. Am. **51**, 1300 (1961)
2.87 F.J.Low: Proc. IEEE **53**, 516 (1965)
2.88 Y.Oka, K.Nagasaka, S.Narita: Japan. J. Appl. Phys. **7**, 611 (1968)
2.89 R.Bachman, H.Kirsch, G.H.Geballe: Rev. Sci. Instr. **41**, 547 (1970)
2.90 P.S.Nayar: Infrared Phys. **14**, 31 (1974)
2.91 K.Shivanandan, D.P.McNutt: Infrared Phys. **15**, 27 (1975)
2.92 G.Dall'Oglio, B.Melchiorri, F.Melchiorri, V.Natale: Infrared Phys. **14**, 347 (1974)
2.93 A.G.Chynoweth: J. Appl. Phys. **27**, 78 (1956)
2.94 J.Cooper: Rev. Sci. Instr. **33**, 92 (1962)
2.95 J.Cooper: J. Sci. Instr. **39**, 467 (1962)
2.96 R.W.Astheimer, F.Schwarz: Appl. Opt. **7**, 1687 (1968)
2.97 H.P.Beerman: IEEE Trans. Electron Devices **ED-16**, 554 (1969)
2.98 H.P.Beerman: Ferroelectrics **2**, 123 (1971)
2.99 A.M.Glass: J. Appl. Phys. **40**, 4699 (1969)
2.100 S.T.Liu, J.D.Heaps, O.N.Tufte: Ferroelectrics **3**, 281 (1972)
2.101 R.B.Maciolek, S.T.Liu: J. Electron. Mater. **2**, 191 (1973)

2.102 S.T.Liu, R.B.Maciolek: *1973 International Electron Device Meeting Technical Digest* (IEEE, New York 1973) p. 259
2.103 S.T.Liu, R.B.Maciolek: J. Electron. Mater. **4**, 91 (1975)
2.104 P.J.Lock: Appl. Phys. Lett. **19**, 390 (1971)
2.105 R.J.Phelan,Jr., R.J.Mahler, A.R.Cook: Appl. Phys. Lett. **19**, 337 (1971)
2.106 E.H.Putley: In *Semiconductors and Semimetals 5, Infrared Detectors*, ed. by R.K.Willardson and A.C.Beer (Academic Press, New York 1970)
2.107 S.T.Liu: to be published in *Ferroelectrics*
2.108 R.W.Astheimer, S.Weiner: Appl. Opt. **3**, 493 (1964)
2.109 N.B.Stevens: In *Semiconductors and Semimetals 5, Infrared Detectors*, ed. by R.K.Willardson and A.C.Beer (Academic Press, New York 1970)
2.110 M.J.E.Golay: Rev. Sci. Instr. **18**, 347 (1947)
2.111 M.J.E.Golay: Rev. Sci. Instr. **20**, 816 (1949)
2.112 D.W.Hill, T.Powell: *Non-Dispersive Infra-Red Gas Analysis in Science, Medicine, and Industry* (Plenum Press, New York 1968)
2.113 R.W.Bené, R.M.Walser: In *Proceedings of the Special Meeting on Unconventional Infrared Detectors, 16 March 1971* (Infrared Information and Analysis Center, Environmental Research Institute of Michigan, Ann Arbor, Michigan 1971)
2.114 R.M.Walser, R.W.Bené, R.E.Caruthers: IEEE Trans. Electron Devices **ED-18**, 309 (1971)
2.115 P.W.Kruse, M.D.Blue, J.H.Garfunkel, W.D.Saur: Infrared Phys. **2**, 53 (1962)
2.116 E.R.Washwell, S.R.Hawkins, K.F.Cuff: In *Proceedings of the Special Meeting on Unconventional Infrared Detectors, 16 March 1971* (Infrared Information and Analysis Center, Environmental Research Institute of Michigan, Ann Arbor, Michigan 1971)
2.117 J.R.McColl: In *Proceedings of the Special Meeting on Unconventional Infrared Detectors, 16 March 1971* (Infrared Information and Analysis Center, Environmental Research Institute of Michigan, Ann Arbor, Michigan 1971)
2.118 R.D.Ennulat, J.L.Fergason: Mol. Cryst. Liq. Cryst. **13**, 149 (1971)
2.119 W.E.Woodmansee: Appl. Opt. **7**, 121 (1968)
2.120 G.Assouline, M.Hareng, E.Leiba: Proc. IEEE **59**, 1355 (1971)
2.121 P.W.Kruse, L.D.McGlauchlin, R.B.McQuistan: op. cit., p. 315ff.
2.122 M.C.Teich: Proc. IEEE **56**, 37 (1968)
2.123 H.Mocker: Appl. Opt. **8**, 677 (1969)
2.124 T.Gilmartin, H.A.Bostick, L.J.Sullivan: *Proc. NEREM Conference* (Nov. 1970) p. 168
2.125 M.C.Teich: In *Semiconductors and Semimetals 5, Infrared Detectors*, ed. by R.K.Willardson and A.C.Beer (Academic Press, New York 1970)
2.126 F.R.Arams, E.W.Sard, B.J.Peyton, F.P.Pace: In *Semiconductors and Semimetals 5, Infrared Detectors*, ed. by R.K.Willardson and A.C.Beer (Academic Press, New York 1970)
2.127 R.J.Keyes, T.M.Quist: In *Semiconductors and Semimetals 5, Infrared Detectors*, ed. by R.K.Willardson and A.C.Beer (Academic Press, New York 1970)
2.128 A.F.Milton: Appl. Opt. **11**, 2311 (1972)
2.129 J.Waner: Appl. Phys. Lett. **12**, 222 (1968)
2.130 W.B.Gandrud, G.D.Boyd: Optics Commun. **1**, 187 (1969)
2.131 G.D.Boyd, W.B.Gandrud, E.Buehler: Appl. Phys. Lett. **18**, 446 (1971)
2.132 D.Hall, A.Yariv, E.Garmire: Appl. Phys. Lett. **17**, 127 (1970)
2.133 R.L.Aggarwal, B.Lax, G.Favrot: Appl. Phys. Lett. **22**, 329 (1973)
2.134 F.W.Anderson, J.M.Powell: Phys. Rev. Lett. **10**, 230 (1963)
2.135 S.Shapiro: In *Proceedings of the Special Meeting on Unconventional Infrared Detectors, 16 March 1971* (Infrared Information and Analysis Center, Environmental Research Institute of Michigan, Ann Arbor, Michigan 1971)
2.136 R.Chiao: Phys. Lett. **33A**, 177 (1970)
2.137 L.O.Hocker, D.R.Sokoloff, V.Daneu, A.Szoke, A.Javan: Appl. Phys. Lett. **12**, 401 (1968)
2.138 D.R.Sokoloff, A.Sanchez, R.M.Osgood, A.Javan: Appl. Phys. Lett. **17**, 257 (1970)
2.139 R.L.Abrams, W.B.Gandrud: In *Proceedings of the Special Meeting on Unconventional Infrared Detectors, 16 March 1971* (Infrared Information and Analysis Center, Environmental Research Institute of Michigan, Ann Arbor, Michigan 1971)

2.140 C.D. Motchenbacher, F.C. Fitchen: *Low-Noise Electronic Design* (Wiley-Interscience, New York 1973)
2.141 K.M. van Vliet: Appl. Opt. **6**, 1145 (1967)
2.142 A. Van der Ziel: *Fluctuation Phenomena in Semiconductors* (Academic Press, New York 1959)
2.143 A. Van der Ziel: In *Noise in Electron Devices*, ed. by L.D. Smullin and H.A. Haus (John Wiley and Sons, New York 1959)
2.144 R.C. Jones: Proc. IRE **47**, 1481 (1959)
2.145 R.E. Burgess: Brit. J. Appl. Phys. **6**, 185 (1955)
2.146 K.M. Van Vliet, J.R. Fassett: In *Fluctuation Phenomena in Solids*, ed. by R.E. Burgess (Academic Press, New York 1965)
2.147 R.C. Jones: In *Advances in Electronics 5*, ed. by L. Marton (Academic Press, New York 1953)
2.148 A. Van der Ziel: Proc. Inst. Radio Engrs. **46**, 1019 (1958)
2.149 K.M. van Vliet: Proc. Inst. Radio Engrs. **46**, 1004 (1958)
2.150 K.M. van Vliet: Phys. Rev. **110**, 50 (1958)
2.151 R.E. Burgess: Brit. J. Appl. Phys. **6**, 185 (1955)
2.152 R.E. Burgess: Proc. Phys. Soc. **B68**, 661 (1955)
2.153 R.E. Burgess: Proc. Phys. Soc. **B69**, 1020 (1956)
2.154 D. Long: Infrared Phys. **7**, 169 (1967)
2.155 A. Van der Ziel: *Noise* (Prentice Hall, New York 1954)
2.156 W.L. Wolfe: op. cit., Table 11-1
2.157 R.C. Jones: J. Opt. Soc. Am. **50**, 1058 (1960)
2.158 S.R. Borrello: Infrared Phys. **12**, 267 (1972)
2.159 P.W. Kruse, L.D. McGlauchlin, R.B. McQuistan: op. cit., Chap. 9
2.160 S.F. Jacobs, M. Sargent, III: Infrared Phys. **10**, 233 (1970)

3. Thermal Detectors

E. H. Putley

With 14 Figures

Herschel discovered the infrared spectrum in 1800 using as a detector a liquid in glass thermometer which is usually taken as the first thermal detector. Recently it has been discovered that thermal detectors have been in use for perhaps millions of years in target locating systems by two families of snake who use the temperature difference between their prey and its surroundings to locate and capture the small animals on which they feed (*Bullock* and *Barrett* [3.1]). The two families of snake are the pit vipers (Crotalidae) and the boa constrictors (Boidae). The pit vipers (which includes rattlesnakes such as the sidewinder) have a pair of facial pits one below each eye which contain their infrared sensing organs while the boas are provided with a linear array along their jaws. It has now been established that the snakes' sensors are in fact thermal detectors (*Harris* and *Gamow* [3.2, 3]). It has been suggested that moths and other insects use infrared. The only example with clear evidence is the bark beetle *Melanophila acuminata*. Thus some insects and snakes developed thermal detectors long before we did.

Most infrared detectors fall into one of two general classes: 1) thermal detectors and 2) photon detectors. In the photon detectors, the incident radiation excites electronic transitions which change the electronic state of the detector. In thermal detectors, the energy of its absorbed radiation raises the temperature of the detecting element. This increase in temperature will cause changes in temperature dependent properties of the detector. Monitoring one of these changes enables the radiation to be detected. The first detectors used by Herschel and the early workers in the infrared were thermal, and the basic types used by them in the 19th century are still in use today.

Newton used a liquid in glass-thermometer to study heat radiation. Rumford and Leslie used a differential gas thermometer. Herschel reverted to the liquid thermometer, but this was soon replaced by the thermopile (*Melloni* [3.4]). Some time later (*Langley* [3.5]) the first bolometers were used. More recently the use of the gas thermometer, in the shape of the *Golay* [3.6] and Luft cells has been reintroduced and is now widely used in spectrometers. Another type of thermal detector now widely used is that utilizing the pyroelectric effect. In addition to these, several other detection processes have been suggested, including thermal expansion and changed dielectric properties with temperature.

Spatially extended thermal detectors are also used to form infrared imaging systems. These have been employed in electron beam tubes, such as the pyroelectric vidicon and in systems with optical readout, such as the

Evapograph. Considerable development is taking place in this area with the object of producing a cheap, high quality, infrared imaging system.

Although most of these types of thermal detector have been in use for many years, they are all used in modern instrumentation. They have not been rendered obsolete by the more recent development of photon detectors because for many types of application use of the appropriate thermal detector is more suitable than a photon detector.

After discussing in more detail the general principles of thermal detectors, the main types of detector will be described and some typical modern applications will be mentioned.

3.1 Basic Principles

The performance of a thermal detector is calculated in two stages. First by consideration of the thermal characteristics of the system the temperature rise produced by the incident radiation is determined. Secondly this temperature rise is used to determine the change in the property which is being used to indicate the signal. The second stage of the calculation gives the output from the detector and hence its responsivity. The first stage of the calculation is common to all thermal detectors, but the details of the second stage will differ for the different types of thermal detectors. In this section the first stage only will be considered while the second stages will be discussed in the sections dealing with the different types of detector.

The simplest representation of the thermal circuit of an infrared detector is shown in Fig. 3.1. The detector is represented by a thermal mass H coupled via a conductance G to a heat sink at a constant temperature T. In the absence of a radiation input the average temperature of the detector will also be T, although it will exhibit a fluctuation about this value. This fluctuation gives rise to a source of detector noise ("temperature noise") which sets the ultimate limit to the minimum signal detected by a perfect thermal detector. When a radiation input is received by the detector, the rise in temperature is found by solving the equation:

$$\eta I = H \left(\frac{d\theta}{dt} \right) + G\theta \qquad (3.1)$$

where I is radiation power incident upon the detector, of which the fraction η is actually absorbed, the fraction $(1-\eta)$ being either reflected from the front surface or transmitted through the detector. The temperature of the detector at the time t is

$$T_D = T + \theta. \qquad (3.2)$$

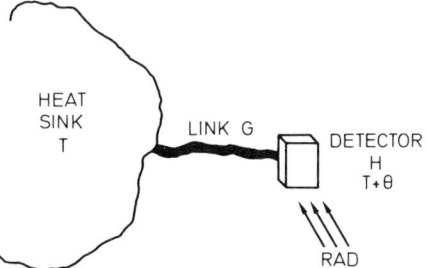

Fig. 3.1. Schematic thermal circuit

I may be time independent but more generally will consist of a time independent component I_0 plus at least one component modulated at an angular frequency ω. Consideration of one modulated component only is sufficiently general for this discussion.

Write

$$I = I_0 + I_\omega e^{j\omega t} \tag{3.3}$$

where $I_\omega \not> I_0$. Solving (3.1) gives for the amplitude θ_ω of the excess temperature component corresponding to I_ω and for its phase difference φ from I_ω

$$\theta_\omega = \eta I_\omega (G^2 + \omega^2 H^2)^{-1/2} \tag{3.4}$$

$$\varphi = \tan^{-1}(\omega H/G). \tag{3.5}$$

Equation (3.4) illustrates several features of thermal detectors. Clearly it is advantageous to make θ_ω as large as possible. To do this G wants to be as small as possible and ω sufficiently low that $\omega H \ll G$. In other words both the thermal capacity of the detector and its thermal coupling to its surroundings want to be as small as possible. These requirements distinguish a thermal detector from an ordinary thermometer. Both require a strongly temperature dependent property and in fact the same effects are used in both types of instrument. The principal differences arise from the need to optimize the interaction of the thermal detector with the incident radiation while reducing as far as possible all other thermal contacts with its surroundings.

Equation (3.4) shows that as ω is increased, the term $\omega^2 H^2$ will eventually exceed G and then θ_ω will fall inversely as ω. A characteristic thermal response time for the detector can therefore be defined as

$$\tau_T = H/G. \tag{3.6}$$

For typical detector design τ_T falls within the range of milliseconds to seconds. This is much longer than the typical response time of a photon detector. For some applications this puts thermal detectors at a disadvantage with respect to photon detectors, but when all the systems tradeoffs are taken into account this disadvantage may not be as great as it would at first sight seem.

The implication of making H as small as possible is that the thermal detector should be as small and of as light a construction as practicable. This is the reason for the fragile and delicate nature of many types of thermal detector. One of the reasons for the large amount of interest in the current developments of the pyroelectric detector is that (as we shall see in Sect. 3.5) it offers a way round this dilemma of achieving both sensitivity and robustness.

Equation (3.4) also shows that G should be made as small as possible. However if we have already made H as small as we can, (3.6) shows that reducing G will increase τ_T, which may be undesirable. Thus it may be necessary to adopt a compromise in the choice of G. The value of G also determines the magnitude of the temperature noise fluctuation [see (3.9)], a small value being required for the highest sensitivity. The minimum possible value of G is that obtained when the only thermal coupling of the detecting element to the heat sink is via radiative exchanges. This limiting value can be estimated from the Stefan-Boltzmann total radiation law. If the thermal detector has a receiving area a of emissivity η then when it is in thermal equilibrium with its surroundings it will radiate a total flux $a\eta\sigma T^4$ where σ is the Stefan-Boltzmann constant. If now the temperature of the detector is increased by a small amount $\theta = dT$, the flux radiated is increased by

$$4a\eta\sigma T^3 dT = 4a\eta\sigma T^3 \theta. \tag{3.7}$$

Hence if the radiative component of the thermal conductance is G_R,

$$G_R = 4a\eta\sigma T^3. \tag{3.8}$$

When the detector is in thermal equilibrium with the heat sink, the rms fluctuation in the power flowing through the thermal conductance into the detector is (*Smith* et al. [3.7])

$$\Delta W_T = (4kT^2 G)^{1/2} \tag{3.9}$$

which will be smallest when G assumes its minimum value, i.e., G_R. Then ΔW_T will be a minimum and its value gives the minimum detectable power for an ideal thermal detector. The minimum detectable signal power P_N is defined as the (rms) signal power incident upon the detector required to equal the (rms) thermal noise power. Hence if the temperature fluctuation associated with G_R is the only source of noise,

$$\eta P_N = \Delta W_T = (16a\eta\sigma kT^5)^{1/2}$$

or

$$P_N = (16a\sigma kT^5/\eta)^{1/2}. \tag{3.10}$$

The definitions of W_T and of P_N assume the amplified noise bandwidth Δf is reduced to 1 Hz. For wider bandwidth Δf the rms noise increases as $(\Delta f)^{1/2}$. Equation (3.10) represents the best performance attainable from a thermal detector. If all the incident radiation is absorbed by the detector, $\eta = 1$.

P_N then becomes

$$P_N \text{ (background limit)} = (16a\sigma kT^5)^{1/2} \tag{3.11}$$

$$= 5.0 \times 10^{-11} \text{ W}.$$

($\sigma = 5.67 \times 10^{-12}$ Jcm^{-2}K^{-4}, $k = 1.38 \times 10^{-23}$ JK^{-1} and assuming $T = 290$ K, $a = 1$ cm^2 and the electronic bandwidth $\Delta f = 1$ Hz). This value is used as a reference to compare the performance of actual detectors. The best room temperature thermal detectors approach within about one order of this value. It will be noted that this calculation has been made without reference to the readout mechanism. It does assume a perfectly noiseless mechanism—the noise level is set by processes outside the detector. An alternative way to perform this calculation is to consider the fluctuations in the background radiation incident upon the detector (*Putley* [3.8]). It would then be found that (3.11) corresponds to the fluctuation associated with the total background radiation incident on the detector from a complete hemispherical field of view.

Thermal detectors are not exclusively operated at room temperature. When cooled to very low temperatures, (3.10) or (3.11) would still represent the ultimate performance if the detector were enclosed completely in a cooled enclosure at the operating temperature. Thus if we were to envisage a detector operating in outer space and cooled to the Universe's background temperature (say 3 K) (3.11) would indicate a limiting sensitivity of

$$P_N(3 \text{ K background}) = 5.5 \times 10^{-16} \text{ W}. \tag{3.12}$$

Bolometers cooled with liquid helium (see Sect. 3.3) have been developed to try to exploit this potential improvement in sensitivity. There is another advantage to be gained from cyrogenic operation. Equation (3.6) shows the desirability of reducing H in order to reduce the thermal time constant. At helium temperature specific heats become very small hence leading to a substantial reduction in H. Although helium cooled bolometers do give a useful improvement in performance, it does not approach that suggested by (3.12). This is because until a complete experiment is conducted at 3 K it is necessary to expose the detector to some radiation from a higher temperature background. Although the field of view may be restricted to a few degrees and cooled filters be used to limit its spectral bandwidth its noise fluctuation will be several orders greater than that of the helium temperature background. *Putley* [3.8] discusses this in more detail with numerical examples showing that the deterioration in a practical case is between three and four orders. Although these remarks have been directed specifically at cooled bolometers they apply in principle to other types of thermal detectors provided that they possess sufficiently large temperature dependent property at low temperature. Since thermoelectric effects tend to zero as T tends to zero thermopiles cannot be used at low temperatures. Similarly, Golay cells would become inoperative. Pyroelectric devices probably could be developed, using a material with a low ferroelectric transition temperature. In fact only cooled bolometers have been developed. These have the advantage that both with superconductors and semiconductors can large temperature variations of resistance be found at suitably low temperatures. Since in most parts of the spectrum cooled photon detectors of high sensitivity are available there seems little point in developing further cooled detectors. The cooled bolometers have a more uniform spectral response than the photon detectors and for this reason they are likely to remain in use.

The performance achieved by any real detector will be inferior to that predicted by (3.11). Comparison with (3.10) shows that unless all the incident radiation is absorbed ($\eta = 1$) the performance will be worse than that of an ideal detector by the factor $\eta^{1/2}$ even in the absence of other sources of noise. The attainment of a value η approaching unity by the proper choice of materials and surface finishes in the construction of detectors calls for care and skill by the designer. A further source of signal loss which does not appear explicitly in the design equations occurs through the need in the majority of cases to encapsulate the detectors in a protective enclosure with a window transparent to the radiation to be detected. Reflection and absorption losses at the window will lead to further degradation in the effective value of η. The need for a window also limits the spectral response of a detector since no material exists which is transparent throughout the whole spectrum. Users of thermal detectors must specify to the manufacturers the spectral range they are working in so that the most suitable windows can be supplied. Table 3.1 summarizes the characteristics of the commoner window material.

Further degradation of performance will arise from the additional noise sources which will always be present. Compare first (3.9) and (3.10). Equation (3.10) assumes that the detecting element can only exchange heat with its surroundings by radiation. This implies that the element must be supported in vacuum by supports of zero thermal conductance and that if electrical contacts are required to extract the output (as they are for all detectors using an electrical temperature dependent property) they also have zero thermal conductance. Since these conditions can never be fully achieved the actual value of G will be greater than G_R and hence P_N will be poorer (greater) than the value determined by (3.10). Contributions to G will come from any gas (conduction and convection) in the encapsulation and from conduction through the supports and electrical connections. The effects of excess thermal conductance on the performance of the pyroelectric detector have been discussed by Logan (*Logan* and *Moore* [3.9], *Logan* [3.10]), to mention one example.

In detectors employing an electrical readout mechanism, electrical noise fluctuations (Johnson noise plus possibly other noise sources such as low frequency contact noise in some cases) must be considered. There will be a Johnson noise source associated with the output impedance of the detector. In thermopiles and bolometers the output impedance is predominantly resistive so that the calculation of the Johnson noise is straightforward. The output impedance of the pyroelectric detector is predominantly capacitative. In this case the resistive component associated with the dielectric loss factor of the

Table 3.1. Some window materials. Most thermal detectors require mounting in a closed encapsulation. Their effective spectral range will then be essentially that of the window. Some window materials have large refractive indices. These materials require blooming to reduce reflection losses. With these materials, the blooming determines the spectral range over which optimum performance is achieved

Material	Useful spectral range (µm)	Refractive index	Notes
KBr	Vis. 30	~1.5	Water soluble and not mechanically strong but are very convenient for laboratory use.
CsI	Vis. 55	~1.7	
BaF$_2$	Vis. 11	~1.4	
Fused silica (SiO$_2$)	Vis. 4.0	~1.4	Fused silica is useful in the near IR but poor at very long wavelengths. Crystal quartz can be used in the near IR, but is a most useful sub mm material.
Crystal quartz (SiO$_2$)	40 – 1000	2.1	
IRTRAN 1 (MgF$_2$)	0.7–7.0	1.3	Pressed powder materials available from Eastman-Kodak.
IRTRAN 2 (ZnS)	1.8–8.0	2.2	
IRTRAN 3 (CaF$_2$)	Vis. → 10.0	1.4	
IRTRAN 4 (ZnSe)	0.7–17	2.4	
IRTRAN 5 (MgO)	0.5–8.0	1.7	
KRS–5	0.7–40	~2.3	Thallium bromide-thallium iodide mixed crystals. Rather brittle liable to plastic deformation. Problems in availability! Blooming important.
Type II diamond	7–1000	2.4	
Si	1.5–13; 30 – >1000	3.4	
Ge	2 –15; 30 – >1000	4.0	Blooming essential, especially in longer wavelength range.

pyroelectric material is responsible for the Johnson noise. Of all the noise sources the Johnson noise is probably the most important. In the higher performance thermal detectors it is often comparable with the temperature fluctuation noise while in the more rugged lower performance types it is usually the dominant noise source.

The final source of noise is that associated with the head amplifier. Usually the output signal at the detector is very small and therefore a low noise amplifier is required. A critical factor in choosing the amplifier is the output impedance of the detector. This can range from a very low value (about an ohm) for thermopiles to an extremely high value ($10^9 - 10^{12}$ ohms) for pyroelectric detectors and some helium cooled bolometers. It is the extreme cases which are the most difficult to deal with. It has been helped in recent years by the development of a range of low noise FETs suitable for both low and high impedance sources (*Putley* [3.11]). Even so situations can still arise when amplifier noise is the dominant noise source. This represents the ultimate limitation to the high frequency performance of pyroelectric detectors and is likely to be the limiting factor with the cooled bolometers.

Having evaluated all the sources of noise the noise equivalent power P_N can best be found by referring all noise sources to the input terminals of the head amplifier (Fig. 3.2) where they can be represented by a number of noise voltage generators ΔV_i in series with the amplifier input. An input signal power P will give rise to a signal voltage V_S at the amplifier input

$$V_S = RP \tag{3.13}$$

where R is the detector's voltage responsivity (volts/watt). The noise generators are combined by summing the squares so that they can be replaced by one equivalent generator V_N where

$$(V_N)^2 = \sum (\Delta V_i)^2 . \tag{3.14}$$

The noise equivalent power P_N is the value of P for which

$$V_S = V_N \tag{3.15}$$

i.e.,

$$RP_N = [\sum (\Delta V_i)^2]^{1/2} . \tag{3.16}$$

To determine P_N, therefore the responsivity has to be measured at signal levels high enough to avoid errors due to the noise sources but low enough to avoid possible nonlinear effects. V_N is usually found by measuring the total noise output from the detector/amplifier combination. It could be calculated from (3.14) when the separate V_i terms have all been measured. Although this is a very instructive exercise for the detector designer, it is not usually necessary or practicable for the user who wants to know the overall performance.

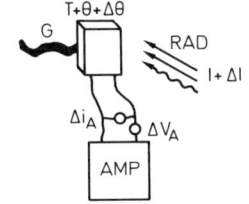
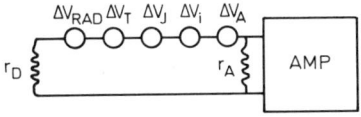

Fig. 3.2. Representation of noise sources

3.2 The Thermopile

The thermopile is one of the oldest infrared detectors, being first described by *Melloni* [3.4] but it is still widely used and in its latest form (the thin film thermopile) is used in space instrumentation.

The basic element in a thermopile is a junction between two dissimilar conductors having a large Seebeck coefficient Θ. To perform efficiently a large electrical conductivity σ is required to minimize Joulean heat loss and a small thermal conductivity K to minimize heat conduction loss between the hot and cold junctions of the thermopile. These requirements are incompatible and we find that in common with other thermoelectric devices (*Goldsmid* [3.12]) the best choice of thermoelectric material is that for which $\sigma\Theta^2/K$ is a maximum and that this occurs for certain heavily doped semiconductors, for example Bi_2Te_3 and related compounds. To make an efficient thermal infrared detector the device must also be an efficient absorber of the incident radiation and must have a small thermal mass to give as short a response time as possible.

In the earlier thermopiles these requirements were met by using fine metallic wires (typically copper-constantan or bismuth-silver) for the elements and attaching the hot junction to a receiver made of thin blackened gold foil. The introduction of semiconductor elements led to an improvement in sensitivity but because the production of fine wire is impracticable and because of the difficulty in making contacts between semiconductors a different form of construction was developed in which the gold foil receiver was also used as a contacting link between the two active elements. Although the sensitivity of the older thermopiles using metal elements is much lower than those using semiconductor elements, the metal elements can be made much more robust and stable so that they are still widely used where a high degree of reliability and of long term stability is required, two examples being industrial radiation pyrometers and ground based meteorological instruments for measuring the radiant intensity of the sun (*Drummond* [3.13]).

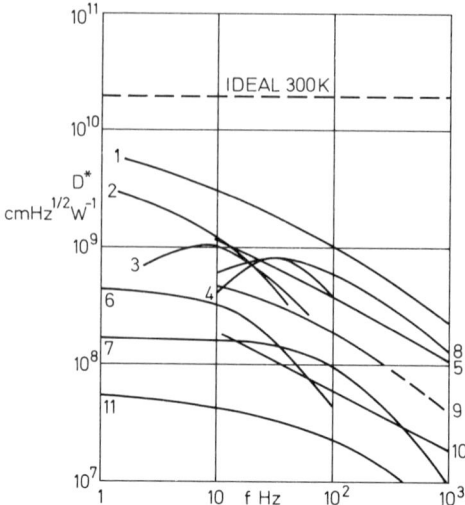

Fig. 3.3. Performance of uncooled thermal detectors
1 Mullard research sample using alanine doped TGS 10 μm thick and area 1.5 × 1.5 mm². Mounted in space qualified encapsulation and performance independently verified
2 Spectroscopic thermopile. Based on *Fellgett* [3.17], *Brown* et al. [3.18], and *Schwarz* [3.19]. Typical receiver area 0.4 mm², time constant 40 ms
3 Golay cell after *Hickey* and *Daniels* [3.20], *Stafsudd* and *Stevens* [3.21], and *Gill* [3.22]
4 TRIAS cell. Space qualified encapsulation (*Chatenier* and *Gauffre* [3.16])
5 Mullard production TGS detector in ruggedized encapsulation. Area 0.5 × 0.5 mm² (after *Baker* et al. [3.23])
6 Evaporated film thermopile (*Stevens* [3.24]). Receiver 0.12 × 0.12 mm², time constant 13 ms
7 Immersed thermistor (*De Waard* and *Weiner* [3.25]). Area of flake 0.1 × 0.1 mm². Time constant 2 ms
8 LiTaO$_3$ pyroelectric detector (*Stokowski* et al. [3.26])
9 SBN pyroelectric detector (*Liu* and *Maciolek* [3.27])
10 Plessey production ceramic pyroelectric detector
11 Thin film bolometer (*Bessonneau* [3.15])

The recent development of thin film techniques has enabled cheap thermopiles to be designed which can be fabricated as complex arrays with good reliability. By using antimony and bismuth for the sensitive elements devices can be produced which have some of the advantages of the metal wire thermopiles but with a higher sensitivity. Table 3.2 summarizes the characteristics of the different types of modern thermopiles. Figure 3.3 which shows the performance of a number of thermal detectors operating at room temperature includes two thermopiles.[1] The performance obtained with the spectroscopic thermopile

---

[1] In Fig. 3.3 the performance is given in terms of D^* which is derived from the noise equivalent power defined in Section 3.1 by the relation

$$D^* = a^{1/2}(\Delta f)^{1/2} P_N^{-1} \, \text{cm}^{1/2} \, \text{Hz}^{1/2} \, \text{W}^{-1}$$

where P_N is the noise equivalent power of a detector of area a measured using an amplifier with noise bandwidth Δf. D^* is useful for comparing detectors of different areas since most (but not all, see Sect. 3.5) noise sources vary as $a^{1/2}$, P_N improving as the area is reduced.

Table 3.2. Uncooled thermal detectors

Type	Responsivity VW^{-1}	NEP or D^* WHz$^{-1/2}$ cm Hz$^{1/2}$ W^{-1}		Response time or freq. resp.	Spectral coverage	Typical area mm^2	Notes/References
		Thermopiles					
Spectroscopic	5	10^9	D^*	10 ms	Visible to about 30 µm	0.5–5.0	see Fig. 3.3 curve 2
Thin film on polymer backing	100	3×10^8	D^*	0.1–10 ms	as above	10^{-2}–1.0	see Fig. 3.3 curve 6 Fast laser detector *Contreras* and *Gaddy* [3.14]
Thin film on heat sink	10^{-6}	10^6	D^*	30 ms	as above		
		Bolometers					
Thermistor	1000	$1.6 \times 10^8 \tau^{1/2}$	D^*	1–10 ms	vis.→40 µm	0.4–4.0	see Fig. 3.3 curve 7
Thin metal oxide film	130	4×10^7	D^*	1 ms	vis.→middle IR	0.5	*Bessonneau* [3.15]
		Pneumatic cells					
Golay		2×10^{-10} NEP		15 ms	vis.→mm (see note)	6	No window available covers whole spectral range With KRS 5 window. *Chatenier* and *Gauffre* [3.16]
TRIAS		3×10^{-10} NEP			0.5–40 µm	3	
		Pyroelectric detectors					
TGS		10^9	D^*	10 Hz	near IR	0.25	see Fig. 3.3 curve 5
		5×10^8		100 Hz	→		
		10^8		1000 Hz	sub-mm		
LiTaO$_3$		6×10^8		10 Hz	as above	1	
SBN doped rare earths		5×10^8		10 Hz	as above	0.05–16	see Fig. 3.3 curve 9
Modified PZT ceramic		10^8		10 Hz		4	see Fig. 3.3 curve 10
		10^7		1000 Hz			
PVF$_2$ film		10^8		10 Hz		80	

(curve 2) has only comparatively recently been bettered by the best attainable pyroelectric detectors. The disadvantages of this type of thermopile are that it is very delicate, has a rather long time constant and because of its very low output impedance needs a specially designed low noise amplifier. A screened step-up input transformer is usually employed with these thermopiles which makes the amplifier rather bulky and expensive. These thermopiles are still widely used in spectrometers. Curve 6 in Fig. 3.3 shows an antimony-bismuth evaporated film thermopile. Although its performance is somewhat lower than that of some other thermal detectors it is a rugged easily manufactured device and can be fabricated either as a single element or in a detector array with over 100 elements. Detectors of this type have been successfully used in a number of space instruments, including some used on interplanetary missions (*Chase* [3.28], *Bender* et al. [3.29]).

3.3 The Bolometer

The bolometer is a resistive element constructed from a material with a large temperature coefficient so that the absorbed radiation produces a large change in resistance.

To operate a bolometer (Fig. 3.4) an accurately controlled bias current i from a suitable source and regulating impedance is passed through the element. If the incident radiation input produces a small change δr in the resistance r of the bolometer then an output voltage

$$V_S = i\delta r \tag{3.17}$$

is generated. If the input radiation produces [cf. (3.4)] an increase θ in the temperature of the element and the temperature coefficient of resistance $(1/r)(dr/dT)$ is α then

$$V_S = i\alpha r\theta \tag{3.18}$$

so that from (3.4) the open circuit output voltage is

$$V_S = \eta I_\omega i\alpha r(G^2 + \omega^2 H^2)^{-1/2} \tag{3.19}$$

corresponding to a voltage responsivity

$$R = \eta i\alpha r(G^2 + \omega^2 H^2)^{-1/2}. \tag{3.20}$$

Equation (3.20) shows the desirability of having large values for α, r, and i as well as small values for G and H. This equation implies that the input impedance of the amplifier is large compared with r. This sets a limit on r so that the resistance cannot be increased independently. Also with very high resistance elements the

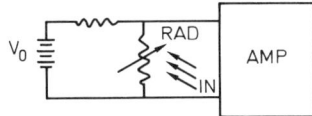

Fig. 3.4. Schematic circuit of bolometer

input capacity of the connecting leads and the amplifier can produce a time constant longer than the thermal time constant. The bias current i cannot be increased independently. When it becomes large Joule heating will raise the temperature of the element and if α is negative destructive thermal run away may occur, but apart from this an increase in noise with increasing i will set an optimum value. The problem of maximizing α then comes to choosing a suitable material. The same basic materials that are used for resistance thermometers are the obvious choices and in fact when constructed in a suitable configuration they have all been used.

In addition to radiation noise and temperature noise associated with the thermal impedance of the element, Johnson noise associated with the resistance r is one of the most important noise sources. With some types of bolometer low frequency current noise is important and is the principal factor limiting i. With room temperature bolometers amplifier noise should not be important but with cyrogenic devices it is usually the dominant noise source, especially when operating with cooled filters to limit the radiation noise to the sub-mm band.

The first bolometer produced by *Langley* in 1880 used a platinum resistance element and later other metals (such as nickel) were used. These metals are still used for resistance thermometers where their high long term stability is an essential requirement. For infrared detectors, however, the older metal film bolometers have been replaced by semiconducting elements which have a much larger temperature coefficient.

The first type of semiconductor bolometer to come into use was the thermistor developed during World War II at Bell Laboratories to provide a simple, reliable but sensitive detector for use both in the then rapidly developing science of infrared spectroscopy and for heat sensing applications. The thermistor bolometer is still one of the most widely used of infrared detectors, although it has been replaced by more sensitive devices for the more demanding applications.

Figure 3.5 illustrates the construction of a thermistor bolometer. Curve 7 of Fig. 3.3 gives the performance of a good thermistor bolometer. Further information is given in Table 3.2. Thermistor bolometers have been used successfully in satellite instrumentation, but are tending to be replaced by pyroelectric detectors. Thermistor bolometers are widely used in burglar alarms and fire detection system. Although, again, pyroelectric devices are replacing them, the possibility of producing cheaper bolometers from evaporated films has received some attention. *Bishop* and *Moore* [3.30] have studied chalcogenide glasses such as $Tl_2SeAs_2Te_3$, while more recently *Moustakas* and *Connell* [3.31] have prepared rf sputtered films of amorphous Ge_xH_{1-x} with a sensitivity within a factor of 5 of the thermistor bolometer.

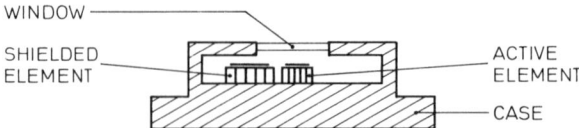

Fig. 3.5. Construction of a thermistor bolometer

With the need for high performance far infrared detectors arising at a time when the longest spectral response of a photoconductive detector was barely 5 μm, attention was given to the development of much higher performance bolometers making use of the fact that at very low temperatures it is possible to obtain much larger relative changes in resistance than near room temperature, the specific heat is much smaller [cf. (3.6)] and that if the detector and its surroundings are cooled to a very low temperature the ultimate sensitivity [see (3.12)] can be orders higher than that for a room temperature device. In practice, for most applications, it is necessary to have an aperture in the enclosure admitting some room temperature background radiation. It is this radiation which determines the ultimate sensitivity of the actual configuration, but with the sizes of apertures which can usually be used the improvement in performance over that of a device operating at room temperature is at least two orders of magnitude. *Putley* [3.8] has derived expressions for the ultimate sensitivity of helium cooled detectors taking into account the effects of room temperature radiation and has shown that the best NEPs which can be achieved in the sub-mm region are within the range $10^{-13} - 10^{-14}$ W, depending upon the spectral filtering and field of view.

The first cooled bolometers used a superconducting element with a transition in the helium temperature region. Problems encountered were that of the temperature stability required to maintain the element at the transition temperature and, since superconductors are very poor absorbers in the far infrared, difficulty in achieving efficient radiation absorption. Figure 3.6 shows a recent design of superconducting bolometer (*Gallinaro* and *Varone* [3.32]). The element consists of an evaporated tin film upon an Al_2O_3 substrate obtained by anodizing the aluminum block. The aluminum block is coupled to the surrounding bath of helium cooled below the λ point by means of the brass rod. The heater wound on the aluminum block is used to raise its temperature to the critical temperature for tin (3.7 K). Because the tin element is in fairly good thermal contact with the aluminum block the time constant of the detector is quite short (~ 3 μs) which is much shorter than that of the earlier designs of superconducting bolometers (e.g., *Martin* and *Bloor* [3.33]) in which a self-supporting film was used. Nevertheless, the sensitivity of this design of detector (NEP $\sim 10^{-13}$ W Hz$^{-1/2}$) is over an order better than that of the earlier one. This at first sight seems inconsistent with the shorter time constant but probably indicates that this structure achieves more efficient absorption of the signal radiation, which for the earlier design was poor. This design of superconducting bolometer overcomes many of the disadvantages of the earlier designs. It should be quite robust and it is for a thermal detector fast. It still has the drawback that accurate temperature control is required (to within 10^{-5} K). In

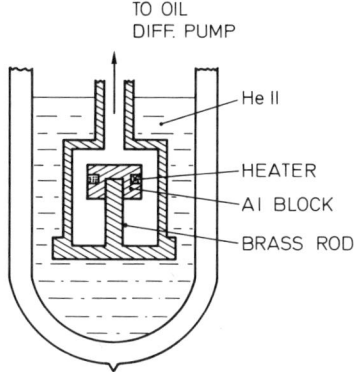

Fig. 3.6. Superconducting bolometer (*Gallinaro* and *Varone* [3.32])

this design this is achieved by controlling the heater current, but a recent suggestion is to use a biasing magnetic field thus varying the superconducting transition temperature to match fluctuations in bath temperature (*Zaitsev* et al. [3.34]).

Cooled semiconductor bolometers are more widely used than the superconducting ones. The semiconducting devices do not require critical temperature control and with the selection of correctly doped samples are better absorbers. The first cooled semiconductor bolometers employed flakes of the carbon composition material used for certain types of carbon resistors and for low temperature carbon thermometers but it was later found that superior performance could be obtained with correctly doped Ge or Si elements (which also are used as low temperature resistance thermometers). Considerable effort has now gone into the design of these bolometers. They are widely used in infrared astronomy where over most of the far infrared spectrum they have a uniform performance comparable in sensitivity with the best photon detectors. Since photon detectors only give their optimum performance over a comparatively narrow spectral band and since for most astronomical applications the relatively long response time of the bolometers is not a disadvantage, it is not surprising that the rapidly growing science of infrared astronomy largely relies on cyrogenic bolometers for its detectors.

Low [3.35] was the first to develop the Ge bolometer operating in liquid helium. Its mode of operation has been discussed in detail by *Zwerdling* et al. [3.36]. Pure Ge does not absorb strongly in the far infrared. If it is lightly doped with suitable impurities it exhibits extrinsic photoconductivity to just beyond 100 µm wavelength. If it is more heavily doped but compensated to maintain a high resistivity (and more important, a high resistance temperature coefficient) the impurities enhance the absorption, especially in the sub-mm region, but the energy absorbed is transferred rapidly to the lattice, raising the temperature of the sample rather than of the existing free carriers as in a photoconductor. The mechanism can be compared with that of the InSb sub-mm detector (*Putley* [3.11]) where impurity absorption is also employed. In the InSb detector the absorbed radiation raises the temperature of the free electrons which are only

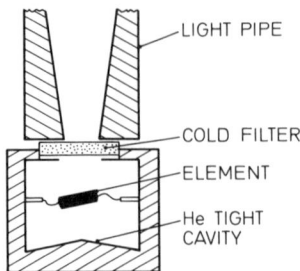

Fig. 3.7. Construction of a cryogenic semiconductor bolometer

loosely coupled to the lattice and it can be thought of as either an electronic bolometer or as a free carrier photoconductor. The Ge device is however a true bolometer but it relies on correct doping with impurities to achieve efficient absorption. This process is less efficient at shorter wavelengths so that for use at 10 μm or less a coating of black may be required. Typical impurity concentrations required are about 10^{16} cm^{-3} Ga with about 10^{15} cm^{-3} In giving p-type conductivity with a compensation ratio of about 0.1. At liquid helium temperatures with these concentrations of impurities an impurity interaction conduction mechanism is responsible for the sample's conductivity leading to an essentially exponential temperature variation of resistance:

$$r = r_0 \exp(AT^{-n}). \tag{3.21}$$

For normal semiconductors $n=1$, but the form of impurity conduction found in bolometer material is better characterized by $n=0.5$ (*Summers* and *Zwerdling* [3.37], *Redfield* [3.38]). From (3.21) the resistance temperature coefficient α is

$$\alpha = (1/r)(dr/dT) = -nAT^{-(n+1)}. \tag{3.22}$$

With $n=1/2$ typical values for A are in the range 20–30 K$^{1/2}$. Equation (3.22) shows that α increases as T is reduced. Since the responsivity is proportional to α and it is desirable to make this as large as possible, operation below 4 K is usually advantageous. There are also other advantages. Equations (3.4) and (3.6) show the desirability of reducing the thermal capacity. Since the specific heat falls as the temperature is reduced, lowering the temperature will again be advantageous. Lowering the temperature will increase the resistance r. If r is not too large this may also be advantageous, but if r rises to such a large value that matching to the amplifier becomes difficult then operation will become difficult and there may be an optimum value for the operating temperature. Reduction below the λ point also reduces low frequency noise from temperature instability and bubbling in the cryostat. In general, it pays to obtain as high a responsivity as possible.

To minimize the thermal capactiy the bolometer elements are made from slices of Ge about 0.2 mm thick. These slices will absorb between 10 and 70% of the incident radiation but by mounting the element in an integrating cavity (Fig. 3.7) practically total absorption can be achieved. The thermal conduction

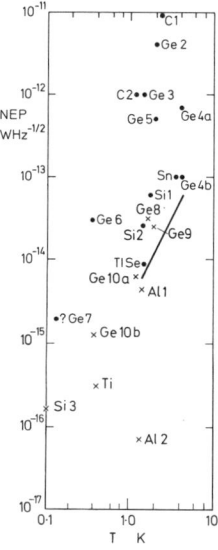

Fig. 3.8. Performance of cooled bolometers NEP of carbon, germanium, silicon, and thallium selenide semiconducting bolometers and tin, aluminum and titanium superconducting bolometers. Details in Table 3.3, composite bolometers in Table 8.1 (p. 303), 50% absorption has been assumed for the composite bolometers. The solid line is *Coron*'s [3.42] estimate of the best attainable performance of a Ge bolometer in the absence of higher temperature background radiation

coupling the element to the surrounding helium bath is usually adjusted by anchoring the bolometer leads to a point in good thermal contact with the bath. The thermal conductance then determines both the time constant (3.7) and the temperature fluctuation noise (3.9). Providing good current contacts are made by the leads, current noise will not be important but Johnson noise must be considered. The largest two noise sources will normally be that associated with the input radiation flux from the room temperature field of view visible to the element and that from the amplifier.

There have been several recent bolometer designs which have succeeded in making the internal noise sources very small. Thus *Drew* and *Sievers* [3.39] have shown that with an element cooled to about 0.5 K in a He^3 cryostat the NEP associated with the temperature and Johnson noise is about 10^{-14} WHz$^{-1/2}$ while *Draine* and *Sievers* [3.40] have shown that using He dilution cooling to 0.1 K this NEP falls to 3×10^{-16} WHz$^{-1/2}$. However this very low value could probably only be exploited in an experiment conducted entirely at cryogenic temperatures, and would certainly require a cryogenically operated low noise amplifier.

Recent improved Ge bolometer designs (*Coron* et al. [3.41–43]) have achieved a performance very close to the theoretical limits. Nevertheless, the use of newer cryogenic bolometers is being considered. *Nayar* and *Hamilton* [3.44] have used a bolometer constructed of thallium selenide. More attention has been given to the use of Si as an alternative to Ge. Si has the advantages that it has a lower specific heat, smaller at low temperatures by a factor of 8; because the impurity centers in Si are more localized than those in Ge a higher concentration is required for impurity interaction, so that the material is easier to prepare and will be a better absorber; and finally because of the highly developed technology of Si, techniques of fabrication, especially making noise free contacts, will be

Table 3.3. Cryogenic bolometers

Material	No. on Fig. 3.8	Operating temp. (K)	Responsivity (VW^{-1})	Noise equivalent power (WHz$^{-1/2}$)	Measuring freq. (Hz)	Response time (s)	Dimensions of element (mm)	References/Notes
Germanium	2	1.7	4.2×10^4	4×10^{-12}	600	2.7×10^{-4}	$1 \times 5 \times 0.25$	Zwerdling et al. [3.36]
Germanium	3	1.5	4×10^4	9×10^{-13}	30	5×10^{-3}	4.5 mm^3	Richards [3.48]
Germanium	4a	4.2	2.5×10^4	7×10^{-13}	80	3×10^{-4}	—	Coron et al. [3.43]. Three part bolometer.
Germanium	4b	4.2	—	1×10^{-13}	20	1×10^{-2}	—	(a) adjusted for large background (cut off filter at 40 μm and entendue 0.1 cm^2. SR) (b) low background.
Germanium	5	2.0	4.5×10^3	5×10^{-13}	200	4×10^{-4}	15 mm$^2 \times 0.12$	Low [3.35]
Germanium	6	0.37	2×10^6	3×10^{-14}	18	1×10^{-2}	$4 \times 2 \times 1$	Drew and Sievers [3.39]
Germanium	7	0.1	3×10^7	See note	33	1×10^{-2}	$3 \times 3 \times 0.5$	Draine and Sievers [3.40] The operational NEP was not measured. Likely to be background limited. Point on Fig. 3.8 is estimate made by comparison with Ge 6 ($\sim 2 \times 10^{-15}$).
Carbon	1	2.1	2.1×10^4	1×10^{-11}	13	1×10^{-2}	20 mm$^2 \times 0.08$	Boyle and Rodgers [3.49]
Carbon	2	1.2	3.5×10^5	1×10^{-12}	15	1×10^{-2}	$1 \times 4 \times 0.03$	Corsi et al. [3.50]
Tin	—	3.7	1.4×10^3	1×10^{-13}	1.25×10^4	3×10^{-6}	10 mm^2	Gallinaro and Varone [3.32]
Silicon	1	1.8	2.8×10^5	6×10^{-14}	10	1×10^{-2}	$5 \times 5 \times 0.4$	Chanin [3.46]
Silicon	2	1.5	1.0×10^6	2.5×10^{-14}	13	1×10^{-2}	—	Kinch [3.45]
Thallium selenide	—	1.5	1.0×10^6	8.3×10^{-15}	32	8×10^{-3}	$5 \times 1 \sim 0.6$	Nayar and Hamilton [3.44]

easier. The results reported for Si bolometers (*Kinch* [3.45], *Chanin* [3.46], *Kunz* [3.47]) show that their performance compares favorably with Ge bolometers.

Figure 3.8 and Table 3.3 give more detailed information on the cryogenic bolometers. In comparing different results it must be remembered that it is not very meaningful to use D^* or other figures of merit. This is because two of the principal noise sources are independent of the detectors' area. Thus the main thermal conductance contributing to the temperature noise term is usually that of the electrical lead and in many cases the amplifier noise is still the largest single noise source.

3.4 The Golay Cell and Related Detectors

Herschel discovered infrared using a mercury thermometer as detector. Before the discovery of the thermopile a more sensitive form of constant pressure gas thermometer was used by *Leslie* and others to study radiant heat. This type of detector went out of use with the introduction of the thermopile and bolometer, but the idea of using a form of gas thermometer was revived by *Hayes* and developed by *Golay* and others just before World War II. The ensuing Golay cell (*Golay* [3.6]) is widely used in laboratory instruments. It is still one of the most sensitive room temperature detectors but its bulkiness, comparative fragility, sensitivity to vibration and rather slow response time limit it to laboratory applications. In the Golay cell (Fig. 3.9) radiation absorbed by a receiver inside a closed capsule of gas (usually xenon for its low thermal conductivity) heats the gas causing its pressure to rise which distorts a flexible membrane on which a mirror is mounted. The movement of the mirror is used to deflect a beam of light shining on a photocell and so producing a change in the photocell current as the output. In modern Golay cells (*Hickey* and *Daniels* [3.20]) the beam of light is provided by a light emitting diode and a solid state photodiode is used to detect it. The reliability and stability of this arrangement is significantly better than that of the earlier Golay cells which used a tungsten filament lamp and a vacuum photocell. Apart from the greater reliability of the modern components, the lower heat dissipation within the device probably increases the reliability of the gas cell itself.

Another method of obtaining an electrical output from the gas cell is to place a fixed conductor near the distorting membrane forming a variable condenser which can be measured with a suitable circuit. This arrangement is used in the ONERA detector (*Chatanier* and *Gauffre* [3.16]) and it is also used in gas analyzers (Luft cells) in which the gas to be analyzed is placed in one cell and the output compared with that from another cell containing a reference sample of gas. An attraction of this type of gas analyzer is that it utilizes the infrared spectral characteristics of the gas without requiring a dispersive element.

Although these various types of pneumatic detector seem awkward compared with modern solid state devices, their importance should not be underestimated. The Golay cell is still the basic detector used in a large amount

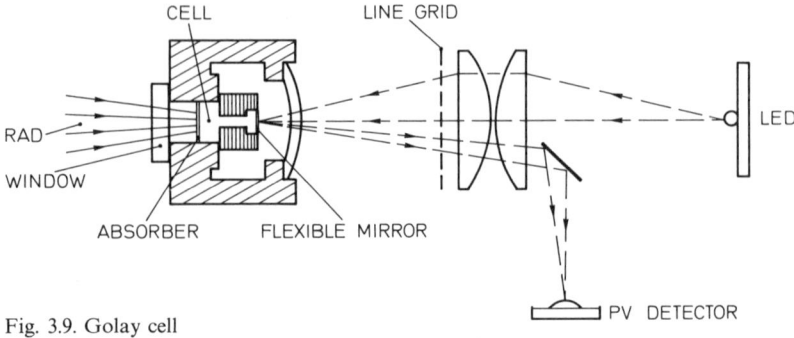

Fig. 3.9. Golay cell

of laboratory instrumentation and the gas analysis technique employing the Luft cell is widely used. One version of this detector has even been qualified for space use (*Chatanier* and *Gauffre* [3.16]). The Golay cell, being used from the visible to wavelengths of several mm has been used over a broader band of the spectrum than any other detector, except possibly the pyroelectric. Table 3.2 and Fig. 3.3 give detailed performance information.

3.5 Pyroelectric Detector

The employment of the pyroelectric effect to detect infrared was first proposed just before World War II. It did not attract a lot of interest until recently. The discovery of new materials with more attractive properties together with a growing requirement for uncooled thermal detectors having a better performance than the thermistor bolometers or the ruggeder types of thermopile but being more robust and suitable for use in industrial or military environments than the high sensitivity thermopiles or Golay cells has resulted in considerable attention to pyroelectric devices making them at the present time the most widely studied of infrared detectors.

A pyroelectric material is one of low enough crystalline symmetry to be able to possess an internal electric dipole moment. Although the external field produced by this dipole will normally be neutralized by an extrinsic charge distribution near the surface of the material, in good pyroelectric material (which are good insulators) this extrinsic charge distribution is relatively stable so that even quite slow changes in the sample's temperature, which produce changes in the internal dipole moment, produce a measurable change in surface charge. Hence if a small capacitor is fabricated by applying a pair of electrodes to the sample the change in temperature and hence the incident thermal radiation can be detected by measuring the charge on the condenser. The magnitude of the pyroelectric effect is such that the sensitivity of the best detectors is comparable with that of the Golay cell or of a high sensitivity thermopile but an element embodying this performance is considerably more robust than the other thermal detectors of comparable sensitivity.

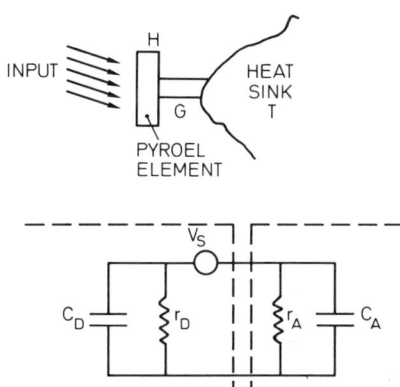

Fig. 3.10. Schematic circuit of pyroelectric detector

The electrical impedance of a pyroelectric detector is almost that of a pure capacitance. Hence an output signal only appears when the input radiation is changing. For maximum output the rate of change of the input radiation should be comparable with the electrical (RC) time constant of the element. Figure 3.10 is the equivalent electrical circuit of a pyroelectric detector (*Putley* [3.11, 51]). Assume that the element receives radiation over an area A normal to the polar axis of the material and that this produces a modulated temperature rise θ_ω (3.4). The corresponding voltage developed across the amplifier input is

$$V = \omega p A \theta_\omega r (1 + \omega^2 \tau_E^2)^{-1/2} \tag{3.23}$$

where p is the pyroelectric coefficient and

$$\tau_E = rC \tag{3.24}$$

is the electrical time constant of the output circuit of the detector (see Fig. 3.10; r and C are the equivalent parallel resistance and capacity, respectively, of the detector and amplifier input). Substituting for θ_ω from (3.4) gives the voltage responsivity

$$R = \eta(\omega p A r / G)(1 + \omega^2 \tau_E^2)^{-1/2}(1 + \omega^2 \tau_T^2)^{-1/2} \tag{3.25}$$

where τ_T the thermal time constant is given by (3.6).

Equation (3.25) shows that at very low frequencies $R \propto \omega$, thus $R \to 0$ as $\omega \to 0$. There will be an intermediate frequency range (assuming $\tau_E \neq \tau_T$) in which R is independent of ω but at high frequencies ($\omega \ll 1/\tau_T, 1/\tau_E$) R will vary as ω^{-1}. Since in practice τ_T and τ_E both usually come within the range $10-0.1$ s, for most applications the high frequency approximation of (3.25) is valid, i.e.,

$$R = \eta p A / \omega H C. \tag{3.26}$$

This equation shows that to achieve a high responsivity a pyroelectric material with a large value of $(p/\varepsilon C')$ is required, where ε is the dielectric constant and C' the volume specific heat of the material. The quantity $(p/\varepsilon C')$ is therefore a useful figure of merit for selecting pyroelectric material, although when noise considerations are included it may be necessary to modify it. It is not possible to define a single pyroelectric figure of merit serving as a guide for all applications.

The principal noise sources are temperature (3.9), Johnson and amplifier noise. Temperature noise sources have been discussed in detail by *Logan* (*Logan* and *Moore* [3.9], *Logan* [3.10]) who has shown that in addition to the radiative conduction, conduction and convection into the ambient gas in the encapsulation and lateral conduction into the surrounds of the element are important. With present materials, however, the Johnson noise under most circumstances is more important. If we assume that the principal contribution to r in Fig. 3.10 comes from the dielectric loss of the pyroelectric material, then the appropriate expressions for the Johnson noise limited noise equivalent power (see (3.16)) is

$$P_N \text{ (Johnson noise)} = \eta^{-1}(4kT)^{1/2}(C'/p)(\omega\varepsilon'\varepsilon_0 \tan\delta)^{1/2}(Ad)^{1/2} \qquad (3.27)$$

where ε' is the real part of the dielectric constant, $\tan\delta$ the dielectric loss factor, ε_0 the free space dielectric constant and d is the thickness of the element. Equation (3.27) emphasizes the need for low dielectric loss material. Finally amplifier noise is important especially with large detectors at high frequencies and small detectors at low frequencies. To illustrate the relative behavior of these noise sources Fig. 3.11 and the accompanying table show for a numerical example, approximating to triglycine sulphate (TGS) how the noise equivalent powers associated with these sources vary with frequency. Study of the expressions for the noise equivalent powers in the table shows the difficulty of defining a material figure of merit.

The principal materials used for pyroelectric detectors are members of the TGS group, lithium tantalate, strontium barium niobate, ceramics members of the lead zirconate titanate (PZT) group and, more recently, films of the polymers polyvinyl fluoride (PVF) and polyvinylidene fluoride (PVF$_2$).

The TGS group includes the isomorphs triglycine selenate and fluoberyllate, the corresponding deuterated compounds, mixed crystals of these and material doped with other amino acids or with metals. These materials are used for the most sensitive detectors and also in the pyroelectric vidicon. They have several disadvantages however, with the result that the refractory oxide materials or the polymer films tend to be preferred where their performance is adequate.

The TGS group are water soluble (the crystals are grown from aqueous solution) and therefore devices made from them must be properly encapsulated. They have relatively low Curie temperatures which restrict their maximum operating temperature. When the pure materials are raised above their Curie temperature they tend to depole and remain depoled upon cooling. Since it can be very inconvenient to have to repole them every time before use (for the

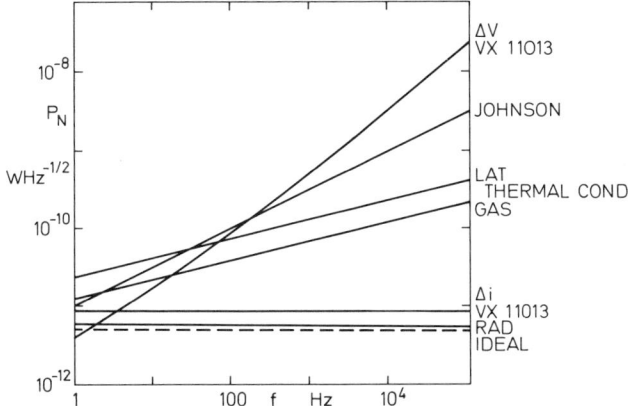

Fig. 3.11. The noise equivalent power of a pyroelectric detector. The frequency dependence of the various noise sources is shown for a detector of area 10^{-2} cm². The following expressions for each source's contribution to the noise equivalent power were used:

$$P_N \text{ (radiation)} = \eta^{-1/2}(16k\sigma T^5)^{1/2} A^{1/2}$$
$$P_N \text{ (gas conduction)} = \eta^{-1}(4kT^2)^{1/2}(2\omega/k_g)^{1/4} K_g^{1/2} A^{1/2}$$
$$P_N \text{ (lateral thermal conduction)} = \eta^{-1}(4kT^2)^{1/2}(2\pi\omega/k_p)^{1/4} K_p^{1/2} A^{1/4} d^{1/2}$$
$$P_N \text{ (Johnson noise)} = \eta^{-1}(4kT)^{1/2}(c'/p)(\omega\varepsilon'\varepsilon_0 \tan\delta)^{1/2}(Ad)^{1/2}$$
$$P_N \text{ (amplifier voltage noise)} = \eta^{-1}(\omega\Delta V)(c'\varepsilon'\varepsilon_0/p)A$$
$$P_N \text{ (amplifier current noise)} = \eta^{-1}(\Delta i)(c'/p)d$$

where k, σ are Boltzmann's and Stefan's constant, respectively, T the absolute temperature, ω the angular frequency, k_g and K_g are respectively the thermal diffusivity and thermal conductivity of the surrounding gas. A is the area of the detecting element and d its thickness. p is the pyroelectric coefficient, c' the volume specific heat, K_p the thermal conductivity, k_p the thermal diffusivity, ε' the real part of the relative dielectric constant, $\tan\delta$ the dielectric loss factor and ε_0 the free space dielectric constant. ΔV and Δi are the equivalent voltage and current noise generators of the amplifier. η is the fraction of the incident radiation absorbed by the detector.
The values assumed for the physical properties were as follows: $p = 3.5 \times 10^{-8}$ C cm^{-2} K^{-1}, $c' = 2.5$ J cm^{-3}, $\varepsilon' = 30$, $\tan\delta = 5 \times 10^{-3}$, $K_p = 6 \times 10^{-3}$ W cm^{-1} K^{-1} ($k_p = K_p c'$), $\eta = 0.8$, $T = 290$ K. The thickness of the element $d = 10$ μm. The amplifier was a Texas Instruments, Ltd VX 11013

detectors, this is normal practice with the pyroelectric vidicon) the introduction of permanently poled material has removed one of the difficulties in using them. It was found (*Lock* [3.52]) that the substitution of a small percentage of the glycine by L-alanine introduced asymmetry in the hysteresis loop which gave the material a preferred poling direction thus eliminating the random poling of adjacent domains typical of the pure material. It was also found that the doped material has somewhat better pyroelectric properties than the pure. It has a slightly higher pyroelectric coefficient and a lower dielectric constant. This is probably a consequence of the doped material being more completely poled than can usually be obtained with pure material where a few domains are permanently locked in opposite directions.

The best detector performance has been obtained with L-alanine doped TGS (see Fig. 3.3 and Table 3.2). Material studies (*Bye* et al. [3.53]) indicate that even better results could be obtained with L-alanine doped mixed crystals of TGS and TGSe, but practical devices using these material do not appear to have been developed yet. For pyroelectric vidicons, where the material requirements are somewhat different, the best results have been reported for the deuterated compounds DTGS and DTGFB (*Nelson* [3.54], *Stupp* [3.55]).

The performance obtained with some other pyroelectric materials is also given in Fig. 3.3 and Table 3.2. These detectors are used where robustness and cheapness are important. Thus they provide simple detectors for high power lasers and may be used to monitor relatively fast wave forms. Earlier it was stated that both the time constants associated with pyroelectric devices are about 1 s. It therefore seems strange at first sight that pyroelectric detectors should be usable at much higher frequencies than other thermal detectors. Referring to (3.27), we see that the Johnson noise limited noise equivalent power deteriorates as $\omega^{1/2}$ due to the increasing dielectric loss with rising frequency. Hence the performance falls off at this rate until at high frequencies the amplifier voltage noise dominates and the performance falls directly as the frequency. This is illustrated in Fig. 3.11. It must be pointed out that the amplifier used in this example was chosen to give the best low frequency performance. It is possible to choose devices with lower voltage noise (but higher current noise) where the best high frequency performance is required (*Putley* [3.11]). From (3.26), if the limiting noise source is the amplifier voltage noise ΔV_A the corresponding noise equivalent power will be

$$P_N = \omega \Delta V_A H C / \eta p A. \tag{3.28}$$

Using the same numerical constants as for Fig. 3.11 but assuming that the amplifier input capacitance cannot be neglected, we find

$$P_N = 1.12 \times 10^{-5} \, \omega \Delta V_A C \, \text{WHz}^{-1/2}.$$

If for example we select a Texas Instruments BF 817 for the amplifier then $\Delta V_A \sim 8 \times 10^{-10}$ VHz$^{-1/2}$ and a reasonable estimate for the total capacity, including both detector and amplifier is ~ 20 pF, we obtain (for $f > 10$ kHz)

$$P_N = 8.6 \times 10^{-15} f.$$

Hence with

$$f = 1 \, \text{MHz}, P_N = 8.6 \times 10^{-9} \, \text{WHz}^{-1/2}.$$

This calculation has assumed the use of TGS, and that if one of the more robust materials is used the performance will be 1 to 2 orders worse. Nevertheless, it still represents an attractive performance for an uncooled detector.

If it is inconvenient to operate with the responsivity varying as $1/f$, then two alternatives are possible. One is to provide a frequency compensating amplifier, as used with television camera tubes. This does not degrade the overall noise

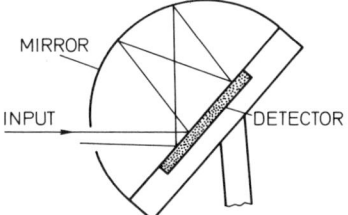

Fig. 3.12. Integrating cavity pyroelectric detector (*Day* et al. [3.57])

performance. The second alternative is to shunt the detector with a low resistance thus reducing the electrical time constant. In this way detectors with response time as short as 1 ps have been demonstrated (*Roundy* et al. [3.56]). This procedure is simple but it has the disadvantage that the Johnson noise from the resistor will degrade the performance.

A growing application for the simple ceramic detectors is in burglar alarms and other security devices. The higher performance devices are becoming more widely used in infrared instrumentation, including satellite radiometers (*Hamilton* et al. [3.57]) and horizon sensors. A recent application is in a special standard radiometer in which one of the metal electrodes is also used as a heating element so that the absolute value of the incident signal can be obtained by a substitution method and a related application is to a cavity detector (Fig. 3.12) where the object is to provide a wide band accurately black detector as a standard for measuring the spectral response of other detectors (*Day* et al. [3.58]). Of future developments, the further use of the plastic materials is one of the most interesting. Further fundamental research is required but there is a significant probability that eventually they will replace most of the materials currently in use.

3.6 Other Types of Thermal Detectors

Several other thermal effects have received some attention but have not received widespread use. Insofar as at frequencies less than 100 Hz the performance of some of the existing detectors approaches close to the ideal limit there seems little advantage in developing a new type of detector. There is still a large discrepancy between the performance of an ideal detector and the best achievable even with a pyroelectric detector at higher frequencies so that new developments here would be worthwhile.

An interesting device (a solid state analogue of the Golay cell?) was developed by *Jones* and *Richards* [3.59] who used an optical lever to amplify the linear expansion of a constantan strip. With an area of $5 \times 0.2\,\text{mm}^2$ the noise equivalent power was $10^{-11}\,\text{WHz}^{-1/2}$, corresponding to a D^* of 10^{10}. The response time was about 0.1 s. The D^* achieved places the performance within a factor of 2 of the ideal limit.

The temperature variation of the dielectric constant has been considered several times (*Moon* and *Steinhardt* [3.60], *Hanel* [3.61], *Maserjian* [3.62]) but the performance attained with dielectric bolometers has not been as high as expected.

Some study has been given to the magnetic analogue of the pyroelectric effect—the pyromagnetic effect (*Walser* et al. [3.63]). At low frequencies this could probably achieve a similar performance to the pyroelectric detector but the requirements for the magnetic field are likely to make higher frequency operation more difficult.

The Nernst effect in suitable semiconductors (InSb–NiSb mixed crystals, *Paul* and *Weiss* [3.64]; Cd_3As_2–InAs, *Goldsmid* et al. [3.65]; Bi–Sb, *Washwell* et al. [3.66]) has also been studied. This has produced a fairly fast rather insensitive detector which could perhaps be developed further.

3.7 The Use of Thermal Detectors in Infrared Imaging Systems

Thermal detectors were used to produce the first infrared imagers. At the present time the majority of imaging systems employ cooled photon detectors. The pyroelectric vidicon is a thermal device now capable of producing images of adequate quality for many purposes and thus providing simpler and cheaper imaging systems. Ultimately an all solid state infrared imager, analogous to the solid state visible TV cameras now being developed, can be expected. This may employ a thermal detection system, but its development has not advanced sufficiently for very reliable prediction of its final form.

Probably the earliest proposed thermal imaging system was that known as the evaporograph which employed a thin film of oil supported on a blackened membrane. The image is focused on the membrane causing differential evaporation of the oil film. The film is illuminated with visible light to produce an interference pattern corresponding to the thermal picture. This device was slow and with poor spatial resolution and it was intermittent in operation because the film had to be replenished from time to time. More recently the Gretag Panicon has employed an oil film sensor. Although the Gretag device was very simple to use it still suffered from the main limitation of the evaporograph (*McDaniel* and *Robinson* [3.67], *Dodel* et al. [3.68]).

The majority of thermal imagers at present in use employ either a single element detector or an array of a relatively small number which is scanned across the scene to build up the image. Some of the first successful imagers of this type employed a single element thermal detector—a thermistor (*Astheimer* and *Wormser* [3.69]) or a pyroelectric detector (*Astheimer* and *Schwarz* [3.70]). The disadvantage of this type of thermal imager was that the time required to scan the scene was too long for real time application.

Another thermal imaging system using a two dimensional sensor is the absorption edge converter (*Hilsum* and *Harding* [3.71]) in which heat absorbed by a selenium film shifts its absorption edge and so modulates a beam of sodium yellow light transmitted through the film. This has a high sensitivity but a poor

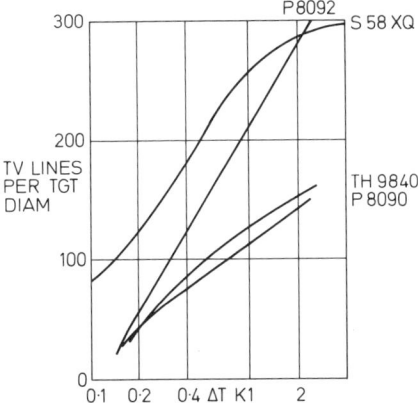

Fig. 3.13. Performance of commercially available pyroelectric vidicons (taken from manufacturers' data sheets)

Fig. 3.14. Thermal picture taken with pyroelectric vidicon similar to P 8092 (Fig. 3.13) using a Rank IRTAL 3 lens

spatial resolution due to thermal spread. Other two dimensional thermal sensors which have been considered include liquid crystals (*Ennulat* and *Fergason* [3.72]), infrared sensitive phosphors (*Urbach* et al. [3.73]) and bolometric layers mounted in a camera tube to utilize electron beam readout. All these devices suffer primarily from poor spatial resolution, although their sensitivity was also rather low.

The most promising thermal imaging system is that using the pyroelectric vidicon. The present development stems from the work of *Hadni* et al. [3.74, 75].

Active development is now taking place in the USA, the UK, France and West Germany. The performance being achieved is now approaching that first predicted (*Le Carvennec* [3.76]). It has been found that pure TGS is not the best material, better results being obtained with DTGS or DTGFB (*Watton* et al. [3.77], *Stupp* [3.55]). The problem of thermal spread responsible for the poor spatial resolution which was the main limitation of the earlier types of thermal camera tube is reduced in the pyroelectric vidicon by the fact that it is necessary to modulate the picture since pyroelectric devices do not produce an output for a steady picture. By modulating the system relatively fast (at a rate corresponding to the television frame time) thermal equilibrium is never established and hence the degradation due to thermal spread is reduced. So far the tubes produced have used thin discs of target material. By reticulating the target into small insulae about 50 μm across the thermal spread can be reduced further (*Stokowski* et al. [3.26], *Yamaka* et al. [3.78]). Figure 3.13 shows typical performance obtainable at the present time from commercially available pyroelectric vidicons while Fig. 3.14 is a photograph obtained using one of the P 8092 tubes (*Watton* et al. [3.79]). With the performance now being attained one can foresee the pyroelectric vidicon making a useful contribution to thermal imaging techniques.

The ultimate in thermal imaging systems will no doubt be some completely solid state system (*Barbe* [3.80]). It is not clear whether pyroelectric (*Steckl* et al. [3.81]) or other thermal sensors can be employed in this, but the possibility has the attraction of appearing to be the only one which does not require any cooling below normal ambient temperatures.

References

3.1 T.H.Bullock, R.Barrett: Comm. in Behav. Biol. **A1**, 19–29 (1968)
3.2 J.F.Harris, R.I.Gamow: Science **172**, 1252–1253 (1971)
3.3 J.F.Harris, R.I.Gamow: Proc. of the 9th Annual Rocky Mountain Bioengineering Symposium and the 10th International ISA Biomedical Science Instrumentation Symposium, Omaha Neb. USA 1–3 May 1972 (ISA BM 72336) **9**, 187–190 (1972)
3.4 M.Melloni*: Ann. Phys. **28**, 371 (1833)
3.5 S.P.Langley*: Nature **25**, 14 (1881)
3.6 M.J.E.Golay: Rev. Sci. Instr. **18**, 357–362 (1947)
3.7 R.A.Smith, F.E.Jones, R.P.Chasmar: *The Detection and Measurement of Infra-Red Radiation*, 2nd Ed. (Oxford University Press, London 1968)
3.8 E.H.Putley: Infrared Phys. **4**, 1–8 (1964)
3.9 R.M.Logan, K.Moore: Infrared Phys. **13**, 37–47 (1973)
3.10 R.M.Logan: Infrared Phys. **13**, 91–98 (1973)
3.11 E.H.Putley: *Semiconductors and Semimetals*, ed. by R.K.Willardson and A.C.Beer, Vol. **12** (Academic Press, New York 1977) pp. 441–449
3.12 H.J.Goldsmid: Thermoelectric Refrigeration (Heywood, London 1964)
3.13 A.J.Drummond: Advan. in Geophys. **14**, 1–52 (1970)
3.14 B.Contreras, O.L.Gaddy: Appl. Phys. Lett. **17**, 450–453 (1970)
3.15 G.Bessonneau: Qual. Rev. Prat. Controle Ind. (France) **15**, 11–14 (1976)

---

* For a recent account of the work of these pioneers see
E.Scott Barr: Infrared Phys. **2**, 67–73 (1962); Infrared Phys. **3**, 195–206 (1963).

3.16 M. Chatanier, G. Gauffre: IEEE Trans. Instr. and Meas. **IM22**, 179–181 (1973)
3.17 P. B. Fellgett: Rad. and Electron Eng. **42**, 476 (1972)
3.18 D. A. H. Brown, R. P. Chasmar, P. B. Fellgett: J. Sci. Instr. **30**, 195–199 (1953)
3.19 E. Schwarz: Research **5**, 407–411 (1952)
3.20 J. R. Hickey, D. B. Daniels: Rev. Sci. Instr. **40**, 732–733 (1969)
3.21 O. Stafsudd, N. B. Stevens: Appl. Opt. **7**, 2320–2322 (1968)
3.22 J. C. Gill: private communication 1958
3.23 G. Baker, D. E. Charlton, P. J. Lock: Rad. and Electron. Eng. **42**, 260–264 (1972)
3.24 N. B. Stevens: *Semiconductors and Semimetals*, ed. by R. K. Willardson and A. C. Beer (Academic Press, New York 1970) **5**, chap. 7, table (1) and figure (15)
3.25 R. de Waard, S. Weiner: Appl. Opt. **6**, 1327–1331 (1967)
3.26 S. E. Stokowski, J. D. Venables, N. E. Byer, T. C. Ensign: Infrared Phys. **16**, 331–334 (1976)
3.27 S. T. Liu, R. B. Maciolek: J. Electron. Mater. **4**, 91–100 (1975)
3.28 S. C. Chase, Jr.: Proc. of the Tech. Prog. Elect. Opt. Systems Design Conf. May 1971, pp. 157–163
3.29 M. L. Bender, P. W. Callaway, S. C. Chase, G. F. Moore, R. D. Ruiz: Appl. Opt. **13**, 2623–2628 (1974)
3.30 S. G. Bishop, W. J. Moore: Appl. Opt. **12**, 80–83 (1973)
3.31 T. D. Moustakas, G. A. N. Connell: J. Appl. Phys. **47**, 1322–1326 (1976)
3.32 G. Gallinaro, R. Varone: Cryogenics **15**, 292–293 (1975)
3.33 D. H. Martin, D. Bloor: Cryogenics **1**, 159–165 (1961)
3.34 G. A. Zaitsev, V. G. Stashkov, I. A. Krebtov: Cryogenics **16**, 440–441 (1976)
3.35 F. J. Low: J. Opt. Soc. Am. **51**, 1300–1304 (1961)
3.36 S. Zwerdling, R. A. Smith, J. P. Theriault: Infrared Phys. **8**, 271–336 (1968)
3.37 C. J. Summers, S. Zwerdling: IEEE Trans. Microwave Theo. and Tech. **MTT-22**, 1009–1013 (1974)
3.38 D. Redfield: Phys. Rev. Lett. **30**, 1319–1322 (1973)
3.39 H. D. Drew, A. J. Sievers: Appl. Opt. **8**, 2067–2071 (1969)
3.40 B. T. Draine, A. J. Sievers: Opt. Comm. **16**, 425–428 (1976)
3.41 N. Coron, G. Dambier, J. Le Blanc: *Infrared Detector Techniques for Space Research*, ed. by V. Manno and J. Ring (D. Reidel Publishing Co, Dordrecht 1972) pp. 121–131
3.42 N. Coron: Infrared Phys. **16**, 411–419 (1976)
3.43 N. Coron, G. Dambier, J. Le Blance, J. P. Moalic: Rev. Sci. Instr. **46**, 492–494 (1975)
3.44 P. S. Nayar, W. O. Hamilton: J. Opt. Soc. Am. **65**, 831–833 (1975)
3.45 M. A. Kinch: J. Appl. Phys. **42**, 5861–5863 (1971)
3.46 G. Chanin: *Infrared Detection Techniques for Space Research*, ed. by V. Manno and J. Ring (D. Reidel, Dordrecht 1972) pp. 114–120
3.47 L. W. Kunz: SPIE **67** Long Wavelength Infrared pp. 37–40 (1975)
3.48 P. L. Richards: *Far Infrared Properties of Solids*, ed. by S. Nudelman and S. S. Mitra (Plenum Press, New York 1970) pp. 103–120
3.49 W. S. Boyle, K. F. Rodgers: J. Opt. Soc. Am. **49**, 66–69 (1959)
3.50 S. Corsi, G. Dall'Oglio, G. Fantoni, F. Melchiorri: Infrared Phys. **13**, 253–272 (1973)
3.51 E. H. Putley: Semiconductors and Semimetals, ed. by R. K. Willardson and A. C. Beer (Academic Press, New York 1970) Vol. 5, 259–285
3.52 P. J. Lock: Appl. Phys. Lett. **19**, 390–391 (1971)
3.53 K. L. Bye, P. W. Whipps, E. T. Keve, M. R. Josey: Ferroelectrics **11**, 525–534 (1976)
3.54 P. D. Nelson: Electron. Lett. **12**, 652 (1976)
3.55 E. H. Stupp: SPIE Technical Symposium East, Reston Virginia 22–25 March 1976, Proc. SPIE Vol. **78**, 23–27
3.56 C. B. Roundy, R. L. Byer, D. W. Phillion, D. J. Kuizenga: Opt. Comm. **10**, 374–377 (1974)
3.57 C. A. Hamilton, R. J. Phelan, Jr., G. W. Day: Opt. Spectra **9**, 37–38 (1975)
3.58 G. W. Day, C. A. Hamilton, K. W. Pyatt: Appl. Opt. **15**, 1865–1868 (1976)
3.59 R. V. Jones, J. C. S. Richards: J. Sci. Instr. **36**, 90–94 (1959)
3.60 P. Moon, R. L. Steinhardt: J. Opt. Soc. Am. **28**, 148–162 (1938)

3.61 R.A.Hanel: J. Opt. Soc. Am. **51**, 220–224 (1960)
3.62 J.Maserjian: Appl. Opt. **9**, 307–315 (1970)
3.63 R.M.Walser, R.W.Bené, R.E.Caruthers: IEEE Trans. Electron Devices **ED-18**, 309–315 (1971)
3.64 B.Paul, H.Weiss: Solid-State Electron. **11**, 979–981 (1968)
3.65 H.J.Goldsmid, N.Savvides, C.Uher: J. Phys. D: Appl. Phys. **5**, 1352–1357 (1972)
3.66 E.R.Washwell, S.R.Hawkins, K.F.Cuff: Appl. Phys. Lett. **17**, 164–166 (1970); Proc. Spec. Meeting on Unconventional Infrared Detectors (University of Michigan, Infrared Information and Analysis Center, Ann Arbor, 1971) pp. 89–111
3.67 G.W.McDaniel, D.Z.Robinson: Appl. Opt. **1**, 311–324 (1962)
3.68 G.Dodel, J.Krautter, H.Häglsperger: Infrared Phys. **16**, 237–242 (1976)
3.69 R.W.Astheimer, E.M.Wormser: J. Opt. Soc. Am. **49**, 184–187 (1959)
3.70 R.W.Astheimer, F.Schwarz: Appl. Opt. **7**, 1687–1695 (1968)
3.71 C.Hilsum, W.R.Harding: Infrared Phys. **1**, 67–93 (1961)
3.72 R.D.Ennulat, J.L.Fergason: Mol. Cryst. Liq. Cryst. **13**, 149–164 (1971)
3.73 F.Urbach, D.Pearlman, H.Hemmendinger: J. Opt. Soc. Am. **36**, 372–381 (1946)
3.74 A.Hadni: J. Phys. **24**, 694 (1963)
3.75 A.Hadni, Y.Henninger, R.Thomas, P.Vergnat, B.Wyncke: J. Phys. **26**, 345–360 (1965)
3.76 F.LeCarvennec: Advan. Electron. Electron. Phys. **28A**, 265–272 (1969)
3.77 R.Watton: Ferroelectrics **10**, 91–98 (1976)
3.78 E.Yamaka, A.Teranishi, K.Nakamura, T.Nagashima: Ferroelectrics **11**, 305–308 (1976)
3.79 R.Watton, D.Burgess, B.Harper: J. Appl. Sci. and Eng. **A2**, 47–63 (1977)
3.80 D.F.Barbe: Proc. IEEE **63**, 38–67 (1975)
3.81 A.J.Steckl, R.D.Nelson, B.T.French, R.A.Gudmundsen, D.Schecater: Proc. IEEE **63**, 67–74 (1975)

4. Photovoltaic and Photoconductive Infrared Detectors

D. Long

With 15 Figures

Photovoltaic and photoconductive effects in solids are widely used for detecting infrared radiation. These detectors offer very high detectivities, although they must often be cooled to achieve such performance. Their performance is high and continues to improve because of the development of highly purified, single-crystal semiconductors as their active materials. However, several different materials appear to be competing for dominance, and it is not clear whether photovoltaic or photoconductive effects should be emphasized. We will attempt to put the situation into better perspective.

Thus we will review recent progress in these detectors, assess their present status, and analyze prospects for their future improvement, emphasizing the relationships of detector performance parameters to semiconductor material parameters and to fundamental limits. There shall be little discussion of detector fabrication technology. Of particular interest shall be the potential performance of the various detector materials if their properties can be optimized, and the comparison of photovoltaic and photoconductive effects in these materials. We will treat only infrared detectors; detectors of optical radiation in the visible region of the electromagnetic spectrum were reviewed recently by *Seib* and *Aukerman* [4.1]. Useful recent general references for this chapter are the reviews of infrared detectors and their applications by *Dimmock* [4.2] and of narrow-gap semiconductors by *Harman* and *Melngailis* [4.3].

The basic theory of photovoltaic and photoconductive detectors shall be presented in Section 4.1 in a unified form convenient for intercomparison of the two effects and of the various detector materials. Then Sections 4.2, 4.3, and 4.4 shall cover photovoltaic, intrinsic photoconductive, and extrinsic photoconductive detectors, respectively, each of these sections including first a subsection in which the general theory of Section 4.1 is specialized to that class of detector, and then a subsection in which specific materials suitable for that class of detector are evaluated in terms of the theory. Finally in Section 4.5 we will draw some conclusions about the status and prospects of photovoltaic and photoconductive infrared detectors. Symbols used in this chapter which are not defined in the text are defined in Table 4.1.

Table 4.1. Definitions of symbols

Symbol	Definition	Symbol	Definition
A	Detector area absorbing photon flux	$\Delta n, \Delta p$	Photoexcited electron, hole concentration
A, A_i (subscript)	Auger lifetime, intrinsic Auger lifetime	N_a, N_d, N_t	Acceptor, donor, trap concentration
b	Electron-hole mobility ratio, thickness of active layer in photovoltaic detector	N_c, N_v	Conduction, valence band density of states
b (subscript)	Background radiation	p	Hole concentration
c	Speed of light	p (subscript)	Quantity in p-type region
D	Carrier diffusion coefficient	P_λ	Power of photon flux
D_λ^*	Spectral detectivity	q	Electronic charge
e (subscript)	Electron	r	Optical reflectivity
E	Energy, electric field	R	Detector resistance
E_F	Fermi level	R (subscript)	Radiative lifetime
E_g	Energy gap	R_r	Direct radiative recombination rate
Δf	Frequency bandwidth	$\mathcal{R}_\lambda$	Spectral responsivity
g, g_{th}	Generation rate of carriers, thermal generation rate	s	Surface recombination velocity
		s (subscript)	Signal
G	Photoconductive gain	t	Photoconductive detector thickness in direction of photon flux
h	Planck's constant		
h (subscript)	Hole	t (subscript)	Concentration in trap
I	Electrical current	T	Absolute temperature
I_n	Noise current	V	Applied voltage
I_p	Photocurrent	w	Photoconductive detector width
I_s	Signal current	W	Space-charge layer width
k_B	Boltzmann's constant	α	Optical absorption coefficient
l	Photoconductive detector length	ε_0	Permittivity of free space
L	Carrier diffusion length	η	Quantum efficiency
m^*	Effective mass of current carrier	κ	Dielectric constant
0 (subscript)	Thermal equilibrium concentration	λ	Photon wavelength
		μ	Carrier mobility
n	Electron concentration	ϱ	Photoconductor resistivity
n (subscript)	Quantity in n-type region	τ	Carrier lifetime
n_i	Intrinsic carrier concentration	ϕ	Photon flux density

4.1 Basic Theory

4.1.1 Direct Photon Detection

Photovoltaic and photoconductive effects result from direct conversion of incident photons into conducting electrons within a material. The two effects differ in the method of sensing the photoexcited electrons electrically. Detectors based on these effects are called *photon* detectors, because they convert photons directly into conducting electrons; no intermediate process is involved, such as the heating of the material by absorption of photons in a *thermal* detector which causes a change of a measurable electrical property.

A photon of sufficient energy can be absorbed by the photon detector material to excite an electron from a nonconducting state into a conducting state, and the "photoexcited" electron can then be observed through its contribution to an electrical current or voltage. To be detected the photon must have an energy $E \gtrsim E_{exc}$, where E_{exc} is the electronic excitation energy. Since

$$E = hc/\lambda, \tag{4.1}$$

where λ is the photon wavelength, a photon detector is sensitive only to photons with $\lambda \lesssim \lambda_{co}$, where λ_{co} is a "cutoff" wavelength given from (4.1) by

$$\lambda_{co} = hc/E_{exc}. \tag{4.2}$$

The excitation can occur either from the valence band to the conduction band to create an electron-hole pair (*intrinsic* excitation) or from a discrete crystal-defect (dopant) energy level to either band to create a conduction electron or hole (*extrinsic* excitation).

4.1.2 Photocurrent, Gain, and Responsivity

If Φ photons/cm^2-s of $\lambda \lesssim \lambda_{co}$ are incident on a photon detector, $\eta\Phi$ are absorbed and converted into photoexcited conducting electrons, where η is the quantum efficiency. Either a photovoltaic or a photoconductive effect is used to convert these photoexcited electrons into a photocurrent. A photovoltaic effect occurs in a material in which there is a space-charge layer; when a photoexcited electron-hole pair enters the layer, the electron and hole are separated by the space-charge field to give a photocurrent. The space-charge layer is formed either by a Schottky barrier, which consists of a metal deposited onto a semiconductor surface, or by a *p-n* junction. A photovoltaic detector (photodiode) can be operated with or without a bias voltage, which can vary the operating point on the current-voltage characteristic. The photoconductive effect simply involves applying a bias voltage across a uniform piece of detector material to generate a photocurrent proportional to the photoexcited electron concentration. In either effect the short-circuit photocurrent is

$$I_p = Aq\eta\Phi G, \tag{4.3}$$

where the *photogain G* represents the number of electrons flowing through the electrical circuit per photon absorbed. The photon flux density Φ may be either "signal" radiation from an object to be detected or radiation from the background. The photogain is determined by properties of the detector, i.e., by which detection effect is used and the material and configuration of the detector. $G = 1$ in a photovoltaic detector of the simple kind (e.g., not an avalanche photodiode), whereas generally $G \neq 1$ in a photoconductor.

If I_p of (4.3) is a photosignal current I_s, then the current responsivity of the detector is

$$\mathscr{R}_\lambda(I_s) = \frac{q\eta\lambda G}{hc}, \tag{4.4}$$

where $\eta = 0$ when $\lambda > \lambda_{co}$ since in general $\Phi = P_\lambda \lambda / Ahc$. Thus if η is relatively constant when $\lambda < \lambda_{co}$, the responsivity of a photon detector ideally has a "sawtooth"-like dependence with a maximum or peak value at the cutoff wavelength λ_{co}. In practice the sawtooth is somewhat rounded or smoothed due to the gradual increase in photon absorption near λ_{co} in the materials used. A smoothed-sawtooth response is characteristic of photon detectors; see for example [4.2].

In principle one could choose for a photon detector any material having an electronic excitation energy satisfying the requirement

$$E_{exc} \leq hc/\lambda_{co}. \tag{4.5}$$

In practice, however, E_{exc} should satisfy the *equality* of (4.5) as closely as possible, both to maximize the operating temperature of the detector and to minimize the amount of background radiation limiting the performance of the detector. Thus near-satisfaction of the equality of (4.5) is the first condition on a photon detector material.

The necessary materials are semiconductors, because they have energy gaps (for photodiodes and intrinsic photoconductors) and impurity ionization energies (for extrinsic photoconductors) corresponding through (4.2) to infrared wavelengths. But the semiconductors used must offer a wide range of excitation energies to satisfy the need for photon detectors optimized for various wavelengths throughout the infrared spectrum. This requirement has led to the use of many different semiconductors as photon detector materials, in contrast, for example, to the field of solid-state electronics where no such requirement exists and silicon is the only semiconductor used for many different devices. Perhaps the most elegant solution to the need for a variety of excitation energies in photon detector materials has been the development of semiconductor alloy systems, in which either the energy gap or an impurity ionization energy can be varied by changing the alloy composition; however, special technological considerations have permitted a few highly developed elemental and compound semiconductors to continue in use or be developed as detector materials for certain wavelength ranges.

4.1.3 Noise Mechanisms

The detectivity of a detector is limited by noise mechanisms. Shot noise is the fundamental mechanism in photovoltaic detectors. The shot noise current is given by [4.4]

$$I_{n,pv}^2 = 2q \left\{ I_p + \frac{I_{sat}}{\beta} \left[\exp(qV/\beta k_B T) + 1 \right] \right\} \Delta f \tag{4.6}$$

for a detector in which the diode current-voltage characteristic is of the form

$$I = I_{sat}[\exp(qV/\beta k_B T) - 1], \tag{4.7}$$

where I_{sat} is the reverse-biased saturation current of the diode. The $I-V$ characteristic of (4.7) is typical for both p-n junctions and Schottky barriers. In a Schottky barrier or in an ideal p-n junction in which only diffusion of minority carriers determines the current, $\beta = 1$. If generation and recombination within the space-charge region of a p-n junction also contribute to the current, $1 < \beta < 2$. Equation (4.7) can be expressed as

$$I = \frac{\beta k_B T}{qR}[\exp(qV/\beta k_B T) - 1], \tag{4.8}$$

since

$$R = -\left(\frac{dI}{dV}\right)^{-1}_{V=0} = \frac{\beta k_B T}{q I_{sat}}, \tag{4.9}$$

where R is the dark resistance of the diode at zero bias voltage; R does not depend upon the background photon flux density Φ_b. Substituting (4.3) and (4.8) into (4.6) and recognizing that $G = 1$ in a photovoltaic detector we get

$$I^2_{n,pv} = 2q\left[q\eta\Phi_b A + \frac{k_B T}{qR}\exp(qV/\beta k_B T) + \frac{k_B T}{qR}\right]\Delta f. \tag{4.10}$$

We have let $\Phi = \Phi_b$ in (4.10) because in any practical application the photon flux contributing to the detector noise will be almost entirely from the background. When $\Phi_b = 0$ and $V = 0$, (4.10) reduces to the well-known expression for the Johnson noise current,

$$I^2_J = \frac{4k_B T}{R}\Delta f. \tag{4.11}$$

Generation-recombination (gr) noise and Johnson noise are the fundamental mechanisms in photoconductive detectors. The total noise current is given by

$$I^2_{n,pc} = 4q\left(q\eta\Phi_b A G^2 + qg_{th}G^2 + \frac{k_B T}{qR}\right)\Delta f. \tag{4.12}$$

The first two terms within the parentheses in (4.12) represent gr noise due to the background photon flux and to thermal equilibrium carriers in the semiconductor, respectively; these terms are derived in Appendix A from the more familiar expression for gr noise. The third term within the parentheses in (4.12) represents the Johnson noise, where R is the photoconductor resistance, which in this case can depend upon the background photon flux density. The photoconductive gain G in (4.12) is generally a function of the bias voltage applied to the photoconductor. At high bias voltages carrier sweepout effects occur which will be discussed later, and the gr noise terms must be multiplied by a numerical factor a, where $1/2 \lesssim a < 1$, but this is only a minor modification.

The expressions (4.10) and (4.12) for the noise currents of photovoltaic and photoconductive detectors are both of the form

$$I_n^2 = uq\left[q\eta\Phi_b AG^2 + I_d(V) + \frac{k_B T}{qR}\right]\Delta f ; \qquad (4.13)$$

for a photovoltaic detector $u=2$ and $G=1$, whereas for a photoconductive detector $u=4$ and G may have a range of values. The $I_d(V)$ term in (4.13) is a voltage-dependent dark current which takes different forms for the two different detector modes.

Additional types of noise may be involved in applications of photon detectors, such as $1/f$ noise and noise of the electronics used in photosignal processing. However, these mechanisms are not fundamental to the detector material. The $1/f$ noise is believed to be caused by electronic transitions involving surface states and/or the electrical contacts to the detector, although no fully satisfactory general theory has been formulated; $1/f$ noise can often be minimized by surface treatments or electrical contacting procedures and is accordingly a technological rather than a fundamental problem. Noise of the electronics can be minimized somewhat, but to realize its potential performance a detector must have properties such that the electronics noise is negligible compared to the fundamental detector noise mechanisms; the detector signal and noise must both be high enough that detector noise is large compared to electronics noise, with the detector still showing a high enough signal/noise ratio. A photoconductive detector may have an advantage over a photovoltaic detector in marginal cases because G can exceed unity in a photoconductor. Electronics noise is indirectly fundamental in the sense that its magnitude determines how large the fundamental detector noise mechanisms must be to dominate and thereby determine detector performance. The reader should keep in mind that in applying the results of this chapter he must always assess the importance of electronics noise to determine whether predicted detector performance can be realized in practice.

4.1.4 Detectivity

The spectral detectivity is given by

$$D_\lambda^* = \frac{\mathcal{R}_\lambda(I_s) A^{1/2} (\Delta f)^{1/2}}{I_n} \qquad (4.14)$$

in general, so that from (4.4) and (4.13) we can write

$$D_\lambda^* = \frac{\eta\lambda}{u^{1/2} hc \sqrt{\eta\Phi_b + \frac{I_d(V)}{q}\frac{1}{G^2 A} + \frac{k_B T}{q^2}\frac{1}{G^2 RA}}} . \qquad (4.15)$$

Equation (4.15) is the general relationship between D_λ^* and detector parameters; it applies to both photovoltaic and photoconductive detectors.

In advanced infrared detector applications it is usually necessary to approach the background noise limited (BLIP) D_λ^*. This objective requires that in the denominator of (4.15).

$$\frac{I_d(V)}{qG^2A} \ll \eta\Phi_b \tag{4.16}$$

and

$$\frac{k_B T}{q^2}\frac{1}{G^2 R A} \ll \eta\Phi_b. \tag{4.17}$$

These are the second and third conditions on the material and design of an infrared photon detector; they are partly interrelated insofar as they involve common parameters. When conditions 1–3 are satisfied the only detector parameter on which D_λ^* depends is η. One generally also wants to achieve the highest possible value of D_λ^* (BLIP) by having the quantum efficiency near its maximum possible value; i.e.,

$$\eta \to 1. \tag{4.18}$$

This is the fourth condition; it also helps inequalities (4.16) and (4.17) to be satisfied.

Summarizing and simplifying (4.6) and (4.16–18), we have four conditions on the material and design of an infrared photon detector:

1) $E_{exc} = hc/\lambda_{co}$; this condition fixes the wavelength response.
2) $\eta G^2 A \gg I_d(V)/q\Phi_b$; this condition usually limits the maximum operating temperature.
3) $\eta G^2 R A \gg k_B T/q^2\Phi_b$; this condition permits nearly BLIP performance at the temperature defined by condition 2.
4) $\eta \to 1$; this condition maximizes the value of D_λ^* (BLIP).

We shall use the above four conditions as a basis for analysis and comparison of the various detector materials and of the two modes, photovoltaic and photoconductive, in the remainder of this chapter. Conditions 1–4 provide a convenient framework for discussing the detector materials, because the parameters on the left sides of the four relationships are directly related to measurable and controllable material properties.

4.1.5 Other Detector Parameters

Achievement of a very high D_λ^* at the desired λ_{co} is usually the primary demand on a photon detector, but other detector performance parameters can be important also. Listed in Table 4.2 are the various detector parameters and the requirements generally placed on them by the most advanced systems.

Table 4.2. Infrared detection systems requirements on detector performance parameters

Detector parameters and properties	System requirements
Wavelengths (λ_{co})	Cover most of infrared spectrum
Detectivity (D_λ^*)	Highest possible (BLIP)
Response time	Very fast in some systems
Operating temperature	High enough for mechanical or electrical cooling
Arrays	$\geq$ 100's of detector elements
Electrical power dissipation	Low per element for large arrays
Cost per element	Low in large arrays
Electronics	Integration with detector array
Ambients	Insensitive to extraneous radiation in some systems

The second most important parameter in many applications is response time. It can be limited by a variety of factors, such as carrier lifetimes within the detector material, the sweepout time of photoexcited carriers, and the RC time constant of the detector and/or its associated circuitry. Detailed analysis of response times goes beyond the scope of this chapter.

Photon detectors must usually be cooled to achieve optimum performance: the longer the cutoff wavelength, the lower must be the detector operating temperature T. There is a fundamental relationship between the temperature of the background viewed by the detector and the lower temperature at which the detector must operate to achieve BLIP performance [4.5]. At several places in this chapter we shall use as a quantitative example a photon detector operating at a temperature $T = 77$ K with a cutoff wavelength $\lambda_{co} = 12.4\,\mu\text{m}$; one can scale the results of this example to other temperatures and cutoff wavelengths by noting that for a given level of detector performance, $T\lambda_{co} \simeq$ constant; i.e., the longer λ_{co} the lower is T while their product remains roughly a constant. This relationship holds because quantities that determine detector performance vary mainly as an exponential of $E_{exc}/k_B T = hc/k_B T \lambda_{co}$.

Related to the problem of detector operating temperature is that of its electrical power dissipation. A photoconductive detector must always carry a primary current I and therefore dissipate $I^2 R$ of electrical power, whereas a photovoltaic detector can be operated without a bias. The trend toward mechanical or electrical cooling, occurring for practical reasons, makes it necessary that a detector dissipate a minimum of electrical power and thereby heat itself as little as possible.

Infrared detector applications often require arrays of many detector elements. The use of arrays has several ramifications. It requires minimum electrical power dissipation in each detector element, so that the entire array can be cooled efficiently. It also requires reasonable uniformity of detector parameters among the elements of an array [4.6]. Another consequence of the increasing sophistication of array technology is the need for lower cost per detector element, so that the cost of a large array not become excessive; thus cost per element can be a sort of "economic" detector parameter just as important as the "technical" detector performance parameters. There is also a trend toward integration of electronics with detector arrays to permit signal processing at the array location, which can affect the choice of detector mode and material.

Finally, some applications require relative insensitivity of the detectors to ambients, such as high storage temperatures, elementary particles, and laser radiation.

4.2 Photovoltaic Detectors

4.2.1 Theory

The important photovoltaic detectors use intrinsic photoexcitation and so are made of semiconductors with energy gaps satisfying condition 1 for the wavelength to be detected. Using the $I_d(V)$ term of (4.10) and letting $G = 1$ for these photovoltaic detectors, condition 2 becomes

$$\eta RA \gg \frac{k_B T}{q^2 \Phi_b} \exp(qV/\beta k_B T). \tag{4.19}$$

Condition 3 then is

$$\eta RA \gg k_B T / q^2 \Phi_b \tag{4.20}$$

because $G = 1$. When V is a large negative (reverse) bias voltage, which is one practical mode of operation of a photovoltaic detector, the right side of (4.19) approaches zero, and the inequality is satisfied for any value of the left side. When the detector is unbiased so that $V = 0$, which is perhaps the most common practical mode of photovoltaic detector operation, (4.19) reduces to (4.20). Therefore, conditions 2 and 3 are satisfied for a photovoltaic detector by achieving high enough values of the RA product and of η to satisfy (4.20) for the background radiation flux density Φ_b in which the detector must operate.

The right side of (4.20) is plotted vs Φ_b in Fig. 4.1 for the common detector operating temperature of 77 K. For example, to have a nearly BLIP $\lambda_{co} = 12.4\,\mu\text{m}$ photovoltaic detector in radiation from a 2π field of view of a 300 K background, one would need $\eta RA \gg 5 \times 10^{-2}$ ohm-cm^2 to satisfy (4.20). The value of ηRA required for a nearly BLIP detector increases with decreasing background flux density according to Fig. 4.1.

The problem of maximizing the detectivity of a photovoltaic detector is thus equivalent to that of maximizing the RA product of the p-n junction or Schottky barrier, as well as maximizing the quantum efficiency η of the device. Photovoltaic detectors are often evaluated by measuring D_λ^* under negligible background radiation and at zero bias voltage, in which case a simple relationship among D_λ^*, η, and RA follows from (4.15):

$$D_\lambda^* = \frac{q\eta\lambda(RA)^{1/2}}{2hc(k_B T)^{1/2}}. \tag{4.21}$$

We shall consider next the theory of RA and η for p-n junctions and Schottky barriers.

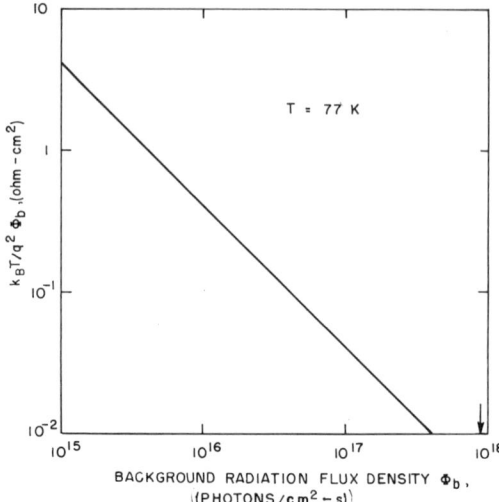

Fig. 4.1. Right side of (4.20) vs background radiation flux density. The arrow at $\Phi_b = 9 \times 10^{17}$ photons/cm²-s represents the flux density absorbed by a $\lambda_{co} = 12.4\,\mu m$ detector from a 2π field of view of a 300 K background; this is the largest flux density ordinarily involved in practice

The general model of a p-n junction photovoltaic detector is shown in Fig. 4.2. *Melngailis* and *Harman* [4.7] have made an almost complete analysis of this model, assuming that the p-n junction current is due only to diffusion of minority carriers. In general, space-charge layer generation-recombination (gr) current can also flow [4.8], but it can be negligible compared to diffusion current in the estimates of potential detector performance of interest here. (See Appendix B to which we will return later when necessary to justify neglecting gr current.) We want now to make one addition to the analysis of [4.7] before proceeding further.

Melngailis and *Harman* calculated the quantum efficiency and the junction saturation current [which gives RA; see (4.9)] for the model of Fig. 4.2, assuming that the back electrical contact ($x = -b$) is a high recombination rate surface at which $\Delta n = 0$. This is the kind of contact one should obtain with a "neutral" metal, i.e., a metal which is not a dopant in the semiconductor to which the contact is made. However, one often makes contact to a semiconductor by using a metal which does act as a dopant of the same type (p or n) as that of the semiconductor material to which the contact is made. If the transition from metal to semiconductor is abrupt enough to permit formation of a space-charge layer between p^+ and p (or n^+ and n) semiconductor material, then the appropriate boundary condition is $d(\Delta n)/dx = 0$, because virtually no minority carrier electron current can flow across such an interface into the heavily doped p^+ material [4.9]. If the transition is gradual, the appropriate boundary condition could still be $\Delta n = 0$. The analysis for a $p^+ - p$ contact is straightforward, and it simply changes the last term in square brackets in (37) of [4.7] to

$$\frac{AqD_e n_{p0}}{L_e} \tanh \frac{b}{L_e},$$

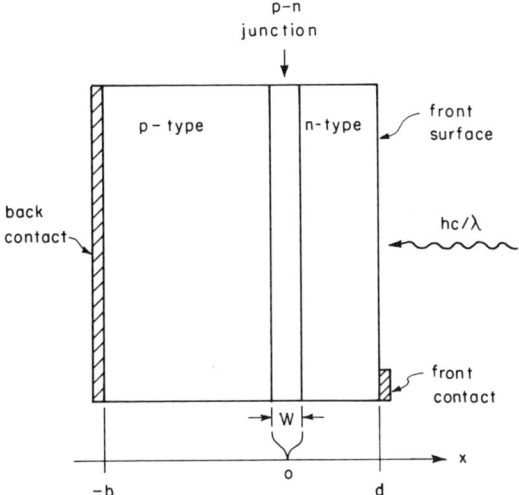

Fig. 4.2. Geometrical model of a *p-n* junction photovoltaic detector

$\tanh b/L_e$ replacing $\coth b/L_e$. Note that the $p^+ - p$ contact gives the same boundary condition as that used by *Melngailis* and *Harman* for the front surface if $s=0$ and if there is no photocurrent ($N=0$ in their notation); i.e., $d(\Delta p)/dx = 0$.

Thus the basic design equations for a *p-n* junction photovoltaic detector are *Melngailis* and *Harman*'s (37) and (38) with the above change in their (37) if a $p^+ - p$ back contact is used. To maximize η we need $r \simeq 0$ (achieved with an antireflection coating), $s \simeq 0$ (achieved with an accumulated (n^+) front surface), and $b/L_h \ll 1$ so that $\cosh b/L_h \simeq 1$; these requirements fortunately also help maximize RA by making the hole current negligible, and they lead from *Melngailis* and *Harman*'s (37) to the following expression for RA:

$$RA = -A\left(\frac{dI}{dV}\right)^{-1}_{V=0} = \frac{k_B T}{q}\left(\frac{qD_e n_{p0}}{L_e}\coth\frac{b}{L_e}\right)^{-1} \quad (4.22)$$

for the $\Delta n = 0$ boundary condition; replace $\coth b/L_e$ by $\tanh b/L_e$ for a $p^+ - p$ back contact. We must then minimize the second term in parentheses in (4.22) to maximize RA. Let us simplify the problem by considering the three possible limiting cases of the two forms,

$$\frac{qD_e n_{p0}}{L_e}\coth\frac{b}{L_e} \quad \text{and} \quad \frac{qD_e n_{p0}}{L_e}\tanh\frac{b}{L_e}.$$

When $b \gg L_e$ both forms reduce to *Case (a)*:

$$qD_e n_{p0}/L_e = qL_e n_{p0}/\tau_e,$$

since $L_e = (D_e \tau_e)^{1/2}$. When $b \ll L_e$ the "coth" form reduces to *Case (b)*:

$$\frac{qD_e n_{p0}}{b} = \frac{qL_e n_{p0}}{\tau_e}\frac{L_e}{b},$$

and the "tanh" form reduces to *Case (c)*:

$$qbn_{p0}/\tau_e.$$

Obviously *Case (c)* gives the smallest current and therefore the highest RA product. Thus we want a p^+-p back contact at a distance $b \ll L_e$ from the junction, provided that the entire device is not then so thin that a significant fraction of the signal radiation is not absorbed by the semiconductor material. Substituting *Case (c)* into (4.22) we get

$$RA = \frac{k_B T}{q^2} \frac{\tau_e}{bn_{p0}} = \frac{k_B T}{q^2} \frac{p_{p0}\tau_e}{bn_i^2} \qquad (4.23)$$

as the highest possible RA product, using the well-known statistical relationship [4.10],

$$p_{p0}n_{p0} = n_i^2. \qquad (4.24)$$

A similar analysis would apply to a detector with the *p*- and *n*-type regions of Fig. 4.2 interchanged; the result would be to replace $p_{p0}\tau_e$ in (4.23) by $n_{n0}\tau_h$. It may not be possible in practice to achieve the kind of electrical contact required for *Case (c)*; if not, then one should use *Case (a)* and calculate RA by replacing b in (4.23) by L_e. Also, gr current may help determine the RA product; the importance of gr current compared to diffusion current can be estimated from the following expression derived by *Sah* et al. [4.8], knowing the carrier lifetime τ in the space-charge region:

$$I_{gr} = qWn_i/2\tau, \qquad (4.25)$$

where W is the junction space-charge layer width.

For comparison later with condition 2 for the photoconductors, we want to express (4.20) in a different form. Substituting (4.23) into (4.20) we get for *Case (c)*:

$$\eta p_{p0}\tau_e/n_i^2 b \gg \Phi_b^{-1}, \qquad (4.26)$$

or for *Case (a)*:

$$\eta p_{p0}\tau_e/n_i^2 L_e \gg \Phi_b^{-1}. \qquad (4.27)$$

Several variations in *p-n* junction photovoltaic detector structure have been investigated, such as heterojunctions in which the wider gap material transmits the incident radiation to the junction and narrower gap material, and junctions illuminated from the back side. However, although possibly improving quantum efficiency, these approaches do not provide performance fundamentally different from that of the model of Fig. 4.2.

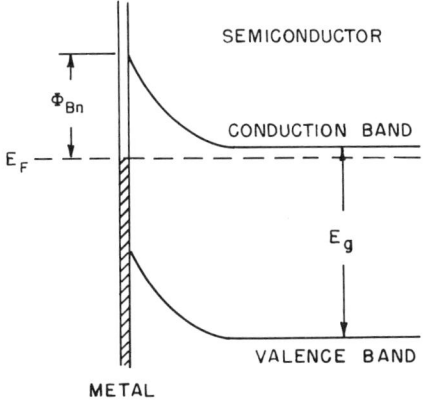

Fig. 4.3. Energy level diagram of a Schottky barrier consisting of a metal on an n-type semiconductor

Another method of realizing a photovoltaic detector is with a Schottky barrier made by depositing a metal onto the surface of a semiconductor. The energy level diagram for a Schottky barrier is shown in Fig. 4.3. Its current-voltage characteristic is [4.10]

$$I = A\left[A^*T^2 \exp\left(\frac{-q\Phi_{Bn}}{k_B T}\right)\right]\left[\exp\left(\frac{qV}{k_B T}\right) - 1\right] \quad (4.28)$$

from which

$$A\left(\frac{dI}{dV}\right)^{-1}_{V=0} = RA = \frac{k_B T}{A^* q T^2} \exp\left(\frac{q\Phi_{Bn}}{k_B T}\right). \quad (4.29)$$

The Richardson constant $A^* = 120$ amp/cm-K^2 for free electrons; A^* will differ from this value by the effective mass ratio m^*/m in a semiconductor, and for nearly all useful semiconductors one has $m^*/m < 1$ and thus $A^* < 120$ amps/cm^2-K^2. One must have a barrier height $\Phi_{Bn} < E_g$ to have a true Schottky barrier; if $\Phi_{Bn} > E_g$, a layer of the semiconductor next to the surface is inverted in type, and there is then a p-n junction within the material. Let us estimate RA with these limitations for a Schottky barrier photovoltaic detector operating at $T = 77$ K; we obtain $RA < 470$ ohm-cm^2 from (4.29). This estimate represents the upper limit of RA achievable with a Schottky barrier. We should keep this result in mind for comparison later with the RA values calculated for p-n junction photovoltaic detectors.

A common additional requirement on a detector is to minimize its response time. The response time of a photovoltaic detector is often determined by its RC time constant [4.7]. Thus for a given RA, one wants to minimize the capacitance per unit area of the junction,

$$\frac{C}{A} = \frac{\kappa \varepsilon_0}{W}. \quad (4.30)$$

To lower C/A one uses as lightly doped material as possible on at least one side of the junction, makes a graded junction, and/or operates the detector under a reverse bias voltage [4.10]. However, there are limits to these approaches, because too wide a W leads to nonnegligible gr current, and the RA product may lower in proportion to the reduction of the doping level. We will consider these tradeoffs further when analyzing detector materials in the next subsection.

4.2.2 Materials

Virtually any semiconductor with an energy gap satisfying condition 1 for the wavelength to be detected is a potential photovoltaic infrared detector material. The remaining three conditions are largely technological for photovoltaic detectors, demanding good enough p-n junction or Schottky barrier technology to yield high RA, and advanced enough coating and surface treatment technology to yield high η. Semiconductor compounds with well developed technologies, such as InSb and InAs, have been exploited as photovoltaic detector materials, but a compound offers optimum performance only at one wavelength. Greater versatility is achieved by using a semiconductor alloy system in which the energy gap, and hence the cutoff wavelength, varies with alloy composition x. The $Pb_{1-x}Sn_xTe$ alloy system has received the most development effort for photovoltaic detectors in recent years [4.3, 7]. The $Hg_{1-x}Cd_xTe$ alloys are being used also but have received much less attention to date as photovoltaic detector materials than $Pb_{1-x}Sn_xTe$. These two alloy systems are the most prominent photovoltaic infrared detector materials under development and will be analyzed in this subsection; their basic properties are summarized in Appendices C and D. Although InSb and InAs are also used as photovoltaic detector materials, they will not be analyzed here; however, as indicated in Appendix D, these two compounds are analogues of the $Hg_{1-x}Cd_xTe$ alloys having the same energy gaps, so that the potential performance of InSb and InAs photovoltaic detectors should be similar to that of the comparable $Hg_{1-x}Cd_xTe$ alloys.

As a numerical example for both alloy systems, we will treat a $\lambda_{co} = 12.4\,\mu m$ detector operating at 77 K. The results of this example should apply approximately to other wavelengths and temperatures satisfying the $\lambda_{co}T \simeq$ constant relationship discussed in Subsection 4.1.5.

Consider $Pb_{1-x}Sn_xTe$ first and refer to Appendix C. For the detector of our example the $Pb_{0.8}Sn_{0.2}Te$ alloy is required to satisfy condition 1; see (4.104). We are interested in the detector performance theoretically possible using these alloys. Accordingly for conditions 2 and 3 we assume material of high enough perfection that Shockley-Read recombination is negligible and only direct radiative recombination occurs. The electron lifetime in the p-type region of Fig. 4.2 is then given by

$$\tau_e = \frac{n_i^2}{R_r p_p}. \tag{4.31}$$

Substituting (4.31) into (4.23) we get for *Case (c)*:

$$RA = \frac{k_B T}{q^2} \frac{1}{bR_r} \tag{4.32}$$

as the maximum possible value of the RA product; this same expression would apply also if the types of the two regions in Fig. 4.2 were reversed.

To calculate RA for $Pb_{1-x}Sn_xTe$ we must estimate how small b can be while still being large enough that nearly all the signal radiation is absorbed. Also, gr current must be negligible; see (4.103) in Appendix B. The requirements on the length parameters are that

$$L_e \gg b \simeq 2.5\alpha^{-1} \gg W, \tag{4.33}$$

since b constitutes most of the thickness of the absorbing detector material; an absorption length of $2.5\alpha^{-1}$ will absorb over 90% of the radiation according to the following general equation for the quantum efficiency [4.11] if the reflectivity $r \simeq 0$:

$$\eta = \frac{(1-r)[1-\exp(-\alpha b)]}{1 - r\exp(-\alpha b)}; \tag{4.34}$$

i.e., $\eta > 0.9$ for this thickness. For $Pb_{1-x}Sn_xTe$, $b \simeq 5 \times 10^{-4}$ cm from Appendix C. Using the general relationship for diffusion length

$$L = D\tau \tag{4.35}$$

and the Einstein relationship

$$qD = \mu k_B T \tag{4.36}$$

as well as equations in Appendix C, we find that for either type of carrier in $\lambda_{co} = 12.4$ μm $Pb_{0.8}Sn_{0.2}Te$ at 77 K, the diffusion length in material limited only by direct radiative recombination is given by

$$L \simeq 5.8 \times 10^{11} p^{-5/6} \text{ cm}. \tag{4.37}$$

The space-charge layer width W of an abrupt p-n junction is given in general by [4.10]

$$W = \sqrt{\frac{2\kappa\varepsilon_0}{q}\left(\frac{n_n + p_p}{n_n p_p}\right) V_{bi}}. \tag{4.38}$$

An abrupt junction gives the smallest possible W, so that we will use it as a "best case" in the spirit of this analysis. If $n_n \ll p_p$, for example, so that the junction is "one-sided",

$$W \simeq \sqrt{\frac{2\kappa\varepsilon_0}{q} \frac{V_{bi}}{n_n}}. \tag{4.39}$$

The "built-in" voltage V_{bi} of the junction depends also on n_n and p_p but only logarithmically [4.10] and for simplicity a good approximation in the cases of interest in this chapter is $V_{bi} \simeq E_g$. The three length parameters of this paragraph are plotted in Fig. 4.4 for the practical range of carrier concentrations in $Pb_{0.8}Sn_{0.2}Te$. Figure 4.4 shows that (4.33) can be satisfied by $Pb_{0.8}Sn_{0.2}Te$ at 77 K when $10^{15} \gtrsim p$ or $n \gtrsim 3 \times 10^{17}$ cm^{-3}.

The RA value for *Case (c)* corresponding to Fig. 4.4b is plotted in Fig. 4.5, using the R_r for $Pb_{0.8}Sn_{0.2}Te$ at 77 K calculated from (4.107) in (4.32). Also plotted in Fig. 4.5 are RA for *Case (a)* using the L of Fig. 4.4, and RA values for both diffusion and gr currents using the carrier lifetime $\tau \simeq 10^{-8}$ s typically observed in presently available $Pb_{1-x}Sn_xTe$ crystals [4.3]. Figure 4.5 thus illustrates the performance potentially achievable by $Pb_{1-x}Sn_xTe$ photovoltaic detectors. If material can be prepared with only direct radiative recombination as a significant lifetime limiting process, then RA values above 100 ohm-cm^2 should be possible. However, with material in which $\tau \simeq 10^{-8}$ s, space-charge generation-recombination current should be important so that $RA < 30$ ohm-cm^2; this conclusion is consistent with results observed experimentally in detectors comparable to our example [4.12]. See also [4.39].

Several close relatives of $Pb_{1-x}Sn_xTe$ have been studied also as photovoltaic detector materials, but they do not have significantly different properties. These variations include $Pb_{1-x}Sn_xSe$ as an alternative alloy system and $Pb_{1-x}Ge_xTe$ with small amounts of Ge to give a material with a slightly wider energy gap than PbTe for 3–5 µm detectors. See Appendix C and [4.3].

Consider $Hg_{1-x}Cd_xTe$ next and refer to Appendix D. Equation (4.108) shows that the $Hg_{0.8}Cd_{0.2}Te$ alloy is required to satisfy condition 1. Again for conditions 2 and 3 we assume material of high enough perfection that Shockley-Read recombination is negligible. In the *p*-type region of Fig. 4.2 only direct radiative recombination is then important, and the electron lifetime is given by (4.31). However, if the types were reversed in Fig. 4.2 to make the region next to the back contact *n*-type, both radiative and Auger recombination would limit the hole lifetime in that region, and

$$\frac{1}{\tau_h} = \frac{1}{\tau_{Ah}} + \frac{1}{\tau_{Rh}} = \frac{n_n^2}{2\tau_{Ai}n_i^2} + \frac{R_r n_n}{n_i^2}. \tag{4.40}$$

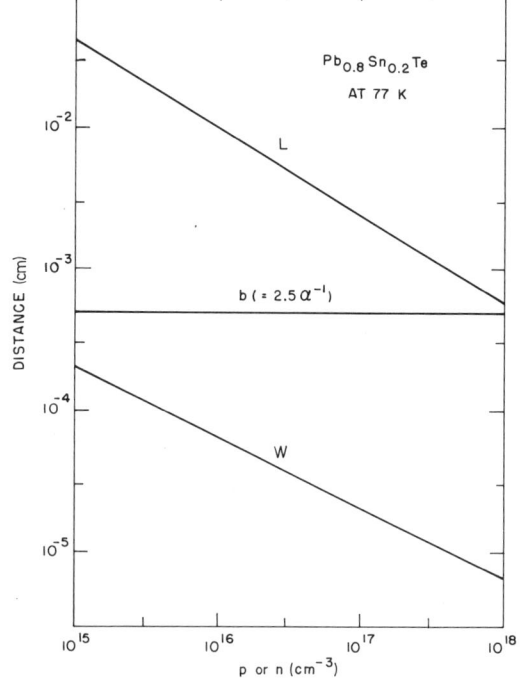

Fig. 4.4. Length parameters vs carrier concentration in a $\lambda_{co} = 12.4\,\mu m$ $Pb_{0.8}Sn_{0.2}Te$ p-n junction photovoltaic detector at 77 K

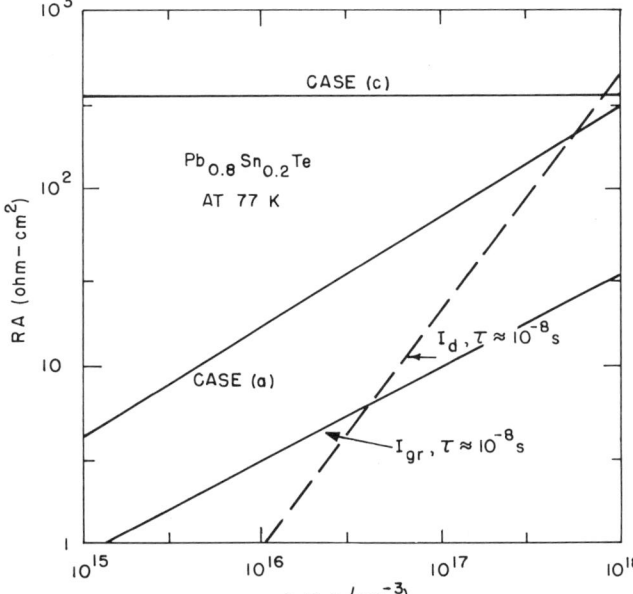

Fig. 4.5. RA product vs carrier concentration for $Pb_{0.8}Sn_{0.2}Te$ p-n junction photovoltaic detector of Fig. 4.4 at 77 K

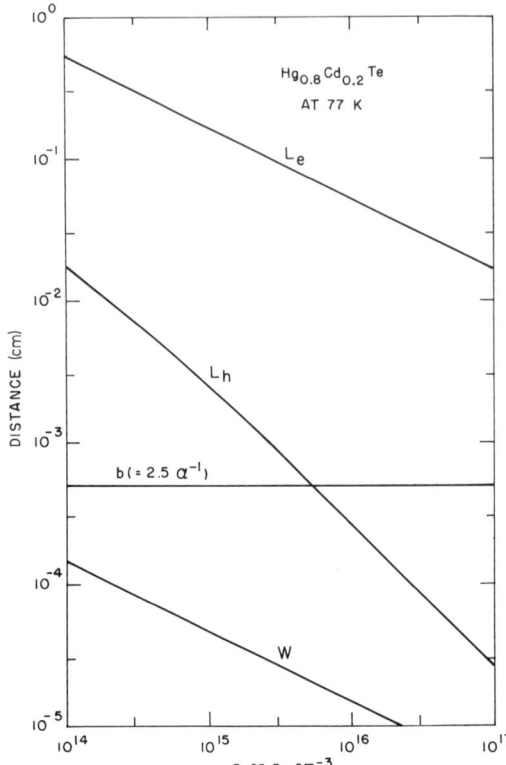

Fig. 4.6. Length parameters vs carrier concentration in a $\lambda_{co} = 12.4\,\mu\text{m}$ $\text{Hg}_{0.8}\text{Cd}_{0.2}\text{Te}$ p-n junction photovoltaic detector at 77 K

Equation (4.32) gives the RA product for *Case (c)* if the region next to the back contact in Fig. 4.2 is p-type, but if it is n-type,

$$RA = \frac{k_B T}{q^2} \frac{1}{b} \left(\frac{n_n}{2\tau_{Ai}} + R_r \right)^{-1}. \qquad (4.41)$$

To calculate RA for $\text{Hg}_{1-x}\text{Cd}_x\text{Te}$ we must again estimate how small b can be while satisfying (4.31); $b \simeq 5 \times 10^{-4}$ cm from Appendix D. Using (4.35) and (4.36) as well as Appendix D and (4.39), we find for $\lambda_{co} = 12.4\,\mu\text{m}$ $\text{Hg}_{0.8}\text{Cd}_{0.2}\text{Te}$ at 77 K the results for the L's, b, and W plotted in Fig. 4.6 for the practical range of carrier concentrations in this alloy system. Figure 4.6 shows that (4.33) can be satisfied for all carrier concentrations of $\text{Hg}_{0.8}\text{Cd}_{0.2}\text{Te}$ at 77 K when the region next to the back contact is p-type, but that it can be satisfied only for $n_n \gtrsim 2 \times 10^{15}$ cm^{-3} when that region is n-type.

The RA values for *Case (c)* corresponding to the b of Fig. 4.6 are plotted in Fig. 4.7, using the R_r and τ_{Ai} values of Appendix D in (4.32) and (4.41). Also plotted in Fig. 4.7 are the RA values for *Case (a)* using the L's of Fig. 4.6. Available n-type $\text{Hg}_{0.8}\text{Cd}_{0.2}\text{Te}$ crystals satisfy our assumption of negligible Shockley-Read recombination [4.13, 14], but it is not known yet how close p-

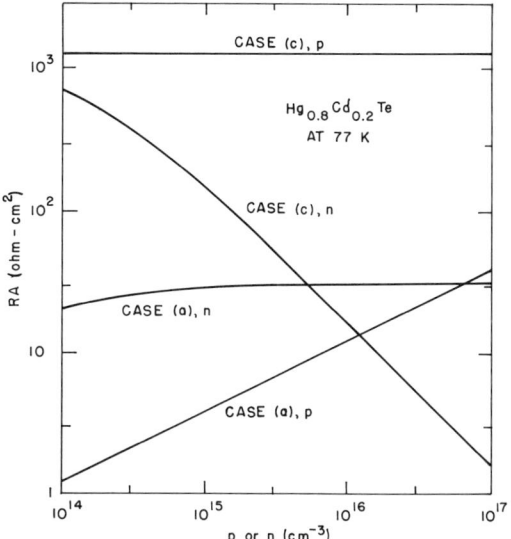

Fig. 4.7. RA product vs carrier concentration for $Hg_{0.8}Cd_{0.2}Te$ p-n junction photovoltaic detector of Fig. 4.6 at 77 K

type $Hg_{0.8}Cd_{0.2}Te$ crystals come to satisfying it. Figure 4.7 thus illustrates the performance potentially achievable by $Hg_{1-x}Cd_xTe$ photovoltaic detectors. Values of RA of $\gtrsim 1$ ohm-cm^2 have been observed recently in $Hg_{1-x}Cd_xTe$ photovoltaic detectors comparable to our example [4.15].

We can conclude from Figs. 4.5 and 4.7 that there is no clear theoretical advantage of one of these two alloy systems over the other with respect to RA, and the same conclusion applies also to η. However, better understanding and improvement of the carrier lifetimes in both types of $Pb_{1-x}Sn_xTe$ and in p-type $Hg_{1-x}Cd_xTe$ are needed to permit realization of the potential photovoltaic detector performance. If Auger recombination is in fact strong in $Pb_{1-x}Sn_xTe$ [4.39], then $Hg_{1-x}Cd_xTe$ with a p-type layer as the active photodiode region would be preferable. There is also the question of whether a back contact like that of *Case (c)* can be made.

The speed of response of a photovoltaic detector is often determined by its RC time constant, as noted in Subsection 4.2.1. The RC time constant is given by the product of RA and (4.30); taking this product and substituting (4.39) for W we get

$$RC \simeq RA \sqrt{\kappa\varepsilon_0 q n_n / 2 V_{bi}} \qquad (4.42)$$

for an abrupt p-n junction, where n_n is the carrier concentration on the less heavily doped side. For a given value of RA a $Hg_{1-x}Cd_xTe$ photovoltaic detector would have a shorter time constant than a $Pb_{1-x}Sn_xTe$ detector, both because the dielectric constant κ of $Hg_{1-x}Cd_xTe$ is lower and because one can use very lightly doped material in the $Hg_{1-x}Cd_xTe$ detector if the region next to the back contact is made n-type.

4.3 Intrinsic Photoconductive Detectors

4.3.1 Theory

For these intrinsic detectors the energy gap of the detector material must correspond to the wavelength to be detected to satisfy condition 1. Condition 2 becomes

$$\eta A/g_{th} \gg \Phi_b^{-1} \tag{4.43}$$

using the $I_d(V)$ term of (4.12). Condition 3 is

$$\eta G^2 RA \gg k_B T/q^2 \Phi_b, \tag{4.44}$$

since $G \neq 1$ in general in photoconductors. The $G^2 RA$ product of an intrinsic photoconductor plays a role analogous to that of the RA product of a photovoltaic detector in condition 3, but condition 2 involves a different inequality for the intrinsic photoconductor. Condition 4 is satisfied by making the detector thick enough and using an anti-reflection coating to achieve a high enough η from (4.34).

In the analysis below we shall consider first the theory underlying (4.43) for an intrinsic photoconductor and then the theory of $G^2 RA$. We shall analyze the simple geometrical model of a photoconductor shown in Fig. 4.8; there is no loss of generality in the results from assuming this regular geometry, and photoconductive detectors nearly always have this configuration anyway.

The simplest model of the material which can account for the major performance features of these photoconductors is the following. The semiconductor is assumed n-type with $n_0 = N_d - N_a$ and $n_0 \gg p_0$, although one could use the same model for p-type material by exchanging n's and p's. This is an *extrinsic* semiconductor while also an *intrinsic* photoconductor. The material contains N_t centers which can trap holes (minority carriers) from the valence band; generally $N_t \ll N_d$. The traps may be defects in the bulk crystal, surface states, or any other such mechanism for immobilizing the minority carriers and making them unavailable for direct recombination. A flux density of $\eta \Phi$ photons/cm^2-s of wavelengths $\lambda \lesssim \lambda_{co}$ is absorbed by the material to excite electrons across the energy gap, creating electron-hole pairs. Some of the photoexcited holes may be trapped at E_t, so that assuming space-charge neutrality (Debye length is very short in these low-resistivity semiconductors),

$$\Delta n = \Delta p + \Delta p_t. \tag{4.45}$$

An untrapped hole can recombine with a conduction band electron by any of three mechanisms: direct radiative, interband Auger, and Shockley-Read recombination.

Most of the applications of these detectors require detection of small signals at relatively high detector operating temperatures and in substantial back-

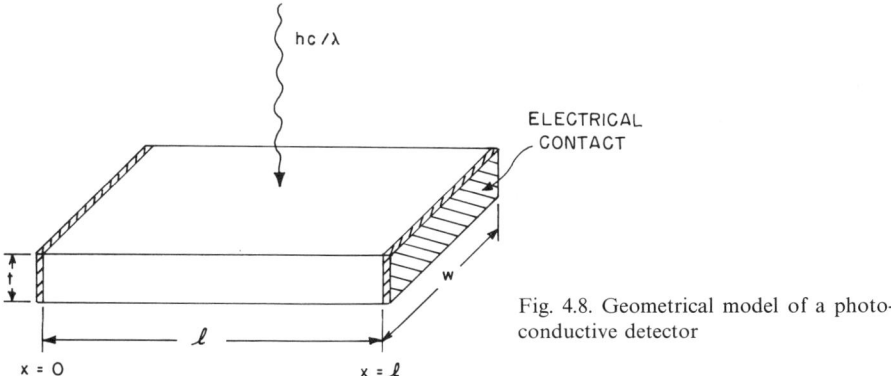

Fig. 4.8. Geometrical model of a photoconductive detector

ground radiation. Under these conditions $\Delta p_b \gg \Delta p_s$ so that $\Delta p \simeq \Delta p_b$. One can show by a procedure like that of *Fan* [4.16] for this model that in fairly intense background radiation the traps are filled and trapping has no effect on the photosignal. Even in dim background radiation if the detector operating temperature is high $\Delta p_t \ll \Delta p$, and trapping is again negligible. In the absence of trapping (4.45) becomes

$$\Delta n = \Delta p. \tag{4.46}$$

Consider condition 2 when (4.46) is satisfied. From the generation-recombination theorem for a two-carrier semiconductor without minority carrier trapping [4.17],

$$p_0 = g_{th}\tau_h/At \tag{4.47}$$

for our n-type model, so that (4.43) can be rewritten as

$$\eta\tau_h/p_0 t = \eta n_0 \tau_h/n_i^2 t \gg \Phi_b^{-1} \tag{4.48}$$

using (4.24). The inequality of (4.48) limits the maximum operating temperature of these detectors, because n_i increases exponentially with temperature while τ_h is either constant or decreases with increasing temperature. Comparing (4.48) with (4.26) and (4.27) we see that condition 2 is similar for the photovoltaic and intrinsic photoconductive detectors, as is expected since they are both based on intrinsic photoexcitation; if the material of both modes of detector were the same, then the only difference would be in the length parameters in (4.26), (4.27), and (4.48).

Let us now develop the theory of G^2RA. The total current of electrons and holes in an intrinsic, two-carrier photoconductor is

$$I = qwt(n\mu_e + p\mu_h)E = qwt(nb + p)\mu_h E, \tag{4.49}$$

and the photocurrent is

$$I_p = qwt(\mu_e \Delta n + \mu_h \Delta p)E = qwt(b\Delta n + \Delta p)\mu_h E, \qquad (4.50)$$

where E is the electric field, and

$$n = n_0 + \Delta n, \qquad (4.51)$$

and

$$p = p_0 + \Delta p. \qquad (4.52)$$

Equating (4.50) to (4.3),

$$G = wt\mu_h E(b\Delta n + \Delta p)/\eta \Phi. \qquad (4.53)$$

From (4.49)

$$RA = El^2 w/I = l^2/qt(nb + p)\mu_h, \qquad (4.54)$$

where n and p are given by (4.51) and (4.52). To calculate $G^2 RA$ we must determine how Δn and Δp depend upon detector material parameters and operating conditions.

An electric field E is applied across the photoconductor to measure the photoconductivity. The field drives electrons toward $x = l$ and holes toward $x = 0$; see Fig. 4.8. The electrical contacts are assumed to be metallic and ohmic, so that photoexcited electron-hole pairs recombine there instantaneously; thus

$$\Delta p = 0 \quad \text{at} \quad x = 0. \qquad (4.55)$$

If the field is strong enough a significant fraction of the holes can be "swept out" of the detector material [4.18, 19], an effect which can be important in the absence of minority carrier trapping; strong trapping prevents the holes from being swept out.

The continuity equation for the minority carrier holes when trapping is negligible is

$$\frac{d\Delta p}{dt} = \frac{\eta \Phi}{t} - \frac{\Delta p}{\tau_h} - \mu_h E \frac{d\Delta p}{dx} \qquad (4.56)$$

neglecting hole diffusion [4.18, 19]. When $d\Delta p/dt = 0$ and $\eta \Phi \neq f(t)$ the (steady-state) solution of (4.56) is

$$\Delta p(x) = \frac{\eta \Phi \tau_h}{t}\left[1 - \exp\left(\frac{-x}{\mu_h \tau_h E}\right)\right] \qquad (4.57)$$

using (4.55). Integrating (4.57) from $x=0$ to $x=l$ and invoking the boundary condition (4.55) we get for the total concentration of photoexcited holes

$$\Delta p = \frac{\eta \Phi \tau_h}{t} \{1 - z_h[1 - \exp(-z_h^{-1})]\} \equiv \frac{\eta \Phi \tau_h}{t} f(z_h), \quad (4.58)$$

where

$$z_h \equiv \mu_h \tau_h E / l. \quad (4.59)$$

Equation (4.58) represents the minority carrier sweepout effect. The total concentration Δp of photoexcited holes maintained in the photoconductor by steady-state irradiation is lower the stronger the electric field, because the field sweeps out some of the holes. The quantity $\tau_h f(z)$ in (4.58) can be thought of as an effective carrier lifetime, reduced from the bulk lifetime τ_h by the sweepout effect.

The speed of response of an intrinsic photoconductive detector is essentially the same as the longest photoexcited carrier lifetime. One can shorten the response time of a detector of this kind by biasing it as far as possible into the sweepout mode, since the effective minority carrier lifetime $\tau f(z)$ is reduced in proportion to the bias field.

Substituting (4.58) into (4.53) and assuming (4.46) we get for the photoconductive gain

$$G_n = (b+1) z_h f(z_h) = \frac{(\mu_e + \mu_h) \tau_h E}{l} f(z_h) \quad (4.60)$$

for our n-type model, where z_h is defined by (4.59). The corresponding expression for an analogous p-type model can be shown to be

$$G_p = (1 + b^{-1}) z_e f(z_e) = \frac{(\mu_e + \mu_h) \tau_e E}{l} f(z_e), \quad (4.61)$$

where

$$z_e \equiv \mu_e \tau_e E / l. \quad (4.62)$$

The $zf(z)$ product is plotted vs z in Fig. 4.9. The saturation of $zf(z)$ when $z > 1$ represents saturation of the photoconductive gain to a value of $G_n = (b+1)/2$ or $G_p = (1 + b^{-1})/2$ when the minority carrier sweepout effect is dominant. Combining (4.60) and (4.48) we get for the n-type model

$$G_n^2 RA = (b+1)^2 [z_h f(z_h)]^2 l^2 / qt(nb+p) \mu_h. \quad (4.63)$$

The corresponding equation for an analogous p-type model can be shown to be

$$G_p^2 RA = (1+b^{-1})^2 [z_e f(z_e)]^2 l^2 / qt(nb+p) \mu_e. \quad (4.64)$$

Theoretically, the minority carrier sweepout effect could be eliminated by having a boundary condition different at $x=0$ from that of (4.61). If the boundary condition were the same as that for the $p^+ - p$ contact proposed in Subsection 4.2.1 i.e., $d(\Delta p)/dx = 0$ for the minority carriers in this n-type

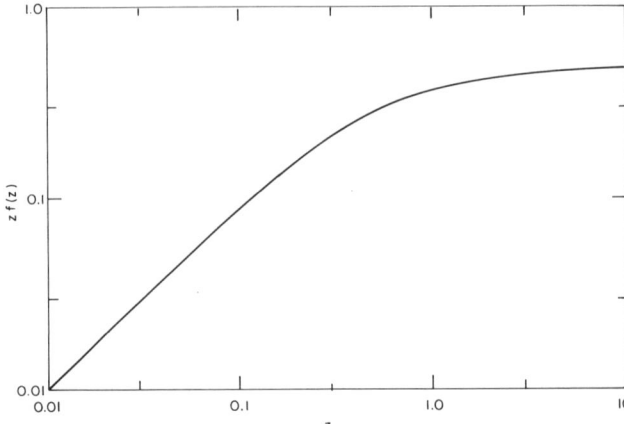

Fig. 4.9. The $zf(z)$ product vs z; see (4.58) and (4.59)

photoconductor, then the holes would be trapped in a potential well between the contacts and not be swept out. This conclusion can be verified by solving (4.56) with this different boundary condition. Again it is not clear that such a contact can be realized in practice.

Another application of these detectors is to detect small signals when background radiation is very dim. It can be shown that in feeble background radiation and at low detector temperatures the minority carrier trapping effect can be important, so that from (4.45)

$$\Delta n \simeq \Delta p_t. \tag{4.65}$$

The trapping lengthens the lifetime of the photoexcited carriers compared to the lifetime in the absence of trapping, and as a consequence it enhances the photoconductive gain relative to the nontrapping gain by the ratio of these lifetimes; an additional gain enhancement can occur also by the trapping preventing minority carrier sweepout. The price paid for the gain enhancement by trapping is a slower detector, since the response time is limited by the longer carrier lifetime.

4.3.2 Materials

We saw in Subsection 4.2.2 that satisfaction of condition 1 was almost sufficient for a semiconductor to be a satisfactory photovoltaic detector material; conditions 2 through 4 placed demands mainly on device design and technology rather than on fundamental properties of the semiconductor material. The situation is more restrictive for intrinsic photoconductive detectors, because condition 3 places specific demands on fundamental material properties which eliminate some classes of semiconductors as satisfactory high-performance detector materials. Let us consider condition 3 in a qualitative way next to determine which materials may be satisfactory. We shall treat conditions 2 and 3 in terms of a quantitative example later.

The general form of condition 3 for an intrinsic photoconductor is (4.44). For a given background radiation flux density and detector operating temperature, the $\eta G^2 RA$ product must be high enough to satisfy (4.44). Condition 3 is expressed in terms of semiconductor material parameters by substituting (4.63) and (4.64) for n-type and p-type material, respectively, into (4.44). It is clear for the n-type model that for given values of $z_h f(z_h)$ and of detector dimensions, one wants a material with the best possible combination of high mobility ratio b and of low majority carrier concentration n and mobility $\mu_e (=b\mu_h)$.

The semiconductors with the zinc blende crystal structure, reviewed in Appendix D, offer very high mobility ratios and can be purified to reach relatively low values of n, but μ_e is quite high; nevertheless, the combination of parameters is such that n-type crystals of these semiconductors are very good intrinsic photoconductor materials. And p-type crystals of these semiconductors can be satisfactory intrinsic photoconductor materials also; although the high mobility ratio works against them [see (4.64)], the hole mobility is low and p can reach rather low values. On the other hand the IV–VI compound semiconductors, reviewed in Appendix C, are unsatisfactory intrinsic photoconductor materials because $b \simeq 1$ and μ_e is high; also it is difficult although not impossible to achieve low n or p.

In other words the intrinsic photoconductor material must provide either a high photoconductive gain if it is of low electrical resistance, or a high enough resistance to compensate for low gain. In the nontrapping regime of detector operation, high photoconductive gain is achieved by having a high ratio of the mobility of the majority carrier to that of the minority carrier; see (4.60). Materials like those in Appendix C could still be useful for intrinsic photoconductors in some applications if they were to show suitable minority carrier trapping, but such effects have not been observed in them.

Thus the zinc blende structure semiconductors can be useful for intrinsic photoconductive detectors. Compounds such as InSb have been used as intrinsic photoconductors [4.20], as well as for photovoltaic detectors, but greater versatility of wavelength response is possible with the $Hg_{1-x}Cd_xTe$ alloy system. The $Hg_{1-x}Cd_xTe$ alloys have received considerable development effort in recent years and are the most prominent intrinsic photoconductor materials; they will be analyzed in this subsection. The development of $Hg_{1-x}Cd_xTe$ has concentrated almost entirely on n-type material since it provides high photoconductive gain; however, p-type $Hg_{1-x}Cd_xTe$ crystals may be useful for intrinsic photoconductive detectors also [4.21].

We shall again use a $\lambda_{co} = 12.4\,\mu m$ detector operating at 77 K as a numerical example. From (4.108) in Appendix D, the $Hg_{0.8}Cd_{0.2}Te$ alloy is required to satisfy condition 1. To determine the best detector performance theoretically possible we ignore Shockley-Read recombination. Then in n-type $Hg_{0.8}Cd_{0.2}Te$ the minority carrier lifetime is given by (4.40) and in p-type by (4.31), assuming that $\Delta n = \Delta p \ll n_0$. This assumption means that the background radiation is not intense enough to modify the semiconductor material parameters significantly; see Appendix E.

Now consider condition 2. For *n*-type $Hg_{0.8}Cd_{0.2}Te$ we can substitute (4.40) into (4.54) to obtain

$$\left(\frac{n_0}{2\tau_{Ai}} + R_r\right) t \ll \eta \Phi_b. \tag{4.66}$$

Both τ_{Ai}^{-1} and R_r increase with temperature approximately as $\exp(-E_g/k_B T)$, so that (4.66) places an upper limit on the detector operating temperature. Using values of τ_{Ai} and R_r from Appendix D and assuming a detector thickness $t = 5 \times 10^{-4}$ cm as in Subsection 4.2.2 to give $\eta > 0.9$, we obtain the curve of the left side of (4.66) vs n_0 plotted in Fig. 4.10 for an operating temperature of 77 K. By comparing Fig. 4.10 with the abscissa of Fig. 4.1, we can see that (4.66) is easily satisfied by the detector of this example when the flux density $\Phi_b > 10^{16}$ photons/cm²-s, and that it is satisfied for lower flux densities by material of lower carrier concentrations. Thus 77 K is a possible operating temperature for this example.

Referring back to (4.26) and (4.27), only if *Case (c)* is achievable can a $Hg_{1-x}Cd_xTe$ photovoltaic detector satisfy condition 2 as well as a $Hg_{1-x}Cd_xTe$ photoconductor, because $b \simeq t$ whereas $L \gg t$. A similar conclusion would apply to a $Pb_{1-x}Sn_xTe$ photovoltaic detector of comparable carrier lifetime, since n_i is roughly the same in these two alloy systems.

When $\Delta n = \Delta p \ll n_0$, (4.63) reduces to

$$G_n^2 RA = (b+1)^2 [z_h f(z_h)]^2 l^2 / q t n_0 \mu_e, \tag{4.67}$$

and (4.54) to

$$RA = l^2 / q t n_0 \mu_e. \tag{4.68}$$

To maximize $G^2 RA$ we must maximize the quantity $z_h f(z_h)$ by applying the highest electric field possible. The limit is usually imposed by the electrical power dissipation P allowable in the detector, where

$$P = E^2 l^2 / R. \tag{4.69}$$

We can write z_h as

$$z_h = \frac{\mu_h \tau_h}{l^2} (PR)^{1/2} = \frac{\mu_h \tau_h}{l} \left(\frac{P}{lw} R\right)^{1/2}, \tag{4.70}$$

where the second equality holds for the common square detector configuration ($l = w$), so that P/lw is the electrical power dissipation per unit dissipating area of the detector. Substituting (4.40) and (4.68) into (4.70) we get

$$z_h = \frac{n_i^2}{bl} \left(\frac{\mu_e}{q t n_0^3}\right)^{1/2} \left(\frac{n_0}{2\tau_{Ai}} + R_r\right)^{-1} \left(\frac{P}{A}\right)^{1/2} \tag{4.71}$$

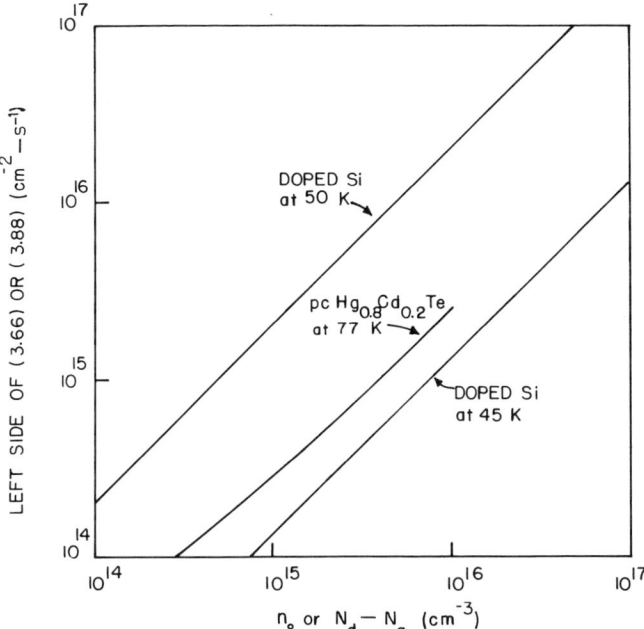

Fig. 4.10. Left sides of (4.66) (for $Hg_{0.8}Cd_{0.2}Te$) and (4.88) (for doped Si) vs concentration (carrier concentration n_0 in n-type $Hg_{0.8}Cd_{0.2}Te$, compensating acceptor concentration N_a in n-type Si). This figure relates to condition 2 for these two photoconductor examples; values of the ordinate must be $\ll \eta \Phi_b$ to satisfy condition 2

for this square n-type $Hg_{0.8}Cd_{0.2}Te$ detector. Using (4.71), Fig. 4.9, and (4.67) we can calculate G^2RA; curves of G^2RA vs carrier concentration n_0 for two representative values of power density are plotted in Fig. 4.11, assuming a square detector with $l = w = 10^{-2}$ cm and thickness $t = 5 \times 10^{-4}$ cm, an electron mobility $\mu_e(77\,\text{K}) = 1.5 \times 10^5$ cm^2/V-s, and the values given in Appendix D for the other parameters. Figure 4.11 illustrates the performance potentially achievable by $Hg_{1-x}Cd_xTe$ intrinsic photoconductive detectors. Experimental results for such detectors are quoted in terms of D_λ^* values in weak background radiation. Plotted in Fig. 4.12 is D_λ^* vs Φ_b for several values of $\eta G^2 RA$ for the example analyzed in this subsection; these curves were calculated from (4.15) assuming condition 2 to be satisfied. Values of D_λ^* as high as 10^{11} cm-Hz$^{1/2}$/watt at 77 K for n-type $Hg_{0.8}Cd_{0.2}Te$ with $n_0 \sim 10^{15}$ cm^{-3} have been reported [4.22], corresponding from Fig. 4.12 to G^2RA products > 1 ohm-cm^2. The curves of Fig. 4.12 can be applied also to (unbiased) photovoltaic detectors by replacing $\eta G^2 RA$ by $RA/2$ and multiplying D_λ^* by $\sqrt{2}$; see (4.10) and (4.15).

The minority carrier sweepout effects have been observed in n-type $Hg_{1-x}Cd_xTe$ by several investigators [4.22, 23]. The speed of response of the photoconductor is improved by biasing into the sweepout mode, as expected, and sweepout is thus a useful effect for controlling detector response time. An

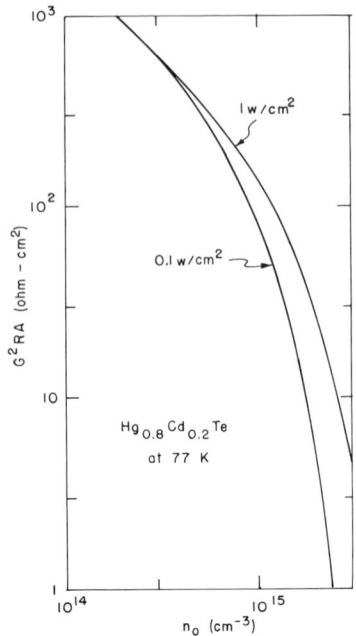

Fig. 4.11. G^2RA product vs carrier concentration for n-type $Hg_{0.8}Cd_{0.2}Te$ intrinsic photoconductive detector at 77 K

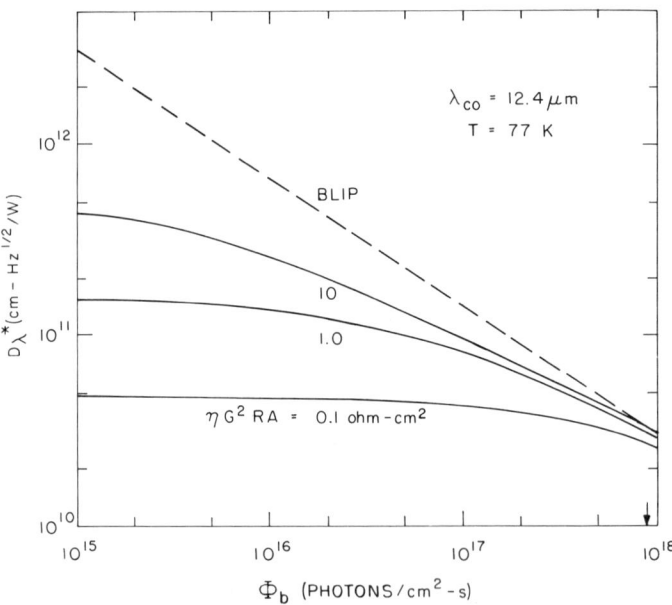

Fig. 4.12. D_λ^* vs Φ_b for several values of $\eta G^2 RA$ calculated from (4.15). The arrow at $\Phi_b = 9 \times 10^{17}$ photons/cm^2-s represents flux absorbed by a 12.4 μm detector from 2π FOV of 300 K background

enhancement of D_λ^* by a factor varying from about $\sqrt{2}$ to 2 is also often observed but not yet fully understood; sweepout can thus be useful for improving D_λ^* as well as reducing detector response time.

We can conclude that the quality of presently available n-type $Hg_{0.8}Cd_{0.2}Te$ crystals with $n_0 \sim 10^{15}$ cm^{-3} is very high for nontrapping detectors, at least as far as D_λ^* is concerned. However, it is still important to achieve $n_0 < 10^{15}$ cm^{-3} more routinely to take advantage of higher detector resistance and lower power dissipation in the sweepout mode as well as the somewhat higher D_λ^*. Although less work has been done to date on $Hg_{1-x}Cd_xTe$ crystals of compositions other than $x \simeq 0.2$ a similar conclusion is likely for them [4.24].

Let us now reconsider the minority carrier trapping case. This case has been studied less extensively than nontrapping detector performance, and no experimental results have been published in the open literature. However, *Beck* and *Broudy* have investigated the trapping effects in detail [4.25] and found them to be more complicated than the predictions of the simple model of Subsection 4.3.1. They have extended the basic theory of *Fan* [4.16] by assuming a continuum of trapping levels in the lower half of the energy gap, and shown that this model can account for observed trapping effects over a wide range of temperature, background radiation, and signal chopping frequency. Very little is known of the nature and concentration of the traps in $Hg_{1-x}Cd_xTe$, so that this is still a major area for research.

4.4 Extrinsic Photoconductive Detectors

4.4.1 Theory

These detectors require a material with an impurity dopant ionization energy corresponding to the wavelength to be detected to satisfy condition 1. Conditions 2 and 3 are given by (4.43) and (4.44), except that here there is only one type of carrier affecting detector performance, and g_{th} in (4.43) represents the generation rate of the thermal equilibrium population of that carrier. Condition 4 is satisfied again by making the detector thick enough and using an antireflection coating.

We shall consider next the theory of condition 2 for an extrinsic photoconductor, then the theory of G^2RA for condition 3, and finally the dependence of η on material parameters. We shall analyze the geometrical model of Fig. 4.8 and assume a simple energy level model of an n-type extrinsic semiconductor consisting of a photoionizable donor level and a compensating acceptor level; properties of a corresponding p-type model would be analogous. This material is *extrinsic* as both a photoconductor and semiconductor.

Sweepout effects can occur in extrinsic photoconductors also [4.26, 27], but we shall not explicitly include them in this model; they are more difficult to understand, but not generally as important in practice as in intrinsic photoconductors. The essential result is that whereas sweepout of carriers can limit the photoconductive gain to 1/2 at modulation frequencies above $\omega = \tau_\varrho^{-1}$, where $\tau_\varrho = \kappa\varepsilon_0\varrho$ is the dielectric relaxation time, the gain can exceed unity at lower frequencies.

Consider condition 2. From the generation-recombination theorem for a one-carrier (n-type) semiconductor [4.17]

$$n_0 = g_{th}\tau_e/At, \tag{4.72}$$

so that (4.43) can be rewritten as

$$\eta\tau_e/n_0 t \gg \Phi_b^{-1}. \tag{4.73}$$

The thermal equilibrium majority-carrier concentration in an n-type semiconductor is given by [4.9]

$$\frac{n_0(n_0 + N_a)}{N_d - N_a - n_0} = \frac{N_c}{g_c}\exp(-E_d/k_B T) \equiv n_1. \tag{4.74}$$

For this model

$$\tau_e = (BN_I)^{-1} = [B(n_0 + N_a)]^{-1}, \tag{4.75}$$

where B is a recombination coefficient and N_I is the concentration of ionized donors [4.28]. Thus (4.73) becomes

$$\eta/Bn_1(N_d - N_a)t \gg \Phi_b^{-1} \tag{4.76}$$

recognizing the need to have $n_0 \ll N_d - N_a$ in any practical application of an extrinsic photoconductor so that most of the uncompensated donors are photoionizable. The inequality of (4.76) limits the maximum operating temperature of these detectors, just as (4.48) does for the intrinsic photoconductors, mainly because of the exponential dependence of n_1 on temperature.

Let us now develop the theory of G^2RA. The expression for the total current for our n-type model is

$$I = qwtn\mu_e E, \tag{4.77}$$

and the photocurrent is

$$I_p = qwt\Delta n\mu_e E; \tag{4.78}$$

n and Δn are related by (4.51). Equating (4.78) to (4.3),

$$G = t\mu_e E \Delta n/\eta \Phi l. \tag{4.79}$$

From (4.77)

$$RA = l^2/qtn\mu_e \simeq l^2/qt\Delta n\mu_e. \tag{4.80}$$

The approximate form of (4.80) is valid when $\Delta n \gg n_0$, which is equivalent to condition 2; see (4.76) and also (4.82) below. To calculate $G^2 RA$ we must determine how Δn depends upon detector material parameters and operating conditions.

The continuity equation for the electrons is

$$\frac{d\Delta n}{dt} = \frac{\eta \Phi}{t} - \frac{\Delta n}{\tau_e}, \qquad (4.81)$$

and its (steady-state) solution is

$$\Delta n = \eta \Phi \tau_e / t. \qquad (4.82)$$

Substituting (4.82) into (4.80) we get

$$RA = l^2 / q\eta \Phi_b \mu_e \tau_e, \qquad (4.83)$$

and we can combine (4.83) with (4.44) to write condition 3 as

$$G^2 l^2 / \mu_e \tau_e \gg k_B T / q. \qquad (4.84)$$

Substituting (4.82) into (4.79) we get

$$G = \mu_e \tau_e E / l \equiv z_e \qquad (4.85)$$

for our n-type model, so that we can write condition 3 in the alternative form

$$z_e l E = \mu_e \tau_e E^2 \gg k_B T / q \qquad (4.86)$$

using (4.85) in (4.84). Both forms of condition 3, (4.84) and (4.86), are independent of the background radiation flux density. The inequality of (4.86) has been termed a "condition for excellence" of an extrinsic photoconductive detector [4.29]; this condition is simply one form of our general condition 3 for infrared photon detectors. Recall that (4.84) and (4.86) implicitly assume condition 2.

Condition 4 requires maximizing the absorption coefficient, which is related to extrinsic material parameters by [4.28]

$$\alpha = \sigma_A (N_d - N_a); \qquad (4.87)$$

σ_A is an absorption or photoionization cross section characteristic of the dopant used as the photoionizable center, and $N_d - N_a$ is the concentration of those centers which are un-ionized and so contain carriers which can be photoexcited into the adjacent band. Achievement of high η is a major problem in these detectors, because extrinsic optical absorption cross sections are relatively low and attainable values of $N_d - N_a$ are limited.

4.4.2 Materials

The first detectors widely used for infrared photon detection in the 8–14 μm wavelength interval were mercury-doped germanium extrinsic photoconductors [4.29]. They were superseded in the late 1960s and early 1970s by intrinsic detectors, especially photoconductive $Hg_{1-x}Cd_xTe$, which offered higher operating temperatures as well as high detectivity in thin devices more compatible with array technology. The Hg-doped Ge detectors are now largely obsolete. However, another consideration has led recently to renewed interest in extrinsic detector materials, specifically doped Si, and this is the desirability of integrating signal-processing electronics with the detector in the same crystal to form a monolithic detector-electronics sensing device. Since Si has by far the most desirable properties and advanced technology of any semiconductor for electronics, the detector should also be made of Si to achieve the best possible monolithic device; the philosophy here is similar to that expressed in an earlier review article [4.30]. Thus we shall analyze only doped Si as an extrinsic photoconductor material because of its potential usefulness in monolithic devices. It is still true in general that extrinsic detectors have disadvantages, compared to intrinsic detectors, such as those which made Hg-doped Ge obsolete, so that there is little incentive for developing extrinsic detector materials other than Si. Studies of Si as an extrinsic photoconductive detector material were pioneered by *Soreff* [4.28].

Condition 1 is not as readily satisfied by doped Si as it is by the alloy semiconductors used in intrinsic photon detectors. The various known dopants can satisfy condition 1 at a number of different wavelengths in the infrared, but none of them is well matched to the important 8–14 μm interval, and possible dopants for the 3–5 μm interval are relatively scarce; see Fig. 4.15 in Appendix F. Thus there is a need for better dopants to satisfy condition 1 more closely for some important detection wavelengths, but the prospects for success of a search for new dopants are uncertain. It remains to be seen whether new doping methods can be developed to satisfy condition 1 better.

Now consider condition 2. We shall use a $\lambda_{co} = 12.4$ μm detector once more as a numerical example, even though there is not yet a known dopant giving exactly this cutoff wavelength. We can rewrite (4.76) as

$$Bn_1(N_d - N_a)t \ll \eta \Phi_b. \tag{4.88}$$

The left side of (4.88) is plotted vs $N_d - N_a$ in Fig. 4.10 for several possible operating temperatures; we have used the values of B and N_c/g_c given in Appendix F, as well as $t = 5 \times 10^{-4}$ cm to be consistent with condition 4 below. By comparing these curves with that for $Hg_{0.8}Cd_{0.2}Te$ in Fig. 4.10 and with the abscissa of Fig. 4.1, we see that this extrinsic Si photoconductive detector requires considerably lower operating temperatures than the intrinsic photoconductor for comparable performance. This result is a well-known disadvantage of an extrinsic photoconductor [4.31]. The curves for a *p*-type example would lie slightly lower than those for the n-type example in Fig. 4.10, because $N_v/g_v < N_c/g_c$.

Let us use (4.84) to evaluate the G^2RA product for these detectors. If sweepout limited the gain to 1/2, then letting $l = 10^{-2}$ cm and using (4.75) and (4.110) for τ_e and μ_e, we find that (4.84) would be satisfied when $N_a \gtrsim 10^{14}$ cm^{-3} and $T \gtrsim 50$ K. However, the dielectric relaxation time would be quite short and the corresponding modulation frequency high for the ranges of material parameters and background flux density of Figs. 4.10 and 4.1, so that at many practical modulation frequencies the gain could become higher. Thus it should usually be possible to satisfy condition 3. Again a p-type example would be similar.

As indicated earlier, condition 4 is not easily satisfied by these detectors. Let us require that $t = 2.5\alpha^{-1}$ so that $\eta > 0.9$, use $\sigma_A(\max) = 10^{-15}$ cm^2 from Appendix F, and assume $(N_d - N_a) = 10^{17}$ cm^{-3} as the maximum value for the shallow dopant of this example at which the ionization energy nearly equals the low doping level value; then $t \simeq 2.5 \times 10^{-2}$ cm is the required detector thickness. We had used this thickness in the above evaluation of condition 2. Thus even in this "best case" estimate, the Si extrinsic photoconductor must be about 50 times thicker than the comparable $Hg_{0.8}Cd_{0.2}Te$ intrinsic photoconductor.

Little data have been published very recently in the open literature on doped Si as an extrinsic photoconductive detector [4.32].

4.5 Summary and Conclusions

We have formulated the theory of infrared photon detectors in terms of four conditions which a detector must satisfy to achieve nearly BLIP performance. These conditions are convenient for intercomparing the materials and effects used in photon detectors.

Condition 1 relates the desired detector wavelength response to the electronic excitation energy needed for that response. It is most readily satisfied by the alloy semiconductors, since they provide continuously variable cutoff wavelengths over a wide spectrum. The $Hg_{1-x}Cd_xTe$ alloy system can cover the greatest wavelength range, < 1 μm to beyond 30 μm.

Condition 2 limits the maximum operating temperature of a photon detector. In (unbiased) photovoltaic detectors this condition is redundant with condition 3. In photoconductive detectors condition 2 requires that the concentration of thermally generated carriers contributing to the gr noise be negligible compared to that of carriers generated by the background radiation. This condition is satisfied best by the $Hg_{1-x}Cd_xTe$ intrinsic photoconductors, i.e., permits the highest operating temperatures for them, but the photovoltaic detectors would be comparable if *Case (c)* could be achieved. Silicon extrinsic photoconductors require lower operating temperatures.

Condition 3 requires that Johnson noise be negligible compared to background noise, and it permits nearly BLIP performance when condition 2 is satisfied. For photovoltaic detectors condition 3 demands a large enough RA product for a given background radiation flux density. Condition 3 for the photoconductive detectors requires a large enough G^2RA product, so that G^2RA for these detectors is analogous to RA for the photovoltaics. The problem in developing photovoltaic detectors mainly involves achieving good enough junction technology to permit high values of RA. For intrinsic photoconductive detectors the corresponding problem is to find and develop semiconductors with the high mobility ratios and/or resistivities needed for high values of G^2RA; the $Hg_{1-x}Cd_xTe$ alloys provide a solution. Photovoltaic detectors may satisfy condition 3 better at low temperatures and in feeble background radiation than nontrapping intrinsic photoconductors, but minority carrier trapping in intrinsic photoconductors can give very high values of G^2RA by gain enhancement. The RA and G^2RA products potentially achievable, respectively, in the known photovoltaic and intrinsic photoconductive materials are not different enough in general to permit a clear choice of one effect and material as the best for photon detectors. Condition 3 is independent of the background radiation flux density in an extrinsic Si photoconductor; it should usually be possible to satisfy.

Condition 4 provides the highest possible BLIP detectivity by requiring that the quantum efficiency approach its maximum value of unity. This condition is easily met by relatively thin photovoltaic and intrinsic photoconductive detectors. However, it is a major problem for extrinsic Si photoconductors, because limited maximum values of dopant concentrations and absorption cross sections give rather low absorption coefficients, requiring undesirably thick detectors for high quantum efficiencies.

Thus the two well-developed semiconductor alloy systems, $Hg_{1-x}Cd_xTe$ and $Pb_{1-x}Sn_xTe$, can satisfy all four conditions quite well. The $Hg_{1-x}Cd_xTe$ system is the more versatile because it can cover a wider wavelength range and is suitable for both photovoltaic and photoconductive detectors, but photovoltaic $Pb_{1-x}Sn_xTe$ detectors are very promising. The most productive approach will be to improve the technology of these alloy systems to bring the detectors closer to the fundamental limits derived in Sections 4.2 and 4.3, rather than to search for new materials, since these alloy systems offer potentially adequate properties. The choice between the photovoltaic and intrinsic photoconductive detectors for an application must usually be made by criteria other than detectivity, such as response time, power dissipation, detector configuration, etc. Doped Si as an extrinsic photoconductor material is a special case; it is of interest for the possibility of integration with Si electronics in monolithic sensor devices, so that there will be considerable development effort devoted to minimizing its deficiencies.

Acknowledgements. The author is indebted to many of his colleagues at the Honeywell Radiation Center and the Honeywell Corporate Research Center for discussions of various subjects in this chapter. Special thanks are due to Dr. *C. H. Li*, Director of Research, for encouraging and supporting the preparation of this chapter, and also to *S. B. Schuldt* for discussions of photoconductivity and carrier lifetimes.

Appendix A: Generation-Recombination Noise [4.19, 32]

We want to derive in a general way the shot-noise-like expression for generation-recombination (gr) noise used in (4.12) in Subsection 4.1.3. The usual expression for the gr noise current in a two-carrier semiconductor [4.17, 18] can be written as

$$I_{gr}^2 = 4I^2 \left(\frac{b+1}{bN_{maj} + N_{min}} \right) \tau \langle (\Delta N)^2 \rangle \Delta f, \tag{4.89}$$

assuming that $\Delta N_{maj} = \Delta N_{min}$ so that $\tau_e = \tau_h = \tau$ and that $b \equiv \mu_{maj}/\mu_{min}$; here N_{maj} and N_{min} are the majority and minority carrier populations (dimensionless), respectively, and μ_{maj} and μ_{min} are their mobilities. Substituting the generation-recombination theorem [4.17],

$$\langle (\Delta N)^2 \rangle = g\tau, \tag{4.90}$$

into (4.89) and noting that $I_{min} = IN_{min}/(bN_{maj} + N_{min})$ we get

$$I_{gr}^2 = 4I_{min}^2 \left(\frac{b+1}{N_{min}} \right)^2 \tau^2 g \Delta f. \tag{4.91}$$

The generation rate g in (4.90) is the total flux generating carriers; i.e., $g = \eta \Phi_b A + g_{th}$ in the terminology of Subsection 4.1.3, where g_{th} is the generation rate of the thermal equilibrium carrier population. Since $I_{min} = qN_{min}\mu_{min}E/l$, (4.91) becomes

$$I_{gr}^2 = 4q^2 g \left[\frac{(\mu_{maj} + \mu_{min})\tau E}{l} \right]^2 \Delta f = 4q^2 g G^2 \Delta f; \tag{4.92}$$

see (4.60) and (4.61) in Subsection 4.3.1 for G. In an extrinsic photoconductor $\mu_{min} = 0$, so that $G = \mu_{maj}\tau_{maj}E/l$. Thus from (4.92)

$$I_{gr}^2 = 4q[q\eta \Phi_b A G^2 + qg_{th} G^2] \tag{4.93}$$

as in (4.12).

In the above derivation we have implicitly assumed that τ represents the effective carrier lifetime including the possiblity of lifetime reduction by sweepout. The theory of minority carrier sweepout is not yet rigorous, so that this assumption may be questionable.

Appendix B: gr Current of a *p-n* Junction when Shockley-Read Recombination is Negligible

We want to determine the relative magnitudes of the diffusion current and the space-charge layer generation-recombination (gr) current in a *p-n* junction for a model in which there are no Shockley-Read recombination centers in the semiconductor material. We shall follow a procedure analogous to that on pp. 1230 and 1231 of *Sah* et al. [4.8] and will use their notation.

The equation for the radiative recombination rate U analogous to *Sah* et al.'s (6) is [4.9]

$$U = \frac{R_r}{n_i^2}(np - n_i^2). \tag{4.94}$$

In the *n*-type region adjacent to the *p-n* junction,

$$np \simeq n_n p_n + n_n(p - p_n) = n_i^2 + n_n(p - p_n), \tag{4.95}$$

so that (4.94) becomes

$$U \simeq \frac{R_r n_n}{n_i^2}(p - p_n). \tag{4.96}$$

Comparing (4.96) with *Sah* et al.'s (8) we get

$$\tau_p = n_i^2 / R_r n_n. \tag{4.97}$$

Substituting (4.97) into their (9) we get

$$J_d = -2qL_0 R_r \tag{4.98}$$

for the diffusion current density in their simple case of $n_n = p_p$.

In the junction space-charge layer

$$np \ll n_i^2, \tag{4.99}$$

so that (4.94) becomes

$$U = -R_r. \tag{4.100}$$

Comparing (4.100) with *Sah* et al.'s (10),

$$\tau_0 = n_i / 2R_r. \tag{4.101}$$

Substituting (4.101) into their (11) we get

$$J_{gr} = -qWR_r \tag{4.102}$$

for the gr current density. The relative magnitudes of the two currents are then given by the ratio of (4.102) to (4.98), which is

$$\frac{J_{gr}}{J_d} = \frac{W}{2L_0} \tag{4.103}$$

for direct radiative recombination.

The equation for interband Auger recombination analogous to *Sah* et al.'s (6) is of the same form as (4.94) [4.33]. Therefore a procedure like that above for radiative recombination would give (4.103) also for Auger recombination.

Appendix C: Properties of IV–VI Semiconductors

These materials include compounds and alloys formed between elements of groups IV and VI of the periodic table. The properties of the narrow-gap semiconductors of this class which are useful in infrared photon detectors have been reviewed in detail very recently by *Harman* and *Melngailis* [4.3], so that we will only summarize here those properties needed in this chapter. The alloy systems $Pb_{1-x}Sn_xTe$, $Pb_{1-x}Sn_xSe$ and $Pb_{1-x}Ge_xTe$ have all been studied as detector materials, but $Pb_{1-x}Sn_xTe$ has received the most emphasis. Accordingly, we will review the properties only of $Pb_{1-x}Sn_xTe$ in this appendix; the other materials are analogous.

The valence and conduction band edges in the $Pb_{1-x}Sn_xTe$ alloys have the same "many-valley" form and nearly equal effective masses [4.3, 34]. The cutoff wavelength at 77 K vs alloy composition in $Pb_{1-x}Sn_xTe$ in the range of compositions normally used for detectors ($0 \lesssim x \lesssim 0.4$) is given by the equation

$$\lambda_{co}(\mu m) = (-0.406x + 0.175)^{-1}, \quad (4.104)$$

derived from the energy gap vs x data reviewed by *Harman* and *Melngailis* [4.3]. The gap widens and λ_{co} shortens with increasing temperature. The $x \cong 0.2$ alloys have been studied most, because they are used for detectors for the important 8–14 μm wavelength range.

A consequence of the similar valence and conduction bands of $Pb_{1-x}Sn_xTe$ is that the electron and hole mobilities are approximately equal ($\mu_e \simeq \mu_h$) for the same temperatures and doping concentrations. *Zoutendyk* [4.35] has measured these mobilities vs carrier concentration in $Pb_{0.8}Sn_{0.2}Te$, and the equation of his 77 K curve is

$$\mu_e, \mu_h (cm^2/V\text{-}s) \simeq 1.16 \times 10^{16} \, n^{-2/3}, p^{-2/3} (cm^{-3}) \quad (4.105)$$

for both mobilities; this is the expected dependence for full statistical degeneracy. See also [4.36, 37] for data on temperature dependence and on mobilities in PbTe.

The formula for n_i in $Pb_{1-x}Sn_xTe$ has been determined by *Harman* and *Melngailis* [4.3] and is

$$n_i(cm^{-3}) = 2.9 \times 10^{15} (TE_g)^{3/2} \exp(-E_g/2k_BT), \quad (4.106)$$

with T in K and E_g in eV. The static dielectric constant $\kappa \simeq 400$ in $Pb_{0.8}Sn_{0.2}Te$ and PbTe, and the index of refraction $n_1 \simeq 6$ [4.7]. The optical absorption coefficient is $\sim 5 \times 10^3$ cm^{-1} at wavelengths close to the intrinsic absorption edge in PbTe [4.38], and this order of magnitude would presumably be observed in $Pb_{1-x}Sn_xTe$ alloys also.

Interband Auger recombination had been thought to be insignificant in $Pb_{1-x}Sn_xTe$ alloys [4.7], but the very recent theory by *Emtage* casts doubt upon that assumption [4.39]. However, in this chapter we have assumed that only radiative recombination need be considered as an intrinsic lifetime-limiting mechanism. *Melngailis* and *Harman* have calculated an expression for the radiative τ_h from which we can determine R_r using (4.31):

$$R_r(\text{cm}^{-3}\text{-s}^{-1}) = 8.2 \times 10^{-9} n_1 n_i^2 T^{-3/2}(\text{K}) E_g^{-1/2}(\text{eV}). \tag{4.107}$$

In (4.107) we have used the relationship $m^*/m = 0.19 E_g(\text{eV})$, which is derivable from the equation at the top of p. 168 and values in the first full paragraph on p. 169 of [4.7].

The Burstein-Moss or "band filling" effect due to statistical degeneracy begins to shift the cutoff wavelength λ_{co} of a $Pb_{1-x}Sn_xTe$ detector to shorter wavelengths as the carrier concentration becomes high [4.34]. Thus the high RA products at high carrier concentrations shown in Fig. 4.5 may not be attainable without an accompanying shift of λ_{co} downward. This effect may be a problem at concentrations above 10^{17} cm^{-3}; the calculation is not attempted here. Tunneling might reduce RA also at high concentrations.

Appendix D: Properties of Zinc Blende Crystal Structure Semiconductors

The materials in this class which are used in infrared photon detectors include the compounds InSb and InAs and the alloy system $Hg_{1-x}Cd_xTe$. The $Hg_{1-x}Cd_xTe$ alloys are the most important of these materials for detectors, and their properties shall be reviewed here. But InSb and InAs are analogous to those $Hg_{1-x}Cd_xTe$ alloys which have comparable energy gaps. Since the most recent comprehensive review of $Hg_{1-x}Cd_xTe$ was published several years ago [4.40], we shall take this opportunity to update it, although still emphasizing the properties needed in this chapter.

The valence band in any of these semiconductors has a maximum at the center of the Brillouin zone with relatively large effective mass, whereas the conduction band minimum at the zone center has a much smaller effective mass [4.3, 34, 40]. Several consequences of this band structure are important in intrinsic photoconductivity. The electron mobility for virtually any scattering mechanism is considerably greater than the hole mobility because of the mass difference; i.e., $\mu_e/\mu_h \equiv b \gg 1$. The conduction band has a low density of states near its minimum, so that n-type material becomes statistically degenerate at relatively low conduction electron concentrations. "Shallow" donor impurities have energy levels merged with the conduction band or very close to it due to the very small conduction band edge effective mass, whereas the levels of "shallow" acceptors are a few hundredths of an eV from the valence band edge. Intrinsic optical absorption coefficients of $\sim 5 \times 10^3$ cm^{-1} are observed near the cutoff wavelength [4.40, 41].

The cutoff wavelength vs alloy composition and temperature in $Hg_{1-x}Cd_xTe$ is given by the equation [4.40]

$$\lambda_{co}(\mu m) = [1.28x - 0.20 + 0.264x^3 + 4.22 \times 10^{-4} T(K)(1 - 2.08x)]^{-1}. \quad (4.108)$$

Nearly all the development of $Hg_{1-x}Cd_xTe$ alloys as intrinsic photoconductor materials to date has involved the composition range $0.18 \gtrsim x \gtrsim 0.4$, corresponding to cutoff wavelengths of $3 \gtrsim \lambda_{co} \gtrsim 30 \mu m$. Alloys of the $x \simeq 0.2$ composition have been emphasized, because they provide detector material for the important 8–14 μm wavelength interval. However, considerable research has been done recently on the higher-x alloys up to $x \simeq 0.4$. The properties of $Hg_{1-x}Cd_xTe$ throughout the $0.18 \gtrsim x \gtrsim 0.4$ range are qualitatively similar; quantitative differences can usually be accounted for by scaling in proportion to the energy gap or to λ_{co}^{-1}.

Single crystals of $Hg_{1-x}Cd_xTe$ are grown by several different methods [4.42]. Almost regardless of the growth method or composition x in the above range, undoped ("pure") crystals which are n-type at low temperatures have an extrinsic electron concentration n_0 near 10^{15} cm^{-3}; this is a relatively low carrier concentration for a semiconductor, and it is one of the major reasons for the success of $Hg_{1-x}Cd_xTe$ as a photoconductive infrared detector material. However, undoped crystals often turn out p-type with $p_0 \sim 10^{17}$ cm^{-3} or have anomalous electrical properties which are not clearly those of either n- or p-type material.

The anomalous crystals have been studied by several investigators [4.43–46]. Their properties are manifested in Hall coefficient R_H vs temperature curves of peculiar shapes; see Fig. 4.13. Sometimes two reversals of sign of R_H are observed: a reversal from n-type in the intrinsic range to p-type at a lower temperature, and then a second reversal back to n-type at a still lower temperature, as in curve 3 of Fig. 4.13. In other samples R_H remains negative at all temperatures but shows a minimum at a temperature not far below the intrinsic range as in curve 2 of Fig. 4.13. The most likely cause of both curve shapes is a thin n-type inversion layer next to the surface of a p-type sample [4.44, 47]. The two Hall curve shapes described above correspond to different relative contributions of the n-type surface layer and p-type bulk; the sample for which R_H remains negative at all temperatures has the greater surface-layer effect. In past work and even in some recent reports, results have been quoted for so-called "compensated" n-type samples having extrinsic electron concentrations $n_0 \sim 10^{14}$ cm^{-3} and mobilities $\mu_e \sim 10^4$ cm^2/volt-s. These values come from assuming that a sample like that of curve 2 in Fig. 4.13 is n-type. However, such samples are undoubtedly usually of the p-type bulk, n-type surface layer variety described in this paragraph. To be truly compensated n-

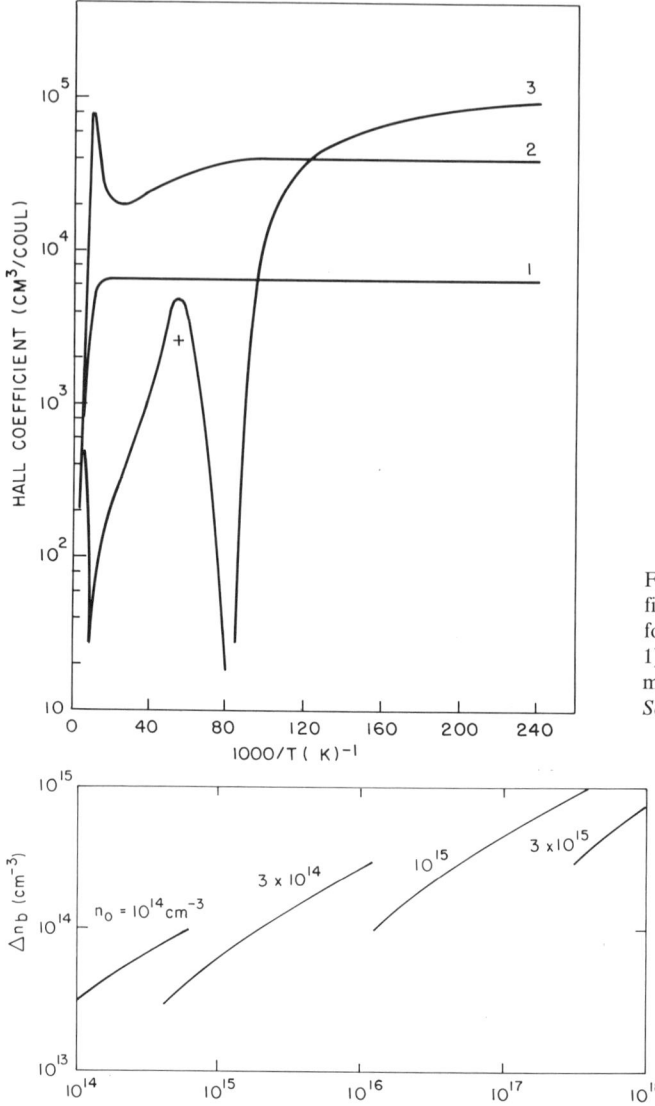

Fig. 4.13. Typical Hall coefficient vs temperature curves for an n-type sample (curve 1) and anomalous p-type samples (curves 2 and 3); after Scott and Hager [4.44]

Fig. 4.14. Background-generated carrier concentration Δn_b vs background radiation flux density Φ_b for representative thermal equilibrium carrier concentrations n_0 in n-type $Hg_{0.8}Cd_{0.2}Te$ at 77 K

type, a sample must have a Hall curve shaped like curve 1 in Fig. 4.14, without the anomalous maximum and minimum of curve 2.

The difference in type between undoped crystals is believed due to differences in the time-temperature history during cooling from the growth temperature, but the details are not yet fully understood. One can often anneal "as-grown" p-

type crystals, however, to convert them to *n*-type with $n_0 \sim 10^{15}$ cm^{-3}. The carrier concentration of $\sim 10^{15}$ cm^{-3} in the *n*-type crystals is probably caused by residual donor impurities, either from the Hg$_{1-x}$Cd$_x$Te starting material or the crystal growth capsule [4.48, 49]. Crystals can also be doped during growth or by diffusion or ion implantation with donor impurities (e.g., In or Al) or acceptor impurities (e.g., Au or Cu), although much needs still to be learned about the properties of dopants in Hg$_{1-x}$Cd$_x$Te.

The electron mobility in Hg$_{1-x}$Cd$_x$Te depends upon x as well as temperature. It is relatively independent of temperature at low temperatures in *n*-type samples with $n_0 \sim 10^{15}$ cm^{-3}, and varies from about 3×10^5 cm^2/volt-s in Hg$_{0.82}$Cd$_{0.18}$Te to about 10^4 cm^2/volt-s in Hg$_{0.6}$Cd$_{0.4}$Te [4.48, 50, 51]. These high mobility values suggest that the donors in such samples are mostly uncompensated by acceptors. The electron mobility in *p*-type material has not been directly measured vs carrier concentration, and it is uncertain how to estimate it. The electron mobility vs n_0 in *n*-type material cannot be used without modification, because the conduction and valence bands are so different; e.g., at a concentration at which an *n*-type sample would be statistically degenerate, a *p*-type sample would be nondegenerate.

Recent data of *Scott* and co-workers [4.52] on electrical properties of *p*-type Hg$_{0.6}$Cd$_{0.4}$Te show that in rather lightly doped material ($p \simeq 10^{15}$–10^{16} cm^{-3}), μ_h (170 K) $\simeq 2 \times 10^2$ cm^2/V-s; since the shape of the valence band should not change much with x, we can use these values in the lower-x example to be considered. Hole mobility data for *n*-type material are scarce, but one can deduce from the minority carrier sweepout data of *Emmons* and *Ashley* [4.53] for Hg$_{0.8}$Cd$_{0.2}$Te at 77 K that $\mu_h \simeq 700$ cm^2/volt-s; since $\mu_e \simeq 2 \times 10^5$ cm^2/volt-s in such material, the mobility ratio $b \simeq 300$, verifying our earlier statement that $b \gg 1$.

The fully *p*-type crystals with $p_0 \sim 10^{17}$ cm^{-3} are of little use as photoconductor material because of their relatively high carrier concentration. Properties of such crystals have been reported by *Elliott* and coworkers [4.54].

The dependence of the intrinsic carrier concentration n_i on composition x and temperature T in Hg$_{1-x}$Cd$_x$Te has been determined for wide ranges of x and T by *Schmit* [4.55]. This subject has been studied also by *Elliott* [4.56] and by *Vérié* [4.57]. *Schmit*'s formula for n_i is

$$n_i(\text{cm}^{-3}) = (8.46 - 2.29x + 0.00342T) \\ \times 10^{14} T^{3/2} E_g^{3/4} \exp(-E_g/2k_B T), \tag{4.109}$$

with T in K and E_g in eV. At 77 K in Hg$_{0.8}$Cd$_{0.2}$Te, $n_i = 5.2 \times 10^{14}$ cm^{-3}.

Carrier recombination in *n*-type Hg$_{0.8}$Cd$_{0.2}$Te has been studied in detail by *Kinch* et al. [4.13] at temperatures high enough for trapping effects to be negligible. They found that the measured lifetime is limited by the interband Auger mechanism in the intrinsic range rather than by direct radiative recombination. The intrinsic Auger lifetime τ_{Ai} vs temperature has been calculated by *Petersen* [4.58] for $Hg_{0.8}Cd_{0.2}Te$ and $Hg_{0.73}Cd_{0.27}Te$ well as by *Buss* for $Hg_{0.805}Cd_{0.195}Te$ (reported in [4.13]); their calculations agree reasonably well, especially considering that the Auger lifetime is a very strong function of the energy gap and so of x. *Kinch* et al. found that Auger recombination was dominant in their best n-type crystals down to the lowest temperature of measurement, 65 K; in other ("compensated") samples Shockley-Read recombination was observed at the low temperatures, but they may have been p-type bulk, n-type surface layer samples like those described above. *Kinch* et al. determined τ_{Ai} quite accurately by comparison of data on n-type $Hg_{0.8}Cd_{0.2}Te$ with theory, and they found $\tau_{Ai} \simeq 10^{-3}$ s at 77 K in $E_g = 0.1$ eV material; see also [4.14]. Radiative recombination can affect the lifetime somewhat at low temperatures. The radiative recombination rate was calculated earlier using the actual band structure of $Hg_{1-x}Cd_xTe$ (nonparabolic conduction band) [4.40], and the recent results of *Kinch* and coworkers agree fairly well. From the recent results of *Kinch* and coworkers we find that $R_r(77 \text{ K}) \simeq 6.5 \times 10^{16}$ cm^{-3}-s^{-1} for $E_g = 0.1$ eV ($\lambda_{co} = 12.4$ μm). Auger recombination should be insignificant in p-type $Hg_{1-x}Cd_xTe$, essentially because of the band structure [4.11]; therefore, only direct radiative recombination need be considered as an intrinsic lifetime-limiting mechanism in p-type material.

The static dielectric constant $\kappa \simeq 20$ in the low-x $Hg_{1-x}Cd_xTe$ alloys of interest here [4.59].

Appendix E: Dependence of Detector Material Parameters on Background Radiation

It is assumed in the examples analyzed in Subsecs. 4.2.2 and 4.3.2 that carriers generated by background radiation are negligible, so that the carrier concentration in the detector material is nearly the same as the thermal equilibrium concentration. This assumption will often be valid in practice, but in cases of relatively intense background radiation and low thermal equilibrium carrier concentration it may not be; this situation may occur especially in n-type $Hg_{1-x}Cd_xTe$ of low carrier concentrations [4.14]. Unfortunately, the rigorous theory of these effect, particularly intrinsic photoconductivity, can become very complicated with inclusion of significant background radiation. For example, the problem is then no longer "mono-molecular", and one must take proper account of the distinction between the incremental and excess carrier lifetimes. (The incremental lifetime is the τ appearing in the gr theorem, whereas the excess carrier lifetime is the τ appearing in expressions such as (4.31) and (4.40)). An analysis of this problem analytically is beyond the scope of this chapter.

Regardless of the theoretical complications, however, it is clear physically that when a detector is to be used in a certain background radiation flux density, it is not helpful to its performance to lower the majority carrier concentration much below the value at which background-generated carriers become comparable. A partial exception to this conclusion may occur when sweepout is important; sweepout complicates the problem even more analytically, but its qualitative effect should be to reduce Δn_b for a given Φ_b and thereby improve detector performance.

Appendix F: Properties of Extrinsic Silicon

Silicon has been studied and applied in devices more than any other semiconductor, and its properties are known in great detail [4.60]. But for this chapter we need review only that information used in Sec. 4.4, which includes properties of dopants, carrier mobilities and lifetimes, and densities of states in the conduction and valence bands.

Impurity dopants in Si can be divided into two categories: those giving "shallow" electronic energy levels and those giving "deep" levels [4.61]. Shallow levels are only a few hundredths of an eV from either band and are provided generally by impurity atoms having one more or less valence electron than the host Si atom. Deep levels are more than 0.1 eV from either band and are provided generally by impurities having greater variations in valence from Si than the shallow impurities.

The energy levels of the shallow dopants can be accounted for by effective mass theory, allowing for a "central cell" correction to the ground state energy [4.62]. In a lightly doped crystal the dopant atoms are sufficiently isolated from each other that their energy levels are discrete, but at higher dopant concentrations the energy levels broaden into "impurity bands" due to overlap of the dopant atom wave functions, and the ionization or ground state energy is reduced [4.63]. Eventually there is so much interaction among the dopant atoms that no bound electronic states are possible, and the dopant states merge with the adjacent band; at this point, the so-called Mott transition [4.64] which occurs at shallow dopant concentrations $\sim 10^{18}\,\text{cm}^{-3}$ in Si, there is no longer a dopant ionization energy and the material cannot be used as an extrinsic photoconductor. For shallow dopants in Si the Mott transition occurs at dopant concentrations well below the solubility limit of the impurity, so that it constitutes the ultimate upper limit to the concentration of impurity centers useful for extrinsic photoconductivity.

The theory of the electronic energy levels of deep dopants in Si is far more difficult and is not yet fully developed. The electrons are much more tightly bound to these impurities than to shallow impurities, and the Mott transition is not observed. Also, the solubilities of these dopants in Si are generally much lower than those of the shallow dopants, so that solubility limits the concentration of a deep dopant which can be achieved in Si for use as an extrinsic photoconductor.

The cutoff wavelengths corresponding to the ground-state ionization energies of impurity dopants in Si are shown in Fig. 4.15 [4.61]; for the shallow

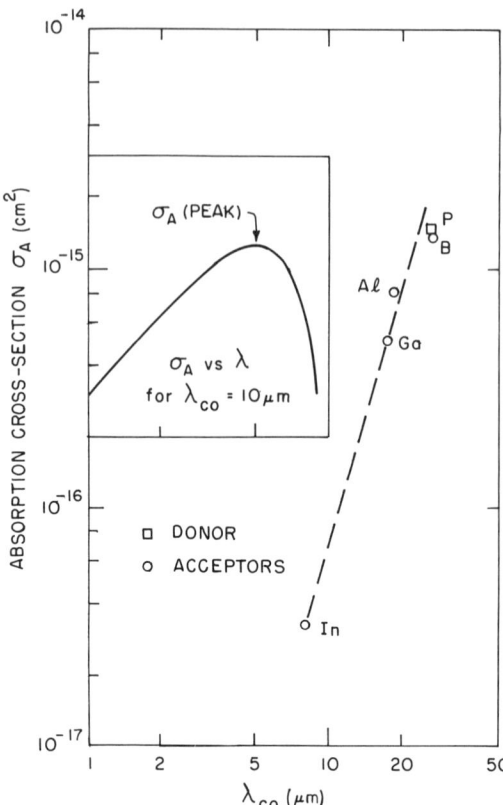

Fig. 4.15. Peak optical absorption cross sections vs cutoff wavelengths of impurity dopants in Si (dashed curve). Solid curve in inset is a typical shape of the σ_A vs λ curve of a dopant [4.65]

dopants these are the cutoff wavelengths for low doping levels and are therefore the shortest possible values. The maximum solid solubilities of all these dopants are given by *Milnes* [4.61].

The optical absorption cross sections σ_A of impurities in Si typically have a dependence on wavelength of the form shown in Fig. 4.15. Maximum or peak values of σ_A for all the impurities in Si for which this parameter is known are also plotted in Fig. 4.15 [4.61].

The carrier mobility is determined mainly by impurity scattering at the low temperatures of extrinsic photoconductor operation. Furthermore, in the relatively uncompensated Si of most interest for detectors, scattering by the neutral majority dopant atoms will dominate. It can be shown that the neutral impurity scattering mobility is given in Si by [4.65]

$$\mu_N(\text{cm}^2/\text{V-s}) = 1.24 \times 10^{22} E_B(\text{eV})/N_N(\text{cm}^{-3}), \tag{4.110}$$

where E_B is the dopant ionization energy and N_N is the concentration of uncompensated, neutral dopant atoms. In most material used in detectors the ionized-impurity scattering mobility will be much smaller [4.66].

Magnitudes of the recombination coefficient B are not accurately known for dopants in Si. As an order of magnitude estimate suitable for our purposes in this chapter we can assume $B \simeq 10^{-6}$ cm^3-s^{-1} [4.67].

The conduction and valence band density of states coefficients in Si are given by [4.68]

$$N_c/g_c = 2.75 \times 10^{15} T^{3/2} \text{ cm}^{-3} \tag{4.111}$$

and [4.69]

$$N_v/g_v = 0.56 \times 10^{15} T^{3/2} \text{ cm}^{-3}. \tag{4.112}$$

For donors in Si the degeneracy factor $g_c = 2$, and for shallow acceptors $g_v = 4$; for deep acceptors (e.g., In) one may have $g_v = 6$ [4.70].

References

4.1 D.H.Seib, L.W.Aukerman: *Advances in Electronics and Electron Physics* **34**, ed. by L.Marton (Academic Press, New York 1973) pp. 95–221
4.2 J.O.Dimmock: J. Electron. Mater. **1**, 255 (1972); see also
 R.B.Emmons, S.R.Hawkins, K.F.Cuff: Optical Engineering **14**, 21 (1975)
 H.Levinstein, J. Mudar: Proc. IEEE **63**, 6 (1975)
4.3 T.C.Harman, I.Melngailis: *Applied Solid State Science* **4**, ed. by R.Wolfe (Academic Press, New York 1974) pp. 1–94; see also
 I.Melngailis: J. Luminescence **7**, 501 (1973)
4.4 G.R.Pruett, R.L.Petritz: Proc. IRE **47**, 1524 (1959)
4.5 W.J.Beyen, B.R.Pagel: Infrared Phys. **6**, 161 (1966)
4.6 D.Long: Infrared Phys. **12**, 115 (1972)
4.7 I.Melngailis, T.C.Harman: *Semiconductors and Semimetals* **5**, ed. by R.K.Willardson and A.C.Beer (Academic Press, New York 1970) pp. 111–174
4.8 C.T.Sah, R.Noyce, W.Shockley: Proc. IRE **45**, 1228 (1957)
4.9 See J.R.Hauser, P.M.Dunbar: Solid-State Electron. **18**, 715 (1975), and references therein. We assume an idealized extreme case of their results in which diffusion current in the p^+ region and space-charge layer gr current of the $p^+ - p$ interface are negligible. One reason for introducing this boundary condition is to emphasize that the $\Delta n = 0$ boundary condition need not always apply
4.10 See, for example, S.M.Sze: *Physics of Semiconductor Devices* (Wiley-Interscience, New York 1969)
4.11 P.W.Kruse, L.D.McGlauchlin, R.B.McQuistan: *Elements of Infrared Technology: Generation, Transmission, and Detection* (Wiley-Interscience, New York 1962)
4.12 W.H.Rolls, D.V.Eddolls: Infrared Phys. **13**, 143 (1973)
 C.C.Wang, S.R.Hampton: Solid-State Electron. **18**, 121 (1975)
 A.M.Andrews, J.T.Longo, J.E.Clarke, E.R.Gertner: Appl. Phys. Lett. **26**, 439 (1975)
 L.H.DeVaux, H.Kimura, M.J.Sheets, F.J.Renda, J.R.Balon, P.S.Chia, A.H.Lockwood: Infrared Phys. **15**, 271 (1975)
 P.S.Chia, J.R.Balon, A.H.Lockwood, D.M.Randall, F.J.Renda, L.H.DeVaux, H.Kimura: Infrared Phys. **15**, 279 (1975)
 P.LoVecchio, M.Jasper, J.T.Cox, M.B.Garber: Infrared Phys. **15**, 295 (1975)
 A.Bradford, E.Wentworth: Infrared Phys. **15**, 303 (1975)
 R.Longshore, M.Jasper, B.Summer, P.LoVecchio: Infrared Phys. **15**, 311 (1975)
 M.R.Johnson, R.A.Chapman, J.S.Wrobel: Infrared Phys. **15**, 317 (1975); see also paper on $Pb_{1-x}Ge_xTe$ by
 G.A.Antcliffe, R.A.Chapman: Appl. Phys. Lett. **26**, 576 (1975)

4.13 M.A.Kinch, M.J.Brau, A.Simmons: J. Appl. Phys. **44**, 1649 (1973)
4.14 M.A.Kinch, S.R.Borrello: Infrared Phys. **15**, 111 (1975)
4.15 G.Fiorito, G.Gasparrini, F.Svelto: Appl. Phys. Lett. **23**, 448 (1973)
J.Marine, C.Motte: Appl. Phys. Lett. **23**, 450 (1973)
T.Koehler, P.J.McNally: Optical Engineering **13**, 312 (1974)
G.Fiorito, G.Gasparrini, F.Svelto: Infrared Phys. **15**, 287 (1975)
J.M.Pawlikowski, P.Becla: Infrared Phys. **15**, 331 (1975)
P.Becla, J.M.Pawlikowski: Infrared Phys. **16**, 457 (1976)
G.Cohen-Solal, A.Zozime, C.Motte, Y.Riant: Infrared Phys. **16**, 555 (1976)
4.16 H.Y.Fan: Phys. Rev. **92**, 1424 (1953)
4.17 K.M.van Vliet: Proc. IRE **46**, 1004 (1958)
4.18 K.M.van Vliet: Appl. Opt. **6**, 1145 (1967)
4.19 R.L.Williams: Infrared Phys. **8**, 337 (1968)
4.20 P.W.Kruse: *Semiconductors and Semimetals* **5**, ed. by R.K.Willardson and A.C.Beer (Academic Press, New York 1970) pp. 15–83
4.21 E.L.Stelzer, D.Long: (unpublished results)
4.22 V.J.Mazurczyk, R.N.Graney, J.B.McCullough: Optical Engineering **13**, 307 (1974)
4.23 R.L.Williams, B.H.Breazeale, C.G.Roberts: Proc. Third International Conf. on Photoconductivity (Pergamon Press, New York 1971) pp. 237–242
M.R.Johnson: J. Appl. Phys. **43**, 3090 (1972)
S.P.Emmons, K.L.Ashley: Appl. Phys. Lett. **20**, 162 (1972)
M.Y.Pines, R.H.Genoud, P.R.Bratt: Proc. IEEE Electron Device Conf. 1974, (IEEE, New York) pp. 456–460
4.24 N.C.Aldrich, J.D.Beck: Appl. Opt. **11**, 2153 (1972)
4.25 J.D.Beck, R.M.Broudy: (unpublished results)
4.26 M.M.Blouke, E.E.Harp, C.R.Jeffus, R.L.Williams: J. Appl. Phys. **43**, 188 (1972); see also references therein
4.27 M.M.Blouke, R.L.Williams: Appl. Phys. Lett. **20**, 25 (1972)
4.28 R.A.Soreff: J. Appl. Phys. **38**, 5201 (1967)
4.29 R.B.Emmons: Infrared Phys. **10**, 63 (1970)
4.30 D.Long: IEEE Trans. Electron Devices **ED-16**, 836 (1969)
4.31 M.M.Blouke, C.B.Burgett, R.L.Williams: Infrared Phys. **13**, 61 (1973)
4.32 M.Y.Pines, R.Baron: Proc. IEEE International Electron Devices Meeting, Washington, DC (1974) pp. 446–450
N.Sclar: Infrared Phys. **16**, 435 (1976)
4.33 See, for example, J.L.Moll: *Physics of Semiconductors* (McGraw-Hill, New York 1964) pp. 101–104
4.34 D.Long: *Energy Bands in Semiconductors* (Wiley-Interscience, New York 1968)
4.35 P.J.A.Zoutendyk: Proc. Semimetals and Narrow-Gap Semiconductors Conf. (Pergamon, New York 1971) p. 421
4.36 J.R.Burke, J.D.Jensen, B.Houston: Proc. Semimetals and Narrow-Gap Semiconductors Conf. (Pergamon, New York 1971) p. 393
4.37 R.S.Allgaier, W.W.Scanlon: Phys. Rev. **111**, 1029 (1958)
4.38 W.W.Scanlon: *Solid State Physics* **9**, ed. by F.Seitz and D.Turnbull (Academic Press, New York 1959) p. 115
4.39 A very recent paper by P.R.Emtage: J. Appl. Phys. **47**, 2565 (1976), predicts theoretically that Auger recombination is strong also in $Pb_{1-x}Sn_xTe$, but the theory has not yet been confirmed by direct experiment; see (4.40) and (4.41) for inclusion of Auger recombination. The strong Auger recombination would explain the short $\sim 10^{-8}$ s carrier lifetimes typically observed in $Pb_{1-x}Sn_xTe$ crystals [4.3]
4.40 D.Long, J.L.Schmit: *Semiconductors and Semimetals* **5**, ed. by R.K.Willardson and A.C.Beer (Academic Press, New York 1970) pp. 175–255
4.41 M.W.Scott: J. Appl. Phys. **40**, 4077 (1969)

4.42 See T.C.Harman: J. Electron. Mater. **1**, 230 (1972), for a report of a recent method and for references to methods developed earlier. Crystals for detectors are still grown often by methods closely related to the modified Bridgeman method developed originally by P.W.Kruse and coworkers in the early 1960s at the Honeywell Corporate Research Center; see P.W.Kruse: Appl. Opt. **4**, 687 (1965)
4.43 C.T.Elliott, I.L.Spain: Solid State Commun. **8**, 2063 (1970)
4.44 W.Scott, R.J.Hager: J. Appl. Phys. **42**, 803 (1971); see also T.T.S.Wong: (thesis, MIT, 1974)
4.45 C.T.Elliot: J. Phys. D: Appl. Phys. **4**, 697 (1971)
4.46 V.V.Ptashinskii, P.S.Kireev: Soviet Phys.—Semiconductors **6**, 1398 (1973)
4.47 G.A.Antcliffe, R.T.Bate, R.A.Reynolds: Proc. Conf. on Physics of Semimetals and Narrow-Gap Semiconductors (Pergamon Press, New York 1971) pp. 499–509
4.48 R.A.Reynolds, M.J.Brau, H.Kraus, R.T.Bate: Proc. Conf. on Physics of Semimetals and Narrow-Gap Semiconductors (Pergamon Press, New York 1971) pp. 511–521
4.49 J.L.Schmit, E.L.Stelzer: (unpublished results)
4.50 J.Stankiewicz, W.Giriat, A.Bienenstock: Phys. Rev. **B4**, 4465 (1971)
4.51 W.Scott: J. Appl. Phys. **43**, 1055 (1972)
4.52 W.Scott, E.L.Stelzer, and R.J.Hager: J. Appl. Phys. **47**, 1408 (1976)
4.53 S.P.Emmons, K.L.Ashley: Appl. Phys. Lett. **20**, 162 (1972)
4.54 C.T.Elliott, I.Melngailis, T.C.Harman, A.G.Foyt: J. Phys. Chem. Solids **33**, 1527 (1972)
4.55 J.L.Schmit: J. Appl. Phys. **41**, 2876 (1970); basic equation given in this paper was subsequently improved by Schmit to give best fit to data; see (4.109)
4.56 C.T.Elliott: J. Phys. D: Appl. Phys. **4**, 697 (1971)
4.57 C.Vérié: Festkörper Probleme X. *Advances in Solid State Phys.* (Pergamon, Braunschweig, W. Germany 1970) pp. 1–19
4.58 P.E.Petersen: J. Appl. Phys. **41**, 3465 (1970)
4.59 D.L.Carter, M.A.Kinch, D.D.Buss: Proc. Semiconductors and Semimetals Conf. (Pergamon, New York 1970) p. 273
4.60 M.L.Schultz: Infrared Phys. **4**, 93 (1964)
4.61 A.F.Milnes: *Deep Impurities in Semiconductors* (Wiley-Interscience, New York 1973); see also J.H.Nevin, H.T.Henderson: J. Appl. Phys. **46**, 2130 (1975) for *Tl*-doped Si
4.62 W.Kohn: *Solid State Physics* **5** (Academic Press, New York 1957) pp. 257–320
4.63 T.F.Lee, T.C.McGill: J. Appl. Phys. **46**, 373 (1975)
4.64 N.F.Mott: Contemp. Phys. **14**, 401 (1973)
4.65 T.C.McGill, R.Baron: Phys. Rev. **B11**, 5208 (1975)
4.66 D.Long: Phys. Rev. **129**, 2464 (1963)
4.67 M.Loewenstein, A.Honig: Phys. Rev. **144**, 781 (1966)
4.68 D.Long, J.Myers: Phys. Rev. **115**, 1119 (1959)
4.69 D.Long, C.D.Motchenbacher, J.Myers: J. Appl. Phys. **30**, 353 (1959)
4.70 J.S.Blakemore, C.E.Sarver: Phys. Rev. **173**, 767 (1968)
R.A.Messenger, J.S.Blakemore: Phys. Rev. **B4**, 1873 (1971)

5. Photoemissive Detectors

H. R. Zwicker

With 22 Figures

5.1 Introduction

5.1.1 Applications and Advantages

There are three general detector applications for which photoemissive devices are uniquely suited, with little competition offered by other forms of detection. These are 1) the detection of low-intensity signals, either weak in total available power or diffuse at the optical focal plane; 2) high-speed detection of low-level signals; and 3) the acquisition of high resolution spatial information (imaging) [5.1–4].

A major advantage of photoemissive (photoelectron-emitting) detectors over other detection devices is the ease with which fast, high gain, low noise amplification can be incorporated within the detector by use of an integral electron multiplier. In several important operational parameters this method of obtaining signal gain outperforms external gain stages which must be used with other low-light-level detectors [5.2, 5] with the possible exception of similar avalanche gain available in some semiconductor devices.

A second major advantage of photoemissive devices is the ease with which uniform, large-area detector surfaces can be fabricated. Such surfaces, while important for simple photomultipliers, are a prerequisite for successful imaging devices, including both direct-view intensifiers and image tubes with scanned electron beam readout [5.3].

The latter, electron-beam-interrogation technique, provides a third major advantage of photoemissive devices, namely, extremely high detector element density which allows high spatial resolution (for imaging) not yet easily obtainable with discrete arrays or integrated semiconductor devices. The advantages of electron-multiplier gain remain available in these scanned devices.

(Other related electron-beam-readout imaging devices with semi-insulating charge-storage surfaces or diode-array retinas are operationally similar to scanned photoemissive sensors and compete with or outperform them in some applications, but these non photoemissive devices are not treated in this chapter [5.6].)

5.1.2 Limitations

The major limitation of photoemissive devices is the restricted range of the spectrum over which response can be obtained [5.7]. In the high energy (UV) region, standard photoemissive tubes do not respond beyond ~0.15 μm owing to transmission limitations of common window materials [5.8]. However, extension of response to 0.12 μm can be obtained with expensive and easily degraded LiF windows, and "windowless" tubes (for example, self-supporting 10 μm thick wafers of GaAs [5.9]) have been used up to x-ray wavelengths. It is also standard practice to obtain UV response in tubes sensitive only to visible light by coating the window with a UV-sensitive downconverter such as sodium salicylate which will fluoresce in a region where the tube is sensitive [5.10]. There is thus no firm upper-energy limit for photoemissive detection, given proper design.

In the low-energy (IR) region of the spectrum, definite materials limitations are encountered which have in the past permitted only marginal response beyond about 1 μm [5.11]. As this chapter shows, however, substantial recent advances have been made in increasing red sensitivity and in extending IR response out to and beyond 1.5 μm [5.12]; indeed this chapter concentrates on these most recent advances into the near-infrared region of the spectrum [5.4, 13–21] where nonphotoemissive detectors previously were the only choice.

The present operating range of high-yield photoemissive devices is thus from less than 0.1 μm to more than 1.1 μm, with an extension to 2 μm a theoretical possibility. Within this spectral range photoemissive devices possess unique capabilities not found in other detectors. Useful photoemissive sensitivities beyond 1.5–2 μm must await significant research breakthroughs, perhaps in the area of externally biased, field-assisted photoemitters [5.22–25].

5.1.3 Types of Photoemissive Surfaces: Classical and NEA

Photoemissive devices can be conveniently divided into two historical (and performance) classes depending upon the material used as the photoemissive layer [5.11]. The first class is the "classical" group [5.26]. Here the photoemitter is a thin evaporated-layer compound containing an alkali metal or metals (almost always including Cs), one or more other metallic elements from group VB of the periodic table (e.g., Sb), and possibly also oxygen and/or silver [5.27]. For ultraviolet applications several other sensor materials are used (e.g., CsI), but such devices become highly specialized and are not discussed further in this chapter [5.28].

The second class of photoemissive device utilizes a photoconductive single-crystal semiconductor substrate with a very thin surface coating consisting of Cs and usually a small amount of oxygen. This second group, the "negative electron affinity" (NEA) class, is the most recent development in applied photoemissive technology; it is the evolution of NEA devices over the past ten years which has dramatically extended high-yield photoemission well into the near IR spectral region.

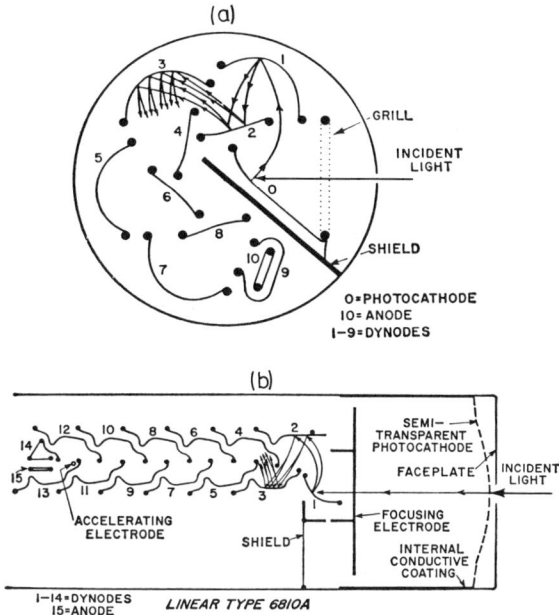

Fig. 5.1. (a) Opaque, or reflection mode, photomultiplier. Electrons are produced at element "O", the photocathode, and electron multiplication gain occurs at each of the dynodes, elements 1–9. The anode (10) collects the resulting electrons [5.153]. (b) One-dimensional sketch of semitransparent, or transmission mode, photomultiplier (adapted from [5.153] and published courtesy RCA Corporation)

5.1.4 RM and TM Modes

Both classical and NEA emitters can be fabricated in two forms [5.29], namely 1) opaque (Fig. 5.1a), where light is incident through the tube envelope directly onto the side of the photocathode from which electron emission occurs, and 2) semitransparent (Fig. 5.1b), where light enters at the rear side of the photoemissive layer and is absorbed throughout its thickness; generated electrons then diffuse through the bulk to the opposite (unilluminated) surface from which they are emitted toward the collecting electrodes. Both forms of emitter provide similar electron collecting efficiency, but the semitransparent structure is presently preferred when imaging or interrogation by a scanning electron beam is used.

It is more difficult to fabricate semitransparent than opaque devices because the thickness of the sensitive layer is an added critical parameter not important in the opaque structure. (Improper thickness in the semitransparent device would limit both the range of useful spectral response and overall quantum efficiency [5.29].) Most new photoemissive surfaces, particularly NEA, are now investigated initially in their opaque ("reflection mode", "RM", or "front illuminated") form, which usually offers the highest initial optical and electron-emissive quantum efficiency, and only after determining optimum thickness and optimum fabrication methods are new surfaces produced in the semitransparent ("transparent", "transmission mode", "TM", or "back illuminated") form. Eventually both forms are offered commercially in several variants, each optimized for selected parameters such as low cost, peak response at a specific wavelength, low thermionic emission (at the expense of decreased sensitivity), or low blemish count. This choice of performance specifications applies particularly to the more completely developed TM and RM devices such as the S–1 or S–20.

5.1.5 Outline of Chapter

Both groups of photosensors (classical and NEA) function after the same physical principles based on the band theory of semiconductors and on electron escape mechanisms from the photoactive layer into the surrounding vacuum. But both suffer from analytical-modeling inadequacies, particularly at their electron emitting surfaces and especially with regard to computation of the size of any electron-emission barrier. The classical devices, discovered by educated accident and fabricated for years by formulas which evolved by trial and error, also suffer (in experimental investigation and in modeling) from their polycrystalline nature and from limited, or short-range, lattice ordering. Although neither class is completely understood physically, the more recent single-crystal NEA photoemitter is substantially more advanced in theoretical modeling than is the classical device (e.g., see [5.30, 31]).

Considered first in this chapter are the physical principles of operation applicable to both classes of photoemissive surface, to the extent to which they are understood. The operation and construction of "classical" devices are then examined in detail, using two extreme cases, (CsSb)[1] and (AgCsO)[1]-S–1, as examples. NEA devices are then considered. These are discussed in somewhat greater depth than classical devices inasmuch as these recently developed detectors are less likely to be familiar to the user. One example of NEA operation, NEA GaAs, is the simplest structure and is discussed in detail for both RM and TM modes. Other IR-sensitive emitters, including those using layers of complicated quaternary compound semiconductor alloys such as InGaAsP, are then briefly summarized. The chapter is concluded with a summary of device-to-device trade-offs in classical and NEA devices.

The emphasis throughout this chapter is on the photoemissive surface itself rather than on the many versions of electron multipliers and tube structure in which devices are offered.

5.2 The Photoemission Process

5.2.1 Fundamentals of Electron Escape Energy

In any stable material, electrons are retained within the substance by electrostatic forces which bind electrons to the positively charged nuclei. An "ionization energy" (renamed the "work function" in a solid and defined below) quantifies

---
[1] For brevity we use parenthesis, e.g., (CsSb), to denote a generic substance without specifying its exact chemical composition by, for example, Cs_3Sb.

the strength of the retentive force, at least on some average. An electron which has energy greater than the "ionization energy" will, if it comes near the surface, have a high probability of escape.

We define this ionization energy to be at the reference level of zero electron volts. The energy E of any bound electron is then measured with respect to this zero or vacuum-level energy, $E_{vac} \equiv 0$. Since electron escape requires $E \gtrsim E_{vac}$, we now consider electron energy values in various materials.

Within a solid there exist a definite number of discrete energy levels in which an electron can reside; all other energy values are "forbidden" [5.32–35] although the number of allowed energy states is always many times greater than the total number of bound electrons actually in the solid.

The equilibrium occupation or "filling" of allowed levels with available electrons is such as to minimize the total energy of the electron population, i.e., electrons will naturally tend to fill the lowest-lying states and leave the remaining upper states empty. At finite temperatures, however, there will always be a small spread in the occupancy of electrons at the highest filled energy levels owing to random thermal motion. Thus some higher-lying states will be filled and some lower-lying states will be empty.

The energy at the middle of this smeared electron distribution is the Fermi energy, E_F; at $E = E_F$ the probability of occupancy is 1/2 (for a nondegenerate state). The occupancy probability falls exponentially with higher energy as $\exp[(E_F - E)/kT]$ at energy E above E_F and rises accordingly for E below E_F. Here k is the Boltzmann constant and T is temperature in Kelvin.

The smallest native value of E_F found for any elemental solid is the 2.1 eV "ionization energy" of metallic Cs ($E_F = -2.1$ eV) [5.36]. The occupancy distribution in Cs thus varies roughly as $\exp[(-2.1 - E)/0.026 \text{ eV}]$, and the probability that an electron has energy greater than the ionization value (i.e., that $E \gtrsim E_{vac}$) is $\gtrsim \exp(-2.1/0.026) \simeq 10^{-35}$ at 300 K. With only about 10^{24} electrons/cm^3 in a realistic material, we may safely assume that "no" electrons within Cs or any room temperature elemental solid have sufficient energy to escape into the surrounding space. (An exception must be made at the very surface, where the 2.1 eV minimum energy above does not hold. Here less tightly bound electrons do exist and can escape. Their number is usually low but this emission does contribute to dark current if the substance is used as a photoemitter.)

There are several processes by which an electron can be given energy sufficiently in excess of its room-temperature thermal-equilibrium value that escape becomes possible. One process is simply to add heat, in which case the exponential factor above can be made large enough that useful electron emission is obtained; thermionic emission from a hot tungsten filament is an example. Of more interest in this chapter is the absorption of incident photon energy by an electron. Useful emission here becomes possible whenever the photon energy absorbed is greater than the ionization energy. In solid Cs, if one electron absorbs a photon of energy ~ 2.1 eV or greater, and if this absorption occurs sufficiently near the surface, then a reasonable probability for electron escape exists. This is the fundamental mechanism of photoemission.

5.2.2 Escape-Energy Parameters for Metals and Semiconductors

The electron "escape energy" from a solid differs substantially from the true ionization energy of an electron in a gas owing to the proximity of lattice ions and nearby localized electrons. In a metal, the "ionization energy" is renamed the *work function*, ϕ, which is defined as the (positive) energy difference between the Fermi level (the "average" energy of those electrons nearest the escape energy) and the minimum free energy of an electron at rest within the vacuum, $E_{vac} \equiv 0$. (For Cs gas the *ionization energy* is 3.7 eV while the *work function* of solid Cs is the 2.1 eV used above [5.36].)

In any *metallic* element or alloy, E_F and thus $\phi \equiv E_{vac} - E_F$ is a materials constant. This is not true, however, in a nonhomogeneous or nonmetallic substance because the Fermi position is not a unique constant of the material. For example, in a *semiconductor* such as GaP, E_F can be made to vary 1–2 eV by "doping", as described below.

The Fermi level (and work function) variation results from the band structure of semiconductors, where uppermost electron levels (those above the tightly bound core states) break into two well-defined nonoverlapping allowed-energy regions separated by a large energy gap of size $E_G (\sim 2\,\text{eV}$ in GaP); the Fermi level can fall anywhere within this gap depending on minute quantities of impurities present, and E_F can vary spatially within the semiconductor if it is placed in contact with other substances.

The variation of the work function and the Fermi level in metals and semiconductors is illustrated in Fig. 5.2. In the metal (Fig. 5.2a), the Fermi level is fixed at the energy where the "N" lowest-lying electron states are filled with the "N" electrons in the (uppermost) band of valence electrons. There are many additional empty electron states above E_F, but these remain empty regardless of the presence of small amounts of foreign contaminants. The position of E_F and the value of ϕ in metals are thus rigorous materials constants.

In the semiconductor (Fig. 5.2b) the Fermi level is located within the forbidden energy bandgap between E_V and E_C. The N states in the valence band (E below E_V) are essentially filled with N valence electrons and the remaining electron states (above E_C) are nearly empty. E_F can be varied from one extreme position to another (from $E_{F_p} \sim E_V$ to $E_{F_n} \sim E_C$) without severely upsetting this balance of filled and empty electron states; the valence band remains essentially filled with the requisite N electrons and the conduction band remains essentially empty unless E_F comes within a few hundredth of an electron volt of either E_V or E_C.

In practice, E_F is adjusted by adding trace quantities of selected impurities— the process of "doping" to obtain *p*-type or *n*- type conduction. The impurities add a very low density of allowed states within the forbidden region of the crystal, and the position of E_F must adjust between E_C and E_V (E_{F_n} to E_{F_p}) to maintain a total of N filled electron states. (For the present discussion the substance remains the same generic semiconductor with and without this $\sim 0.01\%$ doping.) Because the work function is defined as $\phi = E_{vac} - E_F$, its value

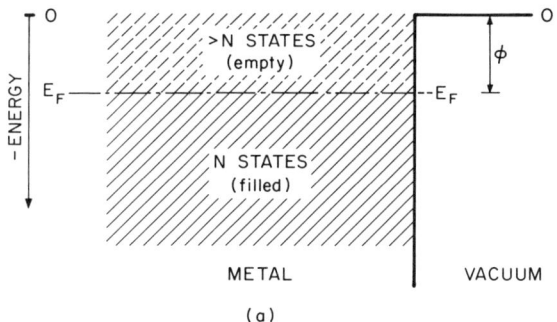

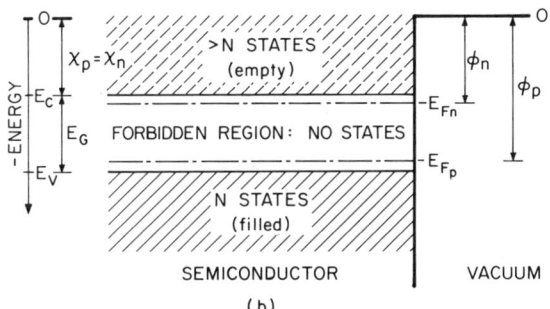

Fig. 5.2. (a) Band structure showing the Fermi level, E_F, and the work function, ϕ, in a metal. (b) Band structure showing Fermi level and work function for both n-type and p-type semiconductors. The forbidden region lies in the gap in allowed states between E_V and E_C and is of width E_G. The Fermi levels E_{F_n} and E_{F_p} vary from n-type to p-type material, and thus also do the work functions, ϕ_n and ϕ_p; the electron affinity, $X = X_n = X_p$, does not

varies along with E_F by $\pm E_G/2$ for any particular semiconductor species; this is illustrated by ϕ_p and ϕ_n in Fig. 5.2b. Obviously the work function is *not* a materials constant for a semiconductor.

In further contrast to metals, there are usually few electrons actually located at the position of E_F within the *forbidden* region in a semiconductor. Thus the concept of the work function as "the energy required to raise an average electron from $E \cong E_F$ to the vacuum energy" has little physical meaning.

A more useful measure of the "ionization energy" in semiconductors is related to the depth of the conduction band edge below E_{vac}. This energy difference is defined as the electron affinity $X \equiv E_{vac} - E_C$, given the alternate symbol E_A by some authors. The electron affinity unlike the work function is a constant of the material, independent of impurity content, as shown in Fig. 5.2b.

5.2.3 Thresholds of Various Materials and Choice of Photoemissive Substances

The photoemissive threshold energy E_{Th} for electron emission in a metal is usually of value $\sim \phi$; in a semiconductor, E_{Th} would be roughly X *if* the density of electrons in the conduction band (above E_C) were sufficiently high to allow photon absorption. This is seldom the case. More often the high density of electrons available to efficiently absorb incident light is in the valence band, near or below $E_V = E_C - E_G$, so that the threshold for photoemission from a semiconductor is in the order of $E_{Th} \cong X + E_G$ [5.37, 38]. This "semiconductor optical ionization energy" is rather constant regardless of doping, and is a more useful quantity than the semiconductor work function for estimating E_{Th}. We now examine photoemissive thresholds and yield limitations in various materials to determine the best candidates for efficient detector cathodes.

Low-Energy-Threshold Materials

In order to obtain useful photoemission at photon energies lower than the threshold of Cs, it would seem necessary simply to find a metallic compound alloy in which the electron work function is lower than the ~ 2 eV elemental minimum (but there is none), or to find a semiconductor where $E_G + X < 2$ eV. This search has led to the discovery of the classical photoemissive compounds. The binary compound with the lowest electron affinity is cesium antimonide, Cs_3Sb, where $X = 0.45$ eV [5.39]. Here $E_G \cong 1.6$ eV for a threshold $E_{Th} \simeq E_G + X$ of about 2 eV. For the more complex substance (NaKCsSb), which is not actually a true compound but is rather (NaKSb) with a surface skin of (NaKCsSb) [5.40], $E_G = 1$ eV and $X = 0.4$ eV (with some variation with surface treatment), which gives (NaKCsSb) the lowest known "classical" threshold of ~ 1.4 eV. (NaKCsSb) is the basis of the ubiquitous S–20 and ERMA-type (extended-red S–20) photocathodes. The S–1 photocathode material (AgCsO) and many NEA emitters have thresholds well below 1.4 eV, but these emitters are not homogenous materials and the simple concept of $E_{Th} = E_G + X$ does not directly apply [5.38]. Thus, in spite of an extensive search for materials with low electron release energy, the minimum threshold remains near 1.4 eV for the simple emitters; longer-wavelength IR-sensitive emitters (e.g., S–1 and NEA) all require a multistep emission process and two or more separate layers to obtain thresholds lower than the native value of any known single material.

Additional Materials Parameters for High Yield

Factors other than the requirement of small "electron binding energy" contribute to high photoemissive efficiency. The most obvious is a high probability that incident photons are actually absorbed, and in a process which excites electrons to a high free energy in the conduction band. This requires both small optical reflectivity and a high absorption constant α. For most metals, the reflectivity (in the visible and near IR region of the spectrum) is very high at 90–99%; metals, including Cs, are thus ruled out as efficient photoemitters [5.41], with the exception of Ag in the S–1 photocathode as discussed below. Semiconductors, on the other hand, generally have moderate 25–50% reflectivities over the range of interest, and absorbed photon energy is also normally taken up by excitation of free electrons; the most common photoemitters therefore are high-α photoconductive semiconductors.

These are two groupings of semiconductors based on their general band structure, namely, "direct" and "indirect" [5.42]. In direct materials, a high, sharply rising absorption constant is found immediately above the bandgap energy, with values of α in the order of 10^4 to 10^5 cm^{-1}. In indirect materials, the absorption at threshold is a more slowly increasing function of energy and α is generally much lower; values $\sim 10^2$ to 10^3 cm^{-1} are common even for photon energies several tenths of an eV above the bandgap. Electron escape depths are, however, much longer in indirect than in direct materials, making net efficiency (proportional to α times escape depth) comparable. Direct emitters (which include most of the III–V compound semiconductors) are, on balance, usually preferable to indirect ones (such as Si and GaP) because they allow the region of photon absorption and electron generation to be kept thin, and for a given thickness they also exhibit a more constant yield above threshold. The common NEA photoemitters are all direct, and classical emitters are also generally direct, with $\alpha \sim 10^4$ to 10^5 cm^{-1} [5.43]. (Their polycrystalline structure does, however, make unambiguous classification and absorption constant measurements difficult.)

Another important materials factor influencing photo yield is the probability that the optically generated electron will travel to the emitting surface without suffering substantial energy loss. Carrier transport efficiency is described by a characteristic transport length L. For "hot" electrons, i.e., those excited high in the conduction band with energy $E \gg E_C$, L is a measure of the mean electron-electron and electron-phonon scattering distance L_S; this is normally in the order of 10^{-2} to 10^{-3} μm (1–10 nm) [5.44]. The amount of energy lost in each collision is small, however, so that an electron can travel farther than one L_S before energy loss prohibits emission, and the energy-loss per collision is thus an additional materials parameter controlling photo yield. This and L_S are also functions of the energy of the traveling electron. All of these factors make analytical treatment of "hot" electron transport complex [5.45], and predicting or controlling L_S nearly impossible.

For "cold" electrons, i.e., those "thermalized" by lattice collisions to energies near the bottom of the conduction band ($E \simeq E_C$), L is the diffusion length L_D; L_D is always longer than L_S, ranging as high as 1–10 μm in direct semiconductors with good lattice structure and free from chemical defects [5.44]. Unlike L_S, however, L_D implies complete loss of carrier energy by recombination at defects or across the bandgap rather than merely fractional loss as for hot electrons. In spite of this, cold electron transport is generally much more efficient than hot electron travel owing to the much longer characteristic length.

As described below, cold electrons can be emitted from NEA surfaces while classical surfaces require hot electrons for emission over a finite energy barrier which exists at the surface. Transport in NEA structures can also be optimized by materials selection, while little can be done to control transport in classical emitters.

Once the electron does arrive near the emitting surface, the probability of its actual escape into the vacuum is the final limitation on yield. Any barrier present at the surface is the primary impediment, but we have partially considered this in describing E_{Th} and in implying that after generation the electron must be transported to the surface without loss of energy below E_{Th} ($\sim \phi$ for metals or

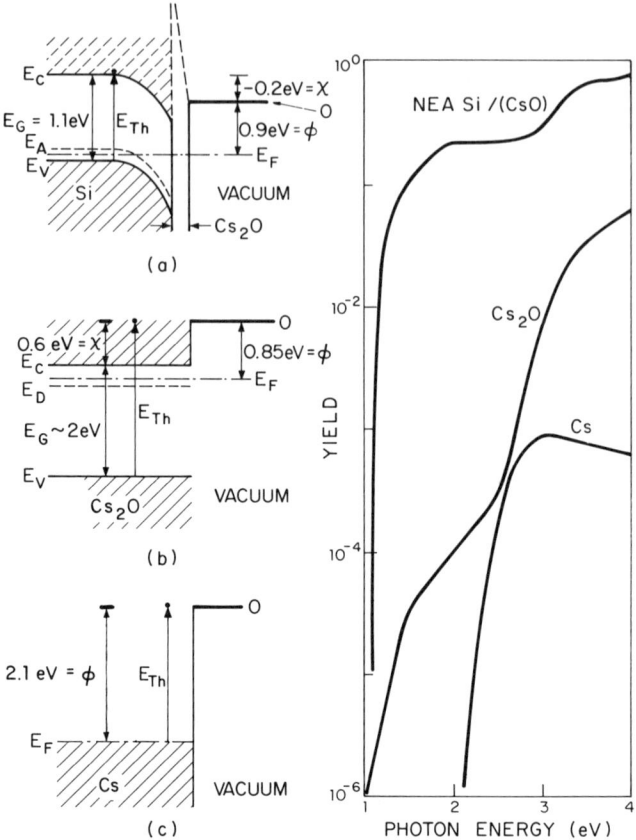

Fig. 5.3a–c. Band structure and photoemissive yield for (a) silicon activated to negative electron affinity; (b) a thin layer of Cs_2O (values are approximate); and (c) metallic Cs. For Si, E_A is the energy of intentionally added p-type dopants, while for Cs_2O, E_D is the energy level of native defects. The relationship between photoemissive threshold energy and the value of $E_G + X$ for the semiconductors, or ϕ for Cs, is clear [5.48–50]

$\sim E_G + X$ for semiconductors). But in addition to the barrier obvious in Figs. 5.2a and b, a further high but thin electron barrier is often found at the very surface of many materials; this barrier may further increase E_{Th} above the values given by the expressions above (see Fig. 5.3a).

Even for electrons with sufficient energy at the surface, the probability of actual emission is always less than unity. Escape varies with the type of material, its crystal orientation, and even with the specific conduction band to which the carrier has been excited. For NEA GaAs, for example, the escape probability from the 111-B (As) face is about 2.5 times larger than that from the 111-A (Ga) face [5.46], and escape from X-states is about twice as probable as escape from Γ-states [5.47]. Little of this kind of detail is known about escape from classical materials. Surface escape factors for both NEA and classical materials are nearly impossible to model, much less predict, and the search for materials with both low electron barriers and high escape probabilities is almost entirely an experimental one.

To recapitulate all factors influencing photo yield, one can conceptualize the photoemission process in two sequential steps. First is the operation of a photoconductor, which includes photoabsorption efficiency plus electron transport efficiency through the bulk of the photoemissive layer to the surface. Second is the surface emission process, which includes all factors enhancing escape once the carrier reaches the surface. In particular this includes rejection of all electrons excited with energy lower than the "ionization energy", i.e., with energy below the surface-barrier energy.

In Fig. 5.3 band models and photoemissive-yield spectral curves (electrons per incident photon) are shown for Cs [5.48], Cs_2O [5.49], and NEA Si/(CsO) [5.50]. For each material the yield is low until the incident photon energy is greater than either the work function ϕ or the semiconductor threshold $E_G + X$.

For Cs, $E_{Th} \sim \phi$, but high metallic reflectivity and very short escape depth (electron-electron scattering) severely limit photoemission. For Si, all the factors which make for efficient emission have been optimized. The electron affinity has been lowered to *less than zero* by a surface treatment plus "band bending" within the bulk as described for NEA photocathodes below. Fairly high yield is found for all photon energies above the (indirect) bandgap of 1.1 eV. The threshold would be steeper for the same semiconductor thickness if Si were a direct material, but the efficiency per *absorbed* photon in Si is high. For Cs_2O, emission is high for energies greater than $E_{Th} = E_G + X \sim 2.6$ eV, but poor absorption (for the assumed thin layer) and inefficient hot electron transport limit emission. This substance is of no value by itself as a photoemitter.

5.3 Classical Photoemissive Surfaces

5.3.1 The (CsSb) Photoemitter

Cesium antimonide as a photoemitter [5.43, 51] has found wide application in commercial photodetection devices. Tubes with (CsSb) on several opaque and semitransparent substrates, and with various window materials, have been given several "S"-numbers including S-4, 5, 11, 13, 17, and 19 [5.52]. The physical properties of this material have also been rather extensively investigated through measurements of, for example, $\dot{\alpha}(E)$, E_G, ϕ, and defect characteristics [5.53]. We have therefore chosen it as the first example of a "classical" photoemissive surface.

The fabrication of sensitive (CsSb) surfaces is rather straightforward, if still something of an art rather than a science [5.39]. Pure Sb is first evaporated *in vacuo* onto the faceplate of a baked-out tube. The faceplate can be transparent glass or opaque metal or it can be glass onto which an electrically conductive but optically transparent thin metal film (used to prevent lateral voltage drop across the high-resistance layer during emission) has been evaporated. For the opaque

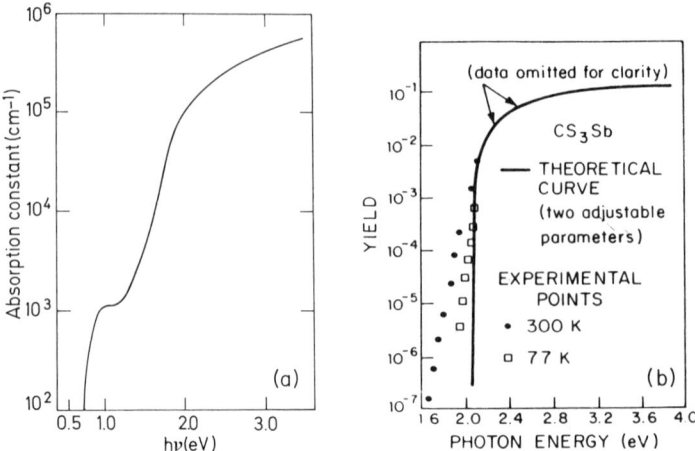

Fig. 5.4. (a) Absorption versus photon energy for (CsSb) at room temperature [5.57]. (b) Spectral yield of (CsSb) at room temperature and with cooling [5.51]

photocathode the thickness of the Sb deposit is not important, while for the transparent cathode the layer is deposited until the white light transmission is about 70% of its initial value [5.54]. The speed of evaporation, the type of chemical Sb source, and the temperature of the tube are among the factors contributing to a properly structured Sb film. Activation then consists of evaporating Cs onto the Sb surface until the white light sensitivity is maximized. The tube, held at a uniform elevated temperature during this evaporation, is carefully cooled to room temperature after the evaporation to prevent nonuniform distillation of Cs on the internal electrodes. After fabrication, the (CsSb) film should be highly stable and, in a semitransparent structure, about 300 Å thick [5.55].

The chemical composition of the (CsSb) film is Cs_3Sb [5.56], with microscopic crystal structure in a DO_3-symmetry cubic lattice. Within these microcrystals Cs_3Sb is a semiconductor with 1.6 eV bandgap. In an optimized surface, excess Sb in the order of $10^{20}/cm^3$ from perfect stoichiometry produces a large density of acceptor levels about 0.5 eV above the valence band; the crystal is therefore p-type with a Fermi level somewhat below 0.5 eV, which produces a moderate conductivity through the film at room temperature.

In Fig. 5.4a and b the absorption constant and the quantum efficiency, respectively, of a "typical" CsSb layer are shown as functions of energy. A schematic model for the band structure of the complete photocathode, derived from comparison of absorption data and emission data, is shown in Fig. 5.5 [5.43, 57]. The electron affinity is about 0.45 eV and the threshold for efficient photoemission is $E_{Th} = E_G + 0.45$ eV, or about 2 eV. The band-bending shown is speculative [5.58].

Operation of CsSb as an efficient photoemitter is straightforward. For incident photon energy less than E_{Th}, electrons generated within the material

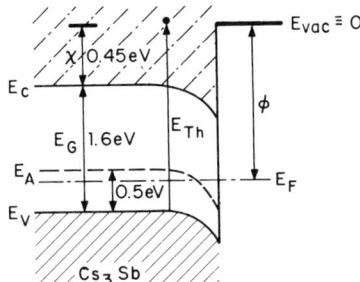

Fig. 5.5. Approximate band structure for a typical (CsSb) photoemissive layer [5.57]

encounter a high barrier at the surface and do not escape. For electrons generated with free energy in excess of the electron affinity ($E_{h\,v} > E_G + X$), escape is determined by the probability of diffusion to the surface prior to energy loss in excess of X (with respect to the bottom of the conduction band). Because the Cs$_3$Sb layer is only some 300 Å thick and the hot electron collision distance is in the order of 150 Å (with a small ~0.05 eV energy loss per collision) [5.51], the quantum efficiency of the layer (per absorbed photon) approaches about one-half at photon energies above ~2.3 eV.

There is a highly extended photoemission threshold for energies *less* than E_{Th}, with measurable emission extending as low as about 1.5 eV (0.69 μm). This emission results from excitation of electrons trapped in the very high density of excess Sb defects within the bulk [5.43]. The related absorption at ~$|E_V + 0.5\,\text{eV}|$ is observed in the 300 K optical data of Fig. 5.4a. This long wavelength "tail" disappears when the tube is cooled owing to emptying of these levels (see Fig. 5.4b, 77 K data).

The CsSb photocathode has the attributes of relatively low cost (owing to the rather simple fabrication techniques) and high stability. The low cost explains the proliferation of structural variants of (CsSb) tubes and of the varieties of glass used for the window. The latter is kept as inexpensive as possible unless ultraviolet response is desired. Other variants include the use of (MnO) as a semitransparent conductive substrate over the glass windows. This substrate has the unexplained (but see [5.58]) advantage of lowering X from that of the glass/CsSb device [5.59], which increases the overall quantum efficiency and extends the response slightly farther into the red. The proliferation of variants indicates that (CsSb) is a well-matured photoemissive substance, with manufacturers offering many versions of a nearly identical tube all with slightly different performance optimizations.

5.3.2 The S–1 (AgCsO) Photocathode

The (CsSb) photocathode is typical of most "classical" photoemissive materials in that it can be modeled as a collection of homogenous semiconductor microcrystals with moderate bandgap energy, low electron affinity, and high

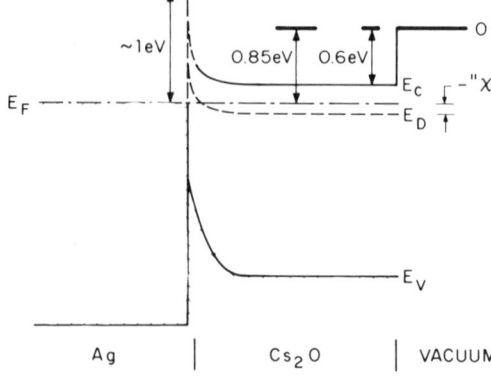

Fig. 5.6. A possible model for a single microcrystal of S–1 (AgCsO) photocathode material. The band structure for the (CsO) layer is as in Fig. 5.3, while that for a coated particle of Ag is similar to Cs in the same figure. A heterojunction barrier occurs between the two materials

photoabsorption. The NEA photocathode discussed in the next section is a complex layered structure including a photoabsorptive substrate material and a separate low work function coating. The S–1 (AgCsO) photocathode [5.60] is the most nearly "NEA" member of the "classical" family in that it includes high X silver particles coated with low X (CsO). Its physical model, although not fully understood, falls midway between the NEA structure and the (CsSb)-like homogenous-semiconductor photoemitter [5.61].

Fabrication of the (AgCsO) surface [5.60] requires deposition of a thin Ag layer followed by oxidation (by glow discharge) and sensitization by the introduction of Cs vapor. There are many variants in the exact fabrication procedure, but in general an S–1 surface is substantially more difficult to form than the (CsSb) surface. Unlike other fairly complex photocathode materials (e.g., NaKCsSb), there are several points in the fabrication process where irreversible harm can be done to the layer. One example is over-cesiation; errors can normally be corrected for other materials but not for (AgCsO). These difficulties, plus low reproducibility, increase the cost of S–1 devices and create a large spread in performance versus cost for commercial offerings.

Modeling the photoemissive surface is highly speculative. A possible band sketch for a single Ag/(CsO) particle (a few tens of Angstroms in size) is shown in Fig. 5.6. The spectral response curve for a "typical" S–1 surface is shown in Fig. 5.7, along with absorption data for a thin Ag film [5.62] and also for Cs_2O [5.49]. The regions of high emission above 3000 Å and below 4000 Å are explained as follows: In the UV range of response (less than 0.3 μm), photoemission is due to absorption within and emission from the (thin) Cs_2O layer [5.49, 63]. The Cs_2O bandgap is about 2 eV and the electron affinity is about 0.6 eV, for a 2.6–3 eV emission threshold (cf. Fig. 5.3). For wavelengths longer than ~0.4 μm, photoemission results from photon absorption within elementary Ag particles; note in Fig. 5.7a the high absorption for a *thin* film. Subsequent electron transport occurs over and/or through the interfacial Ag/Cs_2O barrier and into the low-X Cs_2O layer. To be emitted, these electrons must traverse the CsO while retaining sufficient energy to escape the 0.6 eV Cs_2O/vacuum surface barrier.

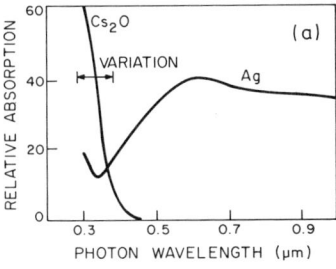
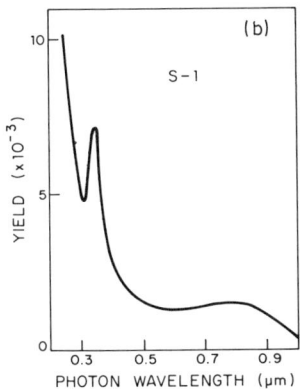

Fig. 5.7. (a) Absorption spectra of thin films of Ag and (CsO) [5.62]. (b) Example spectral response curve for an S–1 (AgCsO) photocathode [5.62]

In the region between 3000 and 4000 Å (the S–1 is seldom used in this range) the emission process is complex and not fully understood [5.49, 63]. Structure at 3200 Å follows a dip in Ag absorption (cf. [5.32]), perhaps evidence of a plasma resonance. Absorption in the (thin) Cs_2O is small here, and what emission does occur is probably the result of absorption within Ag and emission through Cs_2O, as in the longer wavelength region. The source of the peak at 3500 Å is not easily explained, but it also appears in the photoemissive yield curve of Ag/Cs and may be caused by structure within the bands of Ag. Between 3200 and ~ 3000 Å emission is shifting from that of Ag-(CsO) to that of (CsO) alone.

The exact structure in the 3000–4000 Å spectral curve is the result of trade-offs in absorption versus scattering taking place within both (CsO) and Ag. Response depends critically on the relative thickness of these two layers and is highly variable across the surface and from one S–1 device to another. The actual S–1 surface is anything but a perfect layer of single-crystal Ag coated with Cs_2O, so the above picture is highly schematic. The actual emission processes and escape probabilities vary over 50–200 Å wide structural patches of emitting surface; patchiness in the work function also accounts for the absence of sharp threshold structure beyond 0.9 μm [5.64].

Photoemission from (AgCsO) definitely involves a two-step process, with absorption for the more important IR region of the spectrum taking place in one material, Ag, and emission taking place through the low electron affinity Cs_2O covering these particles. In this regard the S–1 is similar to the NEA structure.

5.4 Negative Electron Affinity (NEA) Devices

For both S–1 and NEA operation, an electron-transparent low electron-affinity layer is used to lower the effective electron barrier below the native value in the photoabsorbing substrate. In the NEA device, the electron affinity behaves as if it were actually absent or *negative*, rather than $\sim +0.6\,\text{eV}$ as in the S–1, and in addition the conduction-band discontinuity (1 eV high [5.49] in Fig. 5.6) is low in NEA surfaces and offers less of a limit to emission. (A lower barrier, which can be found in semiconductor/semiconductor, rather than metal/semiconductor, heterojunctions (e.g., [5.65]), is the basis of NEA devices.) We now consider the physics of NEA in detail.

5.4.1 Introduction and Advantages

The negative electron affinity (NEA) photocathode is the most recent and generally the highest-performance photoemissive surface discovered to date [5.4, 11–21]. Its operation is an extension of the same physical principles which apply to all other photoemissive surfaces, differing only in the means of obtaining low electron affinity and in the specific factors which determine photo yield. The principles of operation are so similar to those of classical emitters that the first NEA surface was completely predicted by theoretical extension of classical principles prior to its first experimental fabrication [5.66].

The two major physical differences between NEA and classical emitters are, first, that emitted electrons in NEA surfaces are not "hot", i.e., they have "thermalized" to the bottom of the conduction band, and second, that the value of electron affinity, although genuinely negative, is the *effective* value in the *bulk* and not the true value at the surface [5.15].

Although the principles of NEA emission are a fairly direct extension of classical photoemission theory, the fabrication techniques are quite different and the translation of laboratory success into commercially available photodetection devices has required a very careful marriage of several highly developed technologies, including high vacuum techniques, surface analysis, and compound semiconductor materials growth and purification. All of these techniques are of only very recent vintage.

In this section we discuss first the reflection mode NEA GaAs surface, and second the complicated layered structures required for GaAs transmission mode operation, where GaAs is grown on a separate two-layered semiconductor substrate. Included are principles of operation and the general techniques required for the production of NEA GaAs. The section is concluded with a discussion of other photocathode materials exhibiting longer wavelength response.

NEA photoemissive surfaces are important because they offer at least five substantial advances over classical surfaces. These include 1) increased quantum efficiency in regions of the spectrum where good "classical" surfaces already are available; 2) substantial extension of response into the near infrared region of the spectrum ($\lambda > 1$ micron) where no classical surface with greater than about 0.1% external quantum efficiency has been available; 3) the possibility of extremely good resolution and lag characteristics in image sensors (this resulting from the narrow energy distribution of the emitted "cold" electrons); 4) extremely low room-temperature dark current even for surfaces sensitive beyond 1.1 μm; and 5) the capability of extremely uniform absolute sensitivity over an extended region of the spectrum. Disadvantages include high cost, complexity of fabrication, present small size and uniformity limitations, and, in some cases, lower speed of response and lower long-term durability.

As the discussion of the semitransparent GaAs photocathode will show, there are a tremendous number of trade-offs involved in the selection of materials parameters for NEA emitters, especially in those for imaging applications. As a result, a large number of variants of a seemingly identical device may become available when more devices are offered commercially, and none of these "standard" variants may perform as well as a device tailored to a user's specific application. This is similar to the forced choice of high sensitivity, low dark current, or fully extended infrared response in the case of the S–1 phototube. For NEA devices, however, the choices may be of even greater complexity.

A "general-purpose" TM NEA photoemissive device active to 1.1 μm has yet to be developed. Nevertheless, from a comparison of spectral and performance characteristics of opaque NEA and classical photoemitters, it will be seen that NEA devices may soon permit high-gain room-temperature near-infrared imaging with orders of magnitude increase in performance over previous S–1 and modified S–20 characteristics.

5.4.2 Basic Physics of NEA Operation

NEA photocathodes are photoconductive semiconductors whose surface has been treated to obtain a state of "negative electron affinity". This state is reached when the energy level of an electron at the *bottom* of the conduction band in the *bulk* of the semiconductor is greater than the zero energy level of an electron in the vacuum. Hence an electron excited to the conduction band within the bulk can, if it travels to the activated surface without first recombining, energetically "fall" out of the photocathode into free space.

NEA surfaces differ from classical photoemissive surfaces in that conduction band electrons require no excess thermal energy above E_C to escape. It is a cold electron emission device, while in classical positive electron affinity surfaces a small barrier is present at the surface and either hot electron escape or tunneling of thermalized electrons is required for emission. NEA and classical electron emission are contrasted in Fig. 5.8 [5.67]. Part (a) is a classical emitter, (b) is a *p*-type semiconductor treated to obtain NEA, and (c) is an *n*-type semiconductor similarly treated but without attaining NEA. (Many details of this figure are clarified in later sections.)

As shown for the *p*-type NEA surface (Fig. 5.8b), electrons at the bottom of the conduction band at the *surface* do not have energy greater than E_B, or that required to escape into the vacuum; the very surface is similar to that of a classical, positive electron affinity photoemitter (Fig. 5.8a). But there is a region in the bulk beyond $x = x_{BB}$ where the energy of even a thermalized electron at the *bottom* of the conduction band *exceeds* the vacuum potential. This is the condition of NEA and it does not occur in classical emitters.

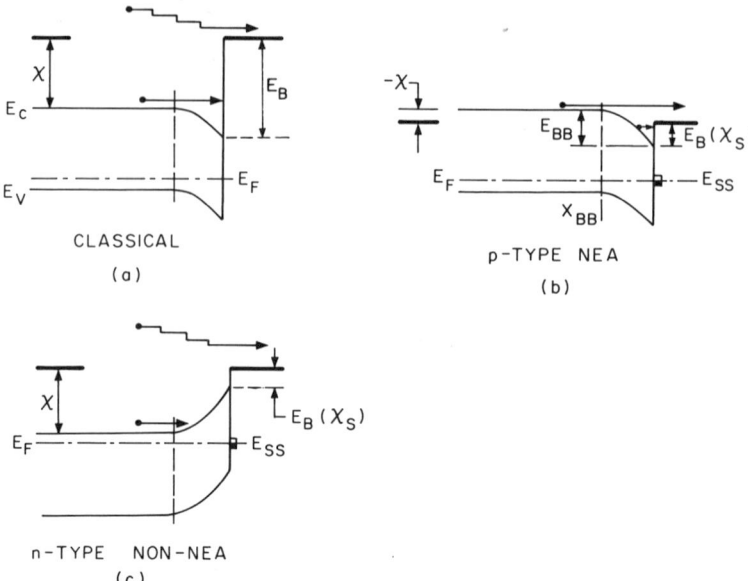

Fig. 5.8a–c. Classical (a), p-type NEA (b), and n-type non-NEA (c) photoemitters. In all portions of the figure E_B is the energy difference between the vacuum level and the value of E_C at the surface and is thus a "surface work function"; the edge of the band-bending region is x_{BB}; the band-bending energy is E_{BB}; and the energy level of assumed surface states is E_{SS}. All of these values are needed for construction of the band diagrams in the figure, and none of the values are obvious without material evolved in the text.—Using the bands resulting from the indicated energy values, bullets indicate electrons with various amounts of energy above and below that required for escape. Only the p-type NEA surface (b) allows escape of cold electrons from the bulk (based on [5.67])

Note in Fig. 5.8b that the bands near the surface are shown bent. This band bending in the p-type photoemitter is important for NEA operation, although not quite so much so as was first thought. (An initial model shows 0.5 eV of band bending for GaAs [5.68], while a more recent one shows only 0.1 eV of bending [5.46].) It is this band bending which allows cold bulk electrons to overcome any small positive barrier which might be present at the coated semiconductor surface. As illustrated in Fig. 5.8b, when the amount of band bending E_{BB} exceeds this barrier energy E_B, zero or negative effective *bulk* electron affinity is obtained in spite of the barrier. As illustrated in Fig. 5.8c, band bending for the n-type semiconductor is in a direction which confines the electrons to the bulk, resulting in a barrier even greater than the positive potential step present at the surface. Because p-type band bending assists in the attainment of negative electron affinity, all NEA emitters are p-type degenerately doped crystals, as in Fig. 5.8b.

A full understanding of the NEA figure involves at least the following three facts relating to cesiated surfaces: 1) most NEA surfaces do exhibit a finite upward discontinuity in the conduction band (a heterojunction energy barrier) at the surface [5.65, 69]; 2) the real value of "χ_s" is independent of the bulk doping of the semiconductor, but is a function of the semiconductor material and of the crystal face; and 3) the value of E_F at the semiconductor surface, and therefore E_C at the surface and ΔE_C from the surface to the bulk ($\Delta E_C \equiv E_{BB}$), is largely determined by the energy of surface states located at E_{SS} [5.68, 70; cf. 5.46]. These states come from the coating material (rather than from the GaAs) and are to some extent independent of the coated material itself. Maintaining a continuous E_F throughout the structure forces the bands to bend until E_F equals the surface state energy at the interface; the Fermi level is said to be "pinned" to E_{SS}. These items, along with factors controlling bulk band bending, are discussed below.

The attainment of NEA requires selection of a surface coating which has both small work function and a small barrier introduced at its interface with the semiconductor active layer. Several coatings have been investigated to date [5.49, 71–74], but thus far the best surface coating in use is "Cs_2O" [5.75], normally of monolayer dimension and deposited on a separate initial monolayer of Cs. A Cs coating by itself can, however, activate GaAs nearly to full NEA, as discussed below.

5.4.3 NEA GaAs

In Fig. 5.9 the development of NEA GaAs is schematically illustrated [5.13]. In Fig. 5.9a the band structure of clean GaAs is shown, with no surface states at the edge of the crystal. The surface barrier (X) is about 4 eV and $E_G = 1.4$ eV, for a photoemissive threshold of 5.4 eV for either p-type or n-type GaAs. (Incident photons with this energy are absorbed within about 10^{-6} cm (100 Å) of the surface and the probability of emission of the resulting hot electrons is high. Pure GaAs is thus a photoemitter for the far UV region of the spectrum if used in very thin TM mode layers or if illuminated from the emitting side where electrons are produced within a few hot electron scattering lengths of the surface.)

Shown also in Fig. 5.9a is Cs metal, while in (b) are shown p- and n-type GaAs after being coated with a monolayer of Cs [5.67]. The work function ϕ is lowered to about 1.4 eV, characteristic of Cs^+ ions on almost any substrate material [5.68, 70]. The Fermi level at the surface is forced to a position $\sim E_{SS}$ owing to pinning by surface states at $E_{SS} \cong E_V + E_G/3$ [5.68], again characteristic of Cs surface states. (In an alternate viewpoint, the value of E_C or E_V at the heterojunction interface is locked to surface states; in either case, specifying the value for an energy at the surface plus specifying the value of E_F within the bulk completely determines the band structures shown in the drawing.)

For p-type material, the Fermi level within the bulk is very near the top of the valence band, or $E_F \cong E_C - E_G$. The electron affinity under these boundary conditions is computed from the identity $\Phi = X + (E_C - E_F)$, where $(E_C - E_F)$ is a positive energy difference $\simeq E_G$; this gives $X \simeq \phi - E_G = 1.4 - 1.4 = 0$ eV. Zero or negative electron affinity is therefore achieved for p-type material. For the n-type material, E_F lies near E_C or $(E_C - E_F) \sim 0$. Here $X = \phi - (E_C - E_F) \sim \phi - 0 = \phi$, and

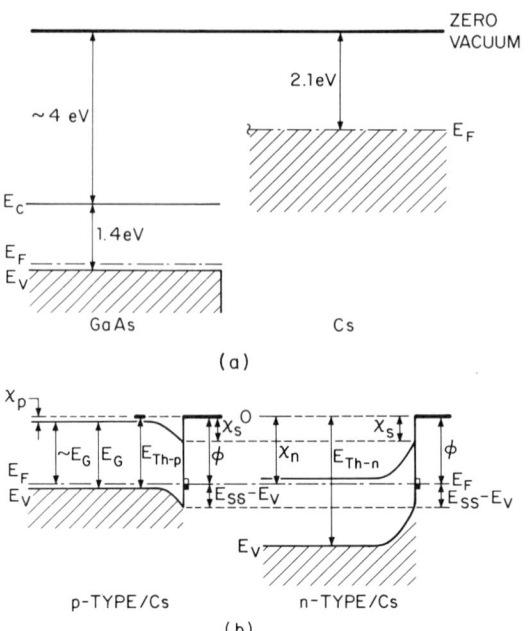

Fig. 5.9a and b. Development of near-NEA GaAs/Cs from the band structure of its constituents. In (a), GaAs and Cs are shown separately; in (b), both p-type and n-type GaAs are shown after coating with a thin layer of Cs. In both sections of part (b), the value of E_F at the surface is pinned to E_{ss}, and the surface values of E_C and E_V are identical for n-type and p-type material. The electron affinity for p-type GaAs/Cs is near zero, while that for an n-type sample is near $E_G = 1.4\,\text{eV}$. The position of E_F in the bulk, determined by dopants of energy not shown, is the only difference between the construction of the two sections. (Although the Fermi level is shown here pinned to E_{ss}, in other models E_V or E_C may be pinned to a surface state energy instead) (adapted from [5.13])

NEA is not reached. In both cases the *real* electron affinity at the *surface* is the same $+0.5\,\text{eV}$; see X_S in both halves of the figure.

Near-zero electron affinity for p-type GaAs results from a fortuitous match between the work function of Cs and the bandgap of GaAs (both at 1.4 eV) for a case where the Fermi level in the bulk is located $\sim E_G$ below the bottom of the conduction band. This was observed prior to fabrication of the first NEA photocathode, p-type GaAs/Cs [5.66]; after construction, the experimentally measured quantum efficiency per absorbed photon was in the order of 0.2 for energies above $\sim 2\,\text{eV}$. A sharp response cutoff at 1.4 eV gave a white light sensitivity of about 500 μA/lm - immediately superior to the classic S-20 surface [5.66].

Further lowering of the surface barrier and some improvement in sensitivity of the activated GaAs surface occur when oxygen and Cs are alternately leaked onto the Cs-covered surface to form (CsO) [5.75]. This X-reduction results qualitatively from the lower electron affinity of the substance Cs_2O, where X-(CsO) ultimately drops to about 0.6 eV for sufficiently thick layers. (See Fig. 5.3,

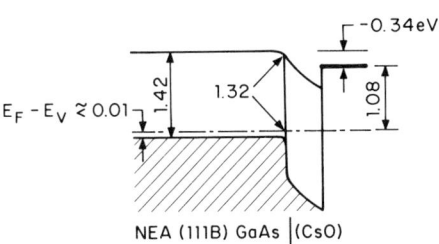

Fig. 5.10. Fully NEA GaAs/(CsO) according to the heterojunction model. For the optimum (CsO) thickness, the work function of (CsO) is 1.08 eV, which gives a bulk GaAs electron affinity of −0.34 eV. The conduction band discontinuity, common in most coated III–V surfaces, is hidden for the 111 B face of GaAs. Note the slight (0.1 eV) band bending, indicating that E_{ss} differs from its value on pure Cs-coated material. The Fermi level at the surface varies also with the choice of activated crystal face [5.46]

and also the dashed curve in Fig. 5.13 below.) The optimum III–V/(CsO) coating thickness is about 8 Å [5.12, 76–80], which gives a (CsO) electron affinity of about 1.1 eV. The white light sensitivity is increased to 2000 µA/lm, with highest response found on the 111B face of the GaAs crystal [5.46].

This structure of the fully NEA GaAs/(Cs:CsO) photoemitter is shown in Fig. 5.10 [5.46]. Note that the GaAs electron affinity is 0.3 eV lower than that found with Cs alone, or $X \cong -0.3$ eV. Further reduction to −0.8 eV would be obtained with a thicker (CsO) layer, but this is neither necessary for NEA nor useful once all energy barriers are overcome.

Band bending within the GaAs in Fig. 5.10 is also shown smaller than that for GaAs/Cs (Fig. 5.9). For the particular model of the NEA GaAs structure used in this figure, a heterojunction barrier between the GaAs and the (CsO) has dropped to about 1.2 eV (measured from the Fermi level), and the conduction band discontinuity has vanished [5.46, 69, 81]. NEA could be achieved even with no band bending. The (CsO) layer in the figure is also shown thick enough to exhibit its own energy band structure. This figure, although adequately explaining NEA GaAs, represents only one of the two primary models for the NEA surface. The alternate model differs substantially in both greater band bending and in the details of the "heterojunction" barrier structure. To understand NEA in other IR photoemitters we now examine the two current models plus the earliest simplified model in more detail.

5.4.4 Modeling the NEA Surface

The Heterojunction and Surface-Dipole Models

The operation of NEA photoemitters can be modeled in the same two major parts as classical emitters, namely, 1) photoconduction (excitation of electrons by photoabsorption in the active region, plus carrier transport to the surface) and 2) escape into the vacuum. The treatment of the bulk photoconductor is straightforward; it involves photoelectron generation and cold electron, diffusion-dominated transport to the "surface" region where the bands begin to bend. Here, and through the surface barrier region, exact transport analysis is difficult owing to 1) field-assisted transport statistics through the bent bands; 2) accompanying "hot" electron scattering which modifies the electron energy distribution between the bulk and the surface barrier; 3) the difficulty of modeling the barrier itself and of treating energy-dependent electron escape over and/or through it; 4) possible field-assisted hot electron transport within the surface layer; and 5) transport over a possible second barrier at *its* surface.

The original simplified model for predicting and computing the quantum efficiency of an NEA photocathode was quite straightforward. It involved determining the photoconductive diffusion current from the bulk to the point $x = x_{BB}$ (see Fig. 5.8) and multiplying this current by a lumped "escape probability" (determined experimentally) to give the yield. Provided that $X \leq 0$, the escape probability is close to unity and the model simply gives a spectral response similar to the absorption curve of GaAs [5.47, 66, 67].

Two less straightforward NEA models have subsequently evolved, these similar to the simple model within the bulk but differing in their treatment beyond x_{BB}. In one model [5.49, 67, 73, 81–85], the GaAs/(CsO) interface is assigned a finite barrier energy determined experimentally. A thermalized (Boltzmann) energy distribution of electrons is then translated, with a uniform loss factor $(1 - T_{BB})$, to this interface. Escape current is computed by determining the number of carriers with energy greater than the barrier height and multiplying this quantity by an average transmission constant T_{BS}. Escape through the surface layer is next assigned a further transmission constant ($T_{ABS} \cdot T_{BV}$), with all of these average transmission constants assumed independent of energy. This model is illustrated in Fig. 5.11. In this, the "heterojunction" model, the (CsO) is assumed to be similar to bulk Cs_2O with its well-defined band structure, defect states (10^{19} donors located 0.25 eV below E_c), and its own surface electron affinity. This additional (CsO)-layer/vacuum electron affinity gives rise to the second, usually ignorable, barrier and transmission factor T_{BV} just at the vacuum surface. The more vital GaAs/(CsO) interface barrier is treated as a normal heterojunction [5.69] in spite of the very small depth of the (CsO) layer (amorphous and ~5–8 Å thick [5.76–80]), but this heterojunction does produce the required discontinuity (interfacial barrier) where the dissimilar semiconductor conduction bands merge [5.49, 65, 69, 86]. (This barrier, rather than the concept of NEA, directly limits the yield of the NEA photocathode.) The heterojunction model was used, a bit simplified, in Fig. 5.10.

The second model is similar but the "barrier" at the semiconductor/surface layer interface is lower and wider. This model is illustrated in Fig. 5.12, [5.12, 15, 87, 88]. Here the bent-band transport factor T_{BB} of Fig. 5.11 is computed explicitly as a function of energy, using hot electron scattering parameters. The probability of escape of the resulting Gaussian (rather than Boltzmann) distribution of electrons is computed at the interface for both tunneling through and for thermal excitation over the barrier. In this, the surface dipole interpretation, the Cs and O are treated as a composite layer of charged ions which produce an electron energy barrier layer of height and width determined by fitting adjustable parameters to experimental data. There are no physical factors T_{ABS} and no second barrier at the vacuum surface to represent the

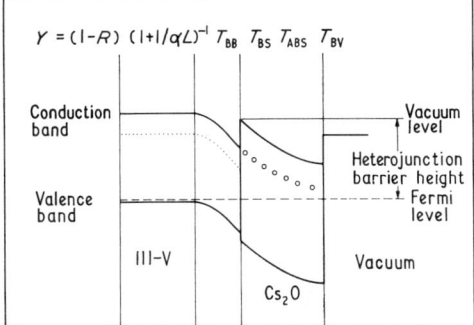

Fig. 5.11. Summary of factors limiting photo yield in the heterojunction model. The optical reflectivity is R; transport factors in the band-bending region are described by T_{BB}, with similar factors T_{BS} at the Semiconductor surface Barrier, T_{ABS} for electron ABSorption in the (CsO) layer, and T_{BV} for loss at the (usually unimportant) Vacuum Barrier [5.82]

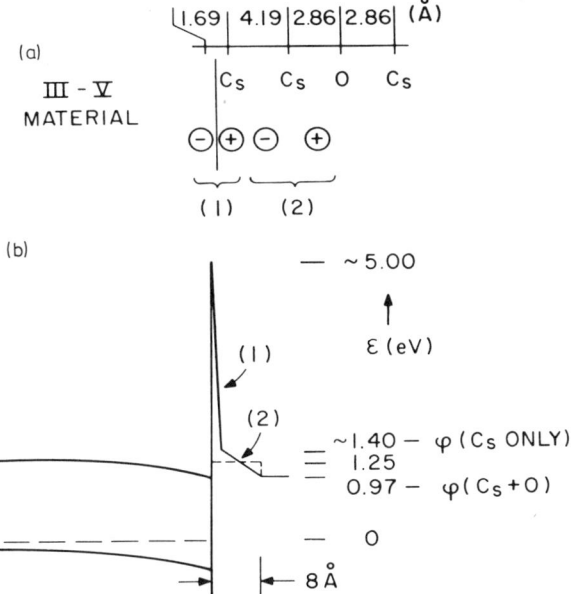

Fig. 5.12a and b. Band structure (b) and physical model (a) for the surface-dipole model of the NEA photoemitter. For analysis, the high but thin potential spike (1) is ignored and the triangular constant ε region is replaced by a constant potential region (2) to simplify calculations. Potential height, after matching to experimental data, is 1.25 eV, while work functions for Cs- and (CsO)-coated GaAs agree well with the results for the heterojunction model. (Zero-reference potential for this figure is the Fermi level in the bulk) [5.12]

electron affinity of the surface layer; the total effect of this surface layer is contained in the wide ionic barrier at the semiconductor surface. Although a triangular potential barrier (constant ε, shown solid in Fig. 5.12) is that more closely called for by the model, for computational simplicity the indicated rectangle is used to match parameters. Numerical values given in the figure are for GaAs; note the similarity to values of Fig. 5.10.

In the heterojunction model the band-bending energy depends on the crystal face, falling as low as 0.1 eV for the (111 B) face as shown in Fig. 5.10. In the surface dipole model, the computed band bending is found to be about the same for all faces and equal to that measured for GaAs–Cs, or ~0.5 eV. Barrier heights are, however, comparable. When all parameters are matched to experimental data, the yield predictions of both models are nearly identical, showing a uniformly high response plateau for photon energies above bandgap. For GaAs the differences between the two models are thus not vital, essentially because nearly all electrons reaching the band-bending region have energy greater than the barrier height in either model. The probability of escape is thus high (as large as 0.5) in spite of the conduction band discontinuity (Figs. 5.11 or 5.12).

5.4.5 Fabrication and Optimization of RM and TM NEA GaAs

Before considering other possible TM or RM IR NEA photoemitters, we discuss in this subsection the fabrication of NEA GaAs and the general factors required to optimize RM mode photocathodes [5.46, 47, 83, 89–91], plus the additional special constraints of TM mode fabrication [5.14, 15, 89, 91, 92–96].

Production of the NEA GaAs Surface

Production of NEA GaAs requires cleaning and cesium-activation procedures [5.97]. All steps are performed under ultrahigh vacuum (oxygen-free) conditions after a system bakeout.

Cleaning is performed by heating above 600 °C until surface contaminants evaporate. In an experimental demountable laboratory system, an additional cleaning by ion bombardment is often performed. The cleaning must produce a surface free of contamination to within 10 % of a monolayer or activation may be impaired [5.98]. The actual activation consists of deposition of about one monolayer of Cs followed by repeated, alternating exposures to oxygen and cesium until the photoemissive response reaches a maximum. Although the surface work function continues to decrease with further deposition of (CsO), this is of no value once full negative electron affinity is reached; for GaAs the emission efficiency actually decreases for Cs_2O layers thicker than the optimum ~1–2 monolayers owing to increased electron loss within a thick surface layer (T_{ABS} of Fig. 5.11 decreases). The point of optimum surface activation is illustrated in Fig. 5.13, where the work function (dashed, reaching 0.6 eV for very thick layers) and the escape probability from the Γ and the X conduction bands of GaAs are shown as functions of (CsO) thickness [5.47]. As noted above, the 111 B face of GaAs offers the highest surface escape probability [5.46] and the smallest heterojunction barrier. (Other III–V compounds are activated by a similar process, although they may require slightly thicker (CsO) layers [5.80] to produce the slightly lower barriers they exhibit [5.81, 82].)

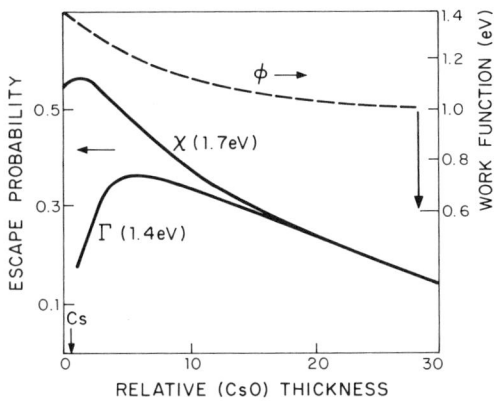

Fig. 5.13. Work function ϕ (dashed curve) and surface escape probability (solid) from the X and Γ conduction bands of GaAs for varying (CsO) coverage [5.47]

RM Mode Materials Selection

Although there are no further variables in the actual activation of GaAs surfaces to NEA, there are several materials parameters trade-offs if high quantum efficiency is to be obtained. The most important variable is the substrate diffusion length L_D which must be at least 1–2 microns for efficient emission. This can be seen in a simplified expression for the yield Y of the opaque (front-illuminated) device (Fig. 5.11), where $Y \cong P(1-R)/[1+1/(\alpha L_D)]$. Here R is the total loss by optical reflection, P is the lumped electron escape probability (transmission) past the beginning of the band-bending region, and α is the optical absorption constant (about 1 micron in GaAs for $h\nu$ above bandgap). High yield requires a high αL_D product, and therefore L_D should be at least 1 μm. This long diffusion length requires excellent crystal quality, which is obtainable for both Si- or Ge-doped [5.71, 99] and Zn-doped [5.47] GaAs photocathode material.

The second important parameter is the p-type doping density [5.46, 90]. The lumped escape probability P includes the hot electron transmission factor T_{BB} through the bent-band region. Any scattered electrons whose energy falls below the barrier are lost. With a 40 Å hot electron scattering length in GaAs, it is desirable that the bent-band region be of comparable dimension. The bent-band length is given by the expression $L_B \cong 35[\Delta V/(N_A/10^{20})]^{1/2}$ Å, where ΔV is the amount of bending in eV. A high doping density, in the order of $5 \times 10^{19}/cm^3$, is therefore desirable if $\Delta V \sim 0.5$ eV. But a high impurity density lowers the carrier lifetime τ for the dominant recombination mechanism (conduction band to acceptor), where $\tau \propto N_A^{-1}$. A compromise doping density must therefore be selected, and $N_A \cong 3-10 \times 10^{18}/cm^3$ is the common choice. Lower values are used where $\Delta V \ll 0.5$ eV, as on the 111 B face of GaAs [5.46].

TM Mode Materials and Fabrication Considerations: The Substrate

For the opaque-mode photocathode there are no further adjustable materials variables beyond L_D and N_A. For semitransparent devices, however, additional parameters remain [5.14, 15, 89, 91–96]. Most obvious is the thickness of the active region. For a very thick absorbing region the majority of electrons is generated very near the nonemitting surface and may not diffuse efficiently to the opposite, activated surface. For too thin an active region the total absorption is small and much incident light completely escapes the cathode. In both thickness extremes efficiency is low. (This also applies, of course, to "classical" cathodes.)

The optimum thickness is determined more by L_D than by α and is generally in the order of $0.75 - 1.5 L_D$ (somewhat longer if response very near the low-α threshold is to be optimized) [5.95]. Because this is only 1–2 µm for (brittle) GaAs, it is clear that the active layer cannot be self-supporting except possibly for very small area devices. Since a supporting substrate must be used between the window and the emitter, photon absorption losses within this substrate, plus carrier losses at the substrate/active layer interface, now become additional factors limiting the performance of semitransparent photocathodes. To minimize loss, a substrate must therefore be chosen with care.

Evaporating GaAs on a glass faceplate, similar to the construction of classical devices, is ruled out owing to the requirement of a high diffusion length, which requires grown-crystal lattice quality. Techniques for processing GaAs on sapphire or glass (including SiO_2) have been investigated, but with one exception [5.100] the published results are disappointing [5.96, 101, 102]. More often successful has been the growth of GaAs on another III–V substrate, always one with a higher bandgap energy. This substrate introduces a short wavelength cutoff (spectral response limit) above which absorption in the inactive substrate, rather than in the GaAs, occurs, and a substrate with the highest possible bandgap is therefore desirable.

Complicating the substrate choice are the requirements of matching the lattice constant of the substrate to that of GaAs [5.14, 15, 93–95, 103, 104] and of choosing a substrate with well-developed technology for large area wafers. Fortunately much of the required technology is currently available owing to similar demands for the fabrication of heterojunction semiconductor lasers.

A close lattice match is required to minimize both structural defects in the grown GaAs and also to limit recombination at interface states between the substrate and the GaAs layer. These defects result from strain introduced by a change in interatomic spacing between the two materials. In general the defect density is equal initially to the density of unmatched bonds, and this is directly related to the difference between the two lattice constants. Dislocations do "grow out of the crystal" if the grown layer is made sufficiently thick, but for photoemitters the growth layer is thin and therefore a large mismatch may propagate unattenuated through the bulk to the surface. This reduces the diffusion constant in the bulk and the image perfection of the surface.

In Fig. 5.14 the lattice parameter a_0 and the bandgap E_G of the most important III–V materials are shown [5.15]. Binary (two-element) compounds are shown as points. Lines connecting these points indicate ternary (three-element) compounds; these have bandgaps and lattice constants intermediate between those of their binary constituents—roughly in proportion to the

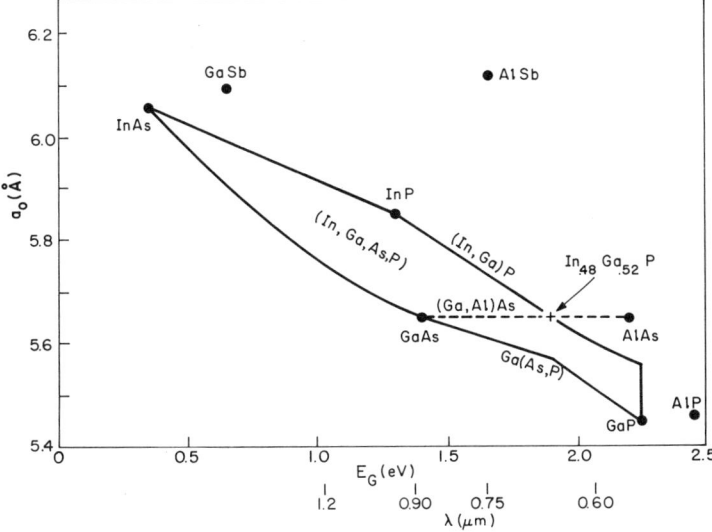

Fig. 5.14. Bandgap energy and lattice constant for binary (points), ternary (connecting lines), and quaternary (region between lines) III–V semiconducting compounds. Breaks in the GaAsP and InGaP lines indicate a transition from indirect to direct band structure [5.15]

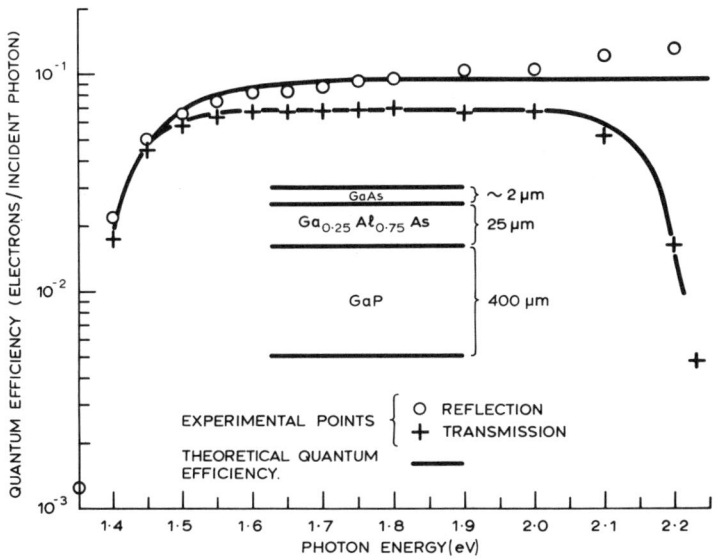

Fig. 5.15. Example multilayer TM GaAs photocathode. The substrate is GaP with a GaAlAs matching layer grown below the GaAs activated layer [5.110]

percentage of each binary in the ternary (e.g. [5.105, 106]). The region enclosed by the ternary connecting lines is the quaternary (four-element) semiconductor InGaAsP[2], which offers continuously varying E_G and a_0 as the In/Ga and As/P ratios are varied [5.107–109].

The obvious choice for a GaAs substrate is AlAs, which has one of the largest bandgaps in the group and a lattice constant within 0.1% of that of GaAs. Unfortunately AlAs is all but unavailable in wafer form, and thus it (or the intermediate ternary AlGaAs) must be custom grown on another more available substrate prior to the GaAs growth. Aluminum-containing compounds are also difficult to grow, owing probably to the formation of aluminum oxide, and a high density of blemishes is often found at the final surface [5.104]. For these technical reasons a better substrate choice is GaP. Here the lattice match is poor, but large, high quality wafers of GaP are widely available. An intermediate layer GaAsP or GaAlAs is often grown between the GaP and the GaAs to minimize this abrupt lattice change.

One example of a complete TM GaP/GaAlAs/GaAs photocathode, along with its RM and TM response, is shown in Fig. 5.15 [5.110]. The ~2 eV high-energy cutoff in the TM data is caused by absorption in the GaP.

TM Mode GaAs Photocathode Models and Performance

To model this now rather complex multilayer photocathode, several authors have derived expressions or computer algorithms which give the yield of transmission mode structures as a function of photon wavelength.

The results of one such calculation are shown as solid curves in Fig. 5.15 [5.110]. Additional generalized curves and closed-form solutions appear in the references [5.14, 15, 89, 92, 93, 95, 96, 101, 104, 110]. The various solutions all agree on the importance of low recombination velocity at the substrate/active layer growth interface and thus on the importance of low lattice mismatch. They also indicate that near threshold the quantum efficiency can be higher in transmission than in the reflection mode owing to one added optical reflection at the vacuum surface [5.95]. From the various separate models have come several versions of an "optimum" structure, but all are similar in general dimensions and structure as outlined above.

AlGaAs- [5.95, 100, 104, 110–114], GaAsP- [5.94, 95, 115, 116], and InGaP-buffered [5.103] cathodes have been fabricated for several variations [5.14, 15] in 1) the exact $Al_xGa_{1-x}As$ or $GaAs_xP_{1-x}$ composition (where x is the mole fraction of the first binary constituent and $(1-x)$ is the remainder); 2) the method of growth (liquid, vapor, or hybrid epitaxy); 3) the thickness of the various layers and their doping; 4) the sequence of growth; and 5) the initial substrate and whether or not it is removed after growth of the intermediate and active layers. A further variant is the use of graded (continuously varying)

---

[2] For brevity we denote generic compounds by simple chemical symbols in this section; the notation $In_xGa_{1-x}As_yP_{1-y}$ is understood.

composition within the interface layer. This grading distributes the lattice mismatch and also introduces a semiuniform electric field which confines generated electrons to the GaAs layer, driving them toward the emitting surface [5.95].

The results to date are preliminary and one may certainly expect further improvements to be reported in the open literature, but at present all of the methods are nearly equally successful. Highest white sensitivity (400 µA/lm) has been achieved with AlGaAs [5.110]. This substrate gives a response window from 6000 to 9000 Å, but early devices suffered from surface defects. Blemishes have been reduced by growth in a highly oxygen-free furnace [5.100] with later cathode sealing to a glass substrate. Sensitivity is high at 390 µA/lm, so the technical impediments to AlGaAs supporting layers do not seem fundamental. Sensitivities for InGaP-matched devices are also nearly identical at 390 µA/lm [5.103], and cutoff is at ~5000–6500 Å. Sensitivities for a GaAsP intermediate layer are only slightly lower (300–350 µA/lm) and the window is only a bit narrower, cutting off near 6500 Å [5.94, 115]. High quality surfaces are easily obtained with this material.

These TM sensitivities are factors of five lower than the best reflection mode GaAs at 2000 µA/lm, but *Allen* [5.92] indicates that while reflection mode performance has nearly reached the theoretical limit of 2400 µA/lm, transmission mode devices may be expected to climb by a factor of three to about 1200 µA/lm. The transmission mode GaAs photocathode is therefore a realistic, if still undeveloped and difficult to fabricate, imaging retina [5.116].

5.4.6 Other NEA IR Photocathodes

We have considered GaAs in detail because it is the best understood (with the exception of NEA Si) and most widely available IR photoemitter, and because GaAs technology provides an excellent platform for extension of NEA further into the IR through use of lower bandgap activated materials. Such work has included InP [5.117], InSbP [5.82], GaAsSb [5.81, 82, 93, 105, 118], InGaAs [5.12, 14, 15, 69, 74, 88, 94, 119–122], InAsP [5.78–80, 82–84, 95, 106], and InGaAsP [5.95, 107–109, 123, 124]. Related work at higher bandgaps has concentrated on the very stable emitter GaAsP [5.125].

The theoretical models developed for GaAs remain valid for these surfaces, and both major models described above have been used to predict the spectral response of all important members of the III–V ternary and quaternary families for many values of material composition (or bandgap). The general predictions of both models are very similar, owing probably to the use of experimental data to adjust theoretical barrier parameters.

The major prediction of both models is the disappointing conclusion that a surface electron affinity which can be made as low as 0.6 eV is *not* a sufficient criterion for efficient NEA operation of small bandgap ($E_G = 1.2$–0.6 eV) materials [5.81, 82]. In one model this results from a rather invariant 1.1 eV

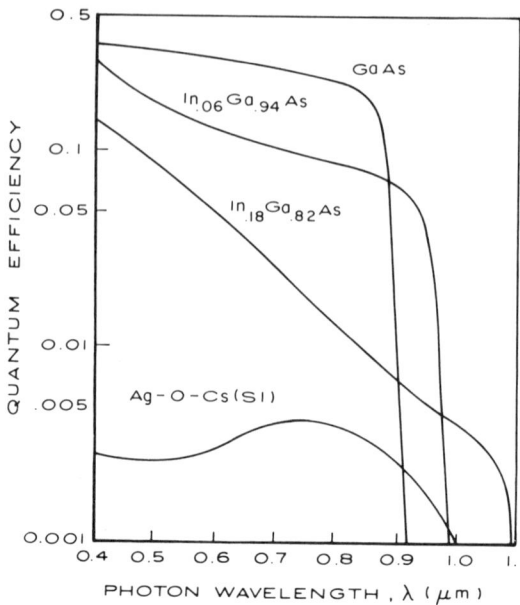

Fig. 5.16. Spectral yield curves for three members of the InGaAs family, including GaAs, illustrating the drop in above-threshold yield as bandgap is lowered to extend IR response. The S–1 AgCsO curve is for reference [5.15]

heterojunction barrier [5.69]; in the second the limitation results from the vanishing probability of lower-energy electrons tunneling through an ionic barrier which remains as high as that found on GaAs [5.12]. In either model this $1.1 - 1.2$ eV interfacial barrier severely limits response beyond 1.1 μm for any substrate material. Some long wavelength response *is* obtainable, but overall yield (even for excitation at > 1.4 eV) falls continuously as E_G is made close to or lower than the barrier height. For highest response, the bandgap of the active layer must be absolutely no lower than necessary to provide usable photo-absorption at the longest wavelength of interest or yield at any energy will be far below that obtained with a higher bulk E_G.

This lower yield as threshold is extended (by lowering E_G) is illustrated by the *crossing* yield curves for several members of the InGaAs family shown in Fig. 5.16. Although lower bandgap materials do allow low energy thresholds beyond even 1.1 μm, a sacrifice in response is made at *all* energies above threshold. Thus no cathode can offer both long threshold response and high overall quantum efficiency.

This sensitivity loss in low bandgap NEA emitters is also illustrated in Fig. 5.17a and b, where threshold escape probability is plotted as a function of bandgap for InGaAs and several groups of III–V compound semiconductors [5.12, 82]. Note the "knee" in each curve near the $1.1-1.25$ eV barrier energy.

The III–V family offering the highest response with the greatest flexibility is the quaternary InGaAsP [5.123]. This material offers the low surface barrier characteristic of InAsP (see Fig. 5.17), a relatively long InP-like diffusion length, the absence of Al, high absorption characteristic of InP, and the broad choice of

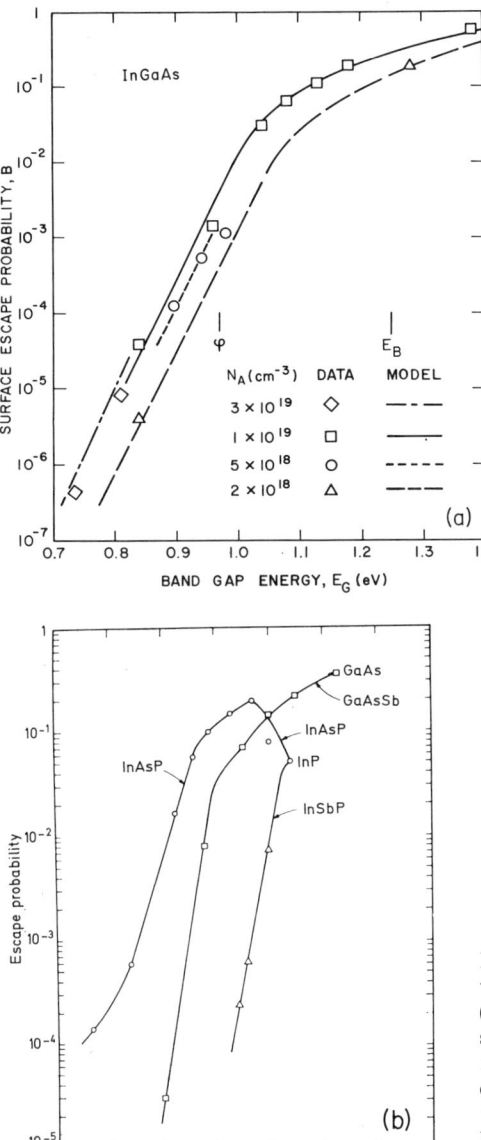

Fig. 5.17a and b. Surface escape probability versus bandgap and semiconductor species for (a) the InGaAs family [5.12] and (b) for several III–V families [5.82]. For the InGaAs family, variation with acceptor doping is explicitly included. Rapid decrease in efficiency for materials of bandgap below a barrier height is obvious

E_G and a_0 discussed above. The related ternary InAsP [5.83] and an alternate ternary InGaAs [5.12] are progressively inferior.

To date the greatest effort in developing IR devices has been in those with response peaked at 1) 8500 Å, for detection of the slightly-below-bandgap GaAs emission and 2) at 1.06 μm, for use in Nd:YAG laser systems. For 8500 Å application, InP, InAsP, InGaAs, and GaAsSb have all been investigated

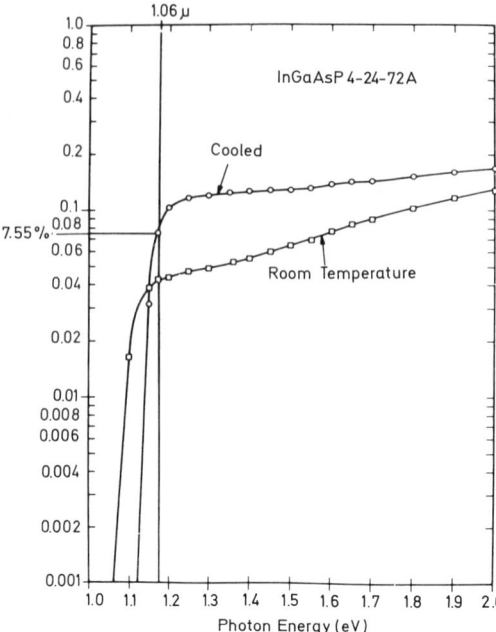

Fig. 5.18. Spectral response for room-temperature and cooled InGaAsP reflection mode photoemitter illustrating high 1.06 μm sensitivity [5.124]

(references p. 177). Here the fabrication technique is similar to that for GaAs. involving growth of an active layer on an intermediate substrate, and results are good even for transmission mode devices (e.g., [5.94]).

For 1.06 μm response, efficiencies as high as 5.4% have been reported for InAsP [5.107], with a slightly lower 3% reported for InGaAs [5.14], both for operation at room temperature. A maximum efficiency of 9% has recently been reported for InGaAsP [5.123]; this is an improvement over the 7.6% efficiency reported for a photocathode cooled to −90 °C (Fig. 5.18) [5.124], which for some time was the best ever achieved. These efficiencies are roughly a factor of 100 higher than the S–1 photocathode response, although the yield of the S–1 device falls off more slowly near threshold. (This is, of course, unimportant for 1.06 μm laser detection.) In the opaque mode, NEA photoemitters can outperform classical devices to beyond 1.6 μm.

Further improvement in either efficiency or threshold of NEA IR photoemitters may require the discovery of a surface-coating material superior to (CsO) [5.13, 65, 73, 85], perhaps one with higher bulk electron affinity but a lower interfacial barrier (say 0.8 eV for each). According to the heterojunction model, the material should be n-type so that full use can be made of the low electron affinity. (For a p-type coating such as Cs_3Sb with $X \sim 0.45$ eV, the effective value of X within the substrate is raised by as much as the bandgap of the coating—see E_F in Fig. 5.11; this yields a higher effective electron affinity in spite of the lower X of the example Cs_3Sb [5.13].) Surface-state pinning position and heterojunction step height are difficult to predict, so that discovery of an improved coating will require a largely experimental effort.

Summarizing IR NEA surfaces, we find that 1) the height of the interfacial discontinuity, rather than X of the surface layer, is the true limit to high IR efficiency; 2) although photoemitters in this general class are all called NEA, those which push the low-energy limit of the concept are not truly NEA but are actually small-positive-affinity devices; and 3) there is a bit of rivalry and uncertainty present in the NEA field (cf. [5.12, 69]).

5.4.7 NEA Silicon

One of the earliest IR-responsive materials activated to NEA was silicon which, with its > 1.1 μm bandgap, seemed to hold great promise for IR operation [5.50]. In spite of its rather small absorption constant (owing to an indirect bandgap), NEA Si has higher yield than comparable InGaAs near ~ 1 μm owing to an extremely long carrier escape depth (long electron lifetime) [5.11]. But Si suffers from one fatal flaw, namely, very high thermionic dark current, in the order of 10^{-9} A/cm^2 at room temperature [5.126]. This is 1–2 orders of magnitude higher than the dark emission from the already poor S–1 surface and 5–7 decades higher than other classical and NEA emitters (see below). Equal emission for n- and p-type bulk indicates that this is entirely a surface-generated current.

Activation of Si to NEA is also much more critical than activation of the III–V compounds. The procedure is similar to that used for the III–Vs, but unlike the amorphous and arbitrarily thick layers found there, the Si/(CsO) layer must be perfect in both structure and composition. For NEA Si, a single layer of Cs must reside on top of "plateaus" of four Si atoms, with oxygen, even though applied last, residing within "caves" below four Si atoms, forming an oxygen sublattice beneath the Si/Cs structure [5.30, 31]. As little as 0.1 % surface contamination can destroy Si NEA, and the (100) face is the only one capable of NEA activation [5.72].

Activated Si has been investigated for purposes other than as a photoemitter [5.127]. These include fabrication of both transmission secondary electron multiplier dynodes [5.128] and cold cathode electron emitters [5.129]. For Si dynodes, the physical strength of thin self-supporting Si "rim" structures and the extremely high multiplier gain would be of great utility in proximity-focussed image tubes, but the high dark current and, in some cases, the slow response time, disallow wide application, at least at room temperature.

In NEA Si cold cathode devices, a planar diode is biased to inject electrons from an n-type substrate to a p-type NEA surface layer, where current is drawn off by an electric field. (Other cold cathodes of different design have been demonstrated in III–V structures [5.130–132].) This efficient emission could be useful in replacing hot (thermionic) cathodes in IR sensitive devices where a low luminence source with small electron energy spread is desirable, but requirements of zero contamination along with electron emission-density restrictions and long-term instability problems have so far prevented practical use.

5.5 Photoemissive Devices: The Photomultiplier

5.5.1 Introduction

A photoemissive detector contains not only the photoemissive sensor layer but also electron detection or gain stages plus a connection to external electronics. In the preceding sections we have limited discussion to the photocathode portion of the photodetector because its quantum efficiency, spectral uniformity, and wavelength limitations are the primary factors limiting the performance of a complete device; advances in the capabilities of the photocathodes have also been most dramatic in recent years.

Some relevant detector device parameters have yet to be considered [5.133, 134]. These are determined in part by the photocathode but also in part by the electrode-portions of the photoemissive device; residual dark current and NEP are, for example, partly characteristics of the photocathode and partly limited by the remaining design of the detector device.

The most vital device specifications, other than dark current, are determined mainly by the internal electron-multiplier amplifier. Specifically, peak output current, signal gain, bandwidth and rise time, and noise figure all are functions of the "dynode" material used for each gain stage and of the geometrical design of the multiplier structure, including the number of stages.

In this section we consider several of these photoemissive device specifications, stressing dark current, the choice of the electron-multiplier dynode material, and the design of the multiplier for optimization of speed or gain. Other device specifications, such as dynode operating voltages, ruggedness and physical size, are best found in the commercial literature [5.2, 3, and, e.g., 5.135–138]. We conclude the chapter with a comparison of spectral response curves for classical and NEA devices.

5.5.2 Photocathode Dark Current

Background

Thermal electron emission from a solid (dark current in a photocathode) can be modeled in various degrees of complexity and full models for either classical or NEA emitters [5.139, 140] require a fair amount of detail. In simplified form we can consider initially the limit of treating the substance as a perfect electron source (an electron "black body"). Here all electrons which are thermally excited to energy greater than the surface barrier (work function or electron affinity) are emitted, and each emitted electron is assumed to be instantly replaced by an electron from the bulk. This is roughly the Richardson model, which applies fairly well to a metal. The dark current computed from this model is an upper limit to emission of a real device.

For a semiconductor we can alternately assume that any thermally excited and emitted electron is *not* instantly replaced owing to diffusion delay from the bulk. The diffusion-limited dark current of this model is substantially lower than emission based on a perfect metal, and computation is more difficult than the Richardson model. For any real photocathode neither model, even if adjusted for nonunity escape probability ("grey body"), is adequate owing to 1) the presence of surface states above the semiconductor valence band, 2) states present within the coating layer, 3) variations in the surface work function, and 4) band bending in the p-type substrate. (A slightly inverted n-type surface region makes electron emission more likely than emission from the bulk.)

Dark Current Values

For a 1.4 eV threshold emitter characteristic of either S–20 or NEA GaAs, the experimentally measured emission is in the order of 10^{-15} to 10^{-16} A/cm^2 at room temperature [5.52], with the slightly lower values found for GaAs. This emission is well above the value computed from the diffusion-limited model even including band bending; the high emission indicates that the thin surface layer of Cs or (CsO) for either photocathode is the actual source of dark current [5.139]. The size of this surface-layer emission is difficult to compute owing to a large number of materials parameters which are unknown for this ill-structured layer.

Emission from most classical and NEA surfaces is in this same $\sim 10^{-14}$–10^{-16} A/cm^2 range, indicating that the surface component dominates the bulk and that it is roughly independent of the substrate material. There is some tendency for higher emission in lower threshold (lower bandgap) emitters, but this is not as pronounced as might be expected from a simple theory for bulk emission over a barrier. The primary exception is the S–1 surface.

For the S–1 Ag/(CsO) photocathode, the low-affinity (CsO) layer covering each Ag particle is much thicker than the (CsO) of NEA structures (tens of Å versus ~ 5 Å); the S–1 dark current is also very high, in the order of 10^{-11} A/cm^2 [5.52], a value which could be expected for a *metal* with a surface barrier of about 1.1 eV. It is unclear whether this emission is dominated by (CsO) emission or by emission from Ag through the (CsO) surface. For a NEA InGaAsP photocathode with a similar 1.1 µm threshold, the surface layer is much thinner while the barrier at the *semiconductor*/surface layer interface is a similar 1 eV; both factors contribute to a lower value of dark current. One manufacturer specifies 2.5×10^{-13} A/cm^2 [5.138], which is much smaller than the S–1 value and only a factor of six higher than S–20 or NEA-GaAs dark emission specified by the same manufacturer. (Others list substantially lower S–20 and GaAs emission, but this may be a difference in specsmanship.) Since some of this dark emission is due to interfacial surface states whose density can be lowered by a more optimum activation technique, one may expect future NEA emitters to drop below the S–1 dark current by still an additional order of magnitude. The NEA surface thus improves the S/N ratio at ~ 1.1 µm by 4–5 decades over the S–1 (including the higher NEA quantum efficiency).

One obvious means of reducing thermal dark emission is to cool the photocathode [5.19, 135, 136, 138, 141–144]. (Another is to reduce either the actual size or the apparent size of the emitting surface; the latter involves "magnetic lenses".) Most classical photoemitters emit well under ten electrons per cm^2-s with cooling to $-20\,°C$. Lower temperatures are required for NEA ($-80°$) and S–1 ($-200°$) emitters. Photocathode cooling has its limitations, however, due to residual tube-structure leakage, high-energy particle interference (from the window, the environment, and extraterrestrial sources), and mechanical stress considerations. More importantly, cooling can raise the resistance of the active layer of most classical devices. This higher resistance causes a lateral cathode voltage drop during emission which requires that emission be reduced to maintain uniformity. For two classical materials (KCsSb) and (CsSb), deep-lying trap levels (cf. Fig. 5.5) lock the Fermi level in a high resistivity position (e.g., see [5.55]) regardless of temperature. This limits obtainable current [5.145] and obviates the use of a cooled tube unless a conducting substrate [5.146] is used. Limitations for other surfaces (S–1 and S–20) are not quite so severe, and with proper choice of tube structural materials, cathode thickness, conducting substrate, and small cathode lateral area, operation of the classical device with cooling is generally advantageous.

For the degenerately doped NEA emitters, cooling is similarly useful, and furthermore there is no resistivity increase with lower temperatures. These emitters are, however, already limited in maximum room-temperature current densities to values about two orders of magnitude lower than classical surfaces to prevent long-term degradation during operation [5.18]. Of course cooling to reduce dark current is more commonly used only at low light levels, and thus the above cathode current limitations, applicable also to the S–1 surface, are not normally a serious problem in cooled tubes.

Along with dark current reduction with cooling, photosensitive thresholds and quantum efficiency can vary with temperature [5.143, 144, 147]. The most important classical IR photocathode, the S–1, either improves or does not exhibit strong effects when cooled. Other common classical surfaces ("ERMA" and to a lesser extent standard S–20) exhibit a small but seldom troublesome threshold energy increase with cooling. Sketchy data on cooled NEA devices indicate slightly higher overall efficiency with a slightly shorter threshold [5.124]. (Because the bandgap of all NEA semiconductor materials increases with lower temperature, a higher energy threshold is expected. For small bandgap emitters, however, a reduction in the interfacial barrier energy may exist, which results in increased yield for photon energies above this slightly higher threshold energy [5.19, 124, 126, 148]; cf. Fig. 5.18.)

5.5.3 The Electron Multiplier

Although photoemissive detectors with no internal electron gain stages are available in the form of photodiodes, most state of the art devices incorporate an "electron multiplier" to boost photocathode output before it is coupled to the external electronics [5.2]. There are many advantages of electron-multiplication gain over gain by external stages, primarily in improved noise figure and faster time response (or larger signal bandwidth).

An electron multiplier consists of several cascaded gain stages or dynodes the first of which receives emitted electrons from the photocathode and the last of which is a collector-anode without gain. The mechanism of gain at each dynode is secondary electron emission [5.149, 150]. The material used for each dynode normally has lower dark current and residual noise than the cathode, limiting photomultiplier noise to either photon noise or residual cathode dark current noise. (In many external gain stage devices, the noise of the first stage is irreducibly larger than the low light level signal noise.)

The bandwidth or anode pulse rise time of an electron multiplier is limited by the spread in electron times of flight (rather than by the flight times per se [5.151]. With proper geometrical and electrostatic design the bandwidth of an electron multiplier can be an order of magnitude greater than that of a similar vacuum tube. A time lag between photon arrival and anode pulse does exist, but this does not affect anode rise time or bandwidth inasmuch as no anode signal is fed back to the input. (Even with gains larger than 10^7 at DC–100 MHz bandwidth, oscillation is not a problem.) Except for the requirements of high voltage, tube-insulating complications, and the large size of multiplier devices, the electron multiplier is a very nearly ideal low noise/high speed amplifier.

Dynode Materials

Secondary emission electron gain is obtained in those materials in which, for each incident high energy electron, several additional dynode electrons are emitted toward the next stage. Using several dynodes in series, each with a gain as low as say three, the gain of a fourteen stage multiplier becomes about 5×10^6 ($\sim 3^{14}$). Dynodes normally have upper current limits which, along with other limits due to defocussing of the traveling electron cloud, anode heating, noise, and space charge effects, combine to limit stable gain to a maximum of about 10^7. For pulsed operation where heating is not a problem, well-designed tubes can be operated at gains of 10^8, although the dynode chain voltage may have to be optimized for such requirements as described in the commercial literature [5.2, 135–137].

A dynode material is chosen primarily from the behavior of its secondary emission ratio δ versus accelerating potential, where δ is the ratio of emitted electrons to incident electrons (for 100% collection efficiency). Example yield data are shown in Fig. 5.19 for several classical compounds. (Good photoemitters are often good secondary electron emitters.) Also shown are data for the new secondary electron emitter, NEA GaP [5.152]. Although there are many materials with electron gains greater than unity, only about three have found wide application in photomultiplier tubes [5.153]. These are (CsSb), which offers very high gain but poor high temperature stability; (AgMgO), which offers nearly as high gain but better stability (with both temperature and with high

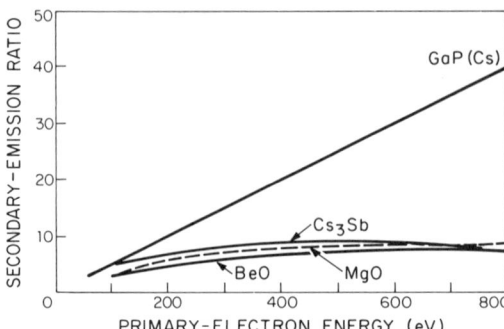

Fig. 5.19. Secondary emission ratios of several classical and one NEA electron multiplier dynode materials for a normal range of acceleration voltages [5.161]

emission density); and (CuBeO), which is similar in gain and also rugged and easily reactivated after exposure to air. (In (CuBeO), (BeO) provides the gain and Cu acts as a conducting substrate, and similarly for (MgO) and Ag in the (AgMgO) dynode.) The high gain properties of (KCsSb) have also been recently recognized [5.154], but this material has not found commercial application owing to competition from the new NEA multiplier materials. (It does occur accidentally, however, during cesiation of tubes with (KCsSb) cathodes and nominally (CsSb) dynodes.)

Certain dynode materials cannot be used in combination with certain photocathodes; thus (CuBeO) is used with the S–1 surface rather than (CsSb). Low dynode photoemission may also be required in some devices, which then prohibits use of high gain but photoemissive (CsSb).

Multiplier Structures and NEA GaP

Both the dynode element structure and the geometry of the dynode chain can take many forms, each with specific advantages [5.155]. A circular arrangement (see Fig. 5.1a) offers simplicity and small size, with little sacrifice in bandwidth or electron collection efficiency. A "Rajchman" recurrently shaped string (sketched two-dimensionally in Fig. 5.1b) offers nearly unlimited addition of stages but this focussing dynode structure is difficult to fabricate. Box-and-grid structures offer high electron-collection efficiencies but have poor response times. Venetian blind structures are compact, have stable gain with power supply fluctuations (as the box-and-grid), and are easily expanded to more stages, but the time response is again poor. (The time response and collection efficiency parameters are rather contradictory inasmuch as limiting spread in time of flight requires a focussing design in which electrons with nontypical paths are rejected at each dynode, while high collection efficiency requires collection of all electrons regardless of velocity or path and thus time of flight.)

Two additional special purpose multiplier designs are also quite common. The first is the crossed-field device [5.138] where a magnetic field is used in conjunction with a specified electrostatic field to force high-energy electrons into long paths and low-velocity electrons into compensating short paths. This minimizes time of flight variations from the photoemissive surface and produces anode rise times of about 0.1 ns, compared with 2 ns for focussed designs and 10 ns for nonfocussed, high efficiency designs. The sacrifice is in bulk, complexity, and lack of control over gain, which must remain fixed (within the limits of dynode deterioration). A second special design is the channel multiplier [5.156, 157] in which a cylinder of high resistivity electron gain material forms a continuous dynode surface. Voltage is applied along its length and multiple reflections of primary and secondary electrons traveling down the tube produce high gain and high collection efficiency. This device sacrifices speed and S/N ratio for structural simplicity. Bundles of these channels are used in imaging devices and a good portion of the image resolution can, with care, be preserved.

To minimize the total operating voltage for a desired gain in standard multipliers, each dynode should be operated near but below the "knee" in its gain-versus-voltage curve, where the gain per volt is highest (see Fig. 5.19). The number of stages is then selected to give the total gain required. For stability with power supply fluctuations, however, operation above the knee (nearer to the "saturation" plateau) is desirable; a higher voltage must be used to obtain the same gain with fewer stages, but better stability and faster response are achieved. Normally 9–14 stages with gains per dynode of 6–3 and 100 V between stages (i.e., above the knee of Fig. 5.19) are used to give 10^6–10^7 gain. As few stages as necessary and as high an operating voltage as possible should be used if time response is to be optimized.

As with NEA and classical photocathode materials, recent developments in NEA [5.128, 152, 158–160] have produced materials with vastly improved gain functions compared with classical compounds (see Fig. 5.19) [5.14, 15, 17, 18, 20, 127]. Gain of NEA GaP increases linearly with applied voltage up to several kV; for single crystals the peak gain is over 150 compared with ~10 for any classical material. With reasonable interstage voltages (500 V/stage and $\delta \simeq 30$) a gain of 10^7 can be obtained with as little as five stages [5.159, 161, 162]. In addition to the obvious simplicity and greatly improved time response (to less than 0.5 ns in focussed designs), this high gain per stage allows, for the first time, unambiguous discrimination between single- and two-electron events [5.161–164]. This results from statistics of the electron gain process [5.148, 165]. Very approximately, the electron multiplier output corresponding to one- and two-electron input events is $G \pm g$ and $2G \pm 2g$, respectively, where G is the nominal total gain and g is a deviation width that varies with δ. For δ smaller than about 10 the factor $\pm g$ is in the order of G, while for δ greater than about 20 the $\pm g$ factor is substantially smaller than G. Thus for low δ the outputs resulting from single- and multiple-electron events statistically overlap, while for high δ as found with GaP the single- and multiple-event amplitudes are well separated and can be discriminated with a pulse height counter. Much single-electron noise can be separated from the multiple-electron signal, allowing a substantial increase in S/N ratio [5.15, 135, 166].

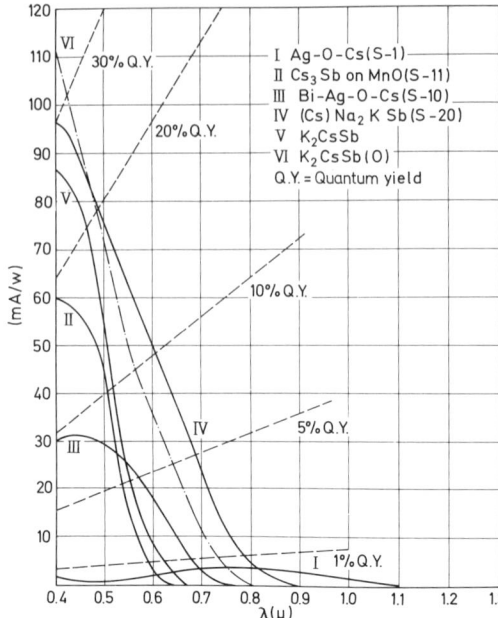

Fig. 5.20. Spectral response of representative classical photocathode surfaces [5.142]

To construct NEA GaP dynodes [5.18, 152], GaP is first deposited by vapor phase or other crystal growth process on substrates shaped to allow focussing. Owing to the high bandgap of GaP (2.3 eV), the cesiated layer required to obtain NEA is not all critical and polycrystalline dynodes can be readily activated to stable NEA across 1–5 µm wide microcrystals. Several commercial photomultipliers offering GaP-dynode options are available [5.137, 138], including devices using GaP only for the most critical first stage [5.161].

5.5.4 Spectral Response Data

Photoemitters are characterized by rather definite upper and lower energy response limits, as stressed throughout this chapter. Spectral sensitivity is commonly plotted in units of amps of cathode emission per watt of illumination. When plotting tube sensitivity versus wavelength, this scale contorts constant quantum efficiency contours into angled lines; sensitivities can be particularly ambiguous near "zero" wavelength. Data for NEA devices are more frequently found plotted directly in units of quantum yield.

To compress the information of a graph into a number, the total sensitivity in amps per lumen for light from a 2854 K tungsten light source (which has peak output at ∼9000 Å) is often specified. Although quoting this number has the advantage of simplicity, it is useless in comparing blue- with IR-sensitive cathodes. Here the use of filtered lumen source data in each spectral region is more meaningful. It should be noted that any "sensitivity number" can be

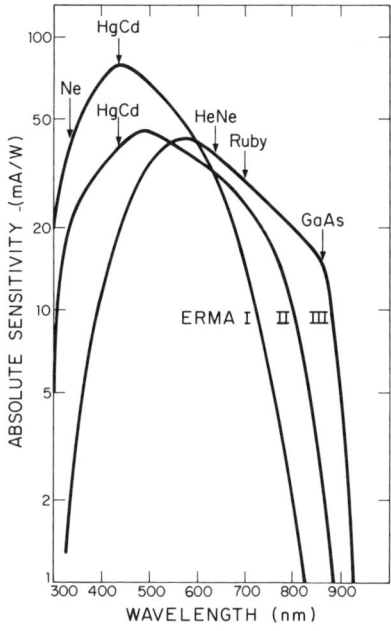

Fig. 5.21. Spectral response of extended-red variants of the S–20 photocathode. Several common laser lines are as indicated ([5.167], published courtesy RCA Corporation)

misleading unless it is used to compare devices with similar spectral response windows.

Excellent summaries of the various response curves available with commercial photomultipliers can be found in commercial sales publications and review articles [5.11] and we will not repeat such data in detail here. One response summary for classical photoemitters is given in Fig. 5.20 [5.11]. Note that the maximum yield is in the order of 0.3 electron per incident photon, with the higher efficiencies and broader response generally obtained with more complex materials. Note also that the IR-sensitive S–1 surface has by far the poorest quantum efficiency over the visible spectrum. Summaries of the general advantages and applications of each generic surface are given in [5.1, 11].

Example curves for one of the more important generic surfaces, the S–20 (NaKCsSb) and its extended red modifications, are illustrated in Fig. 5.21 [5.167]. Each variant is slightly better adapted to, for example, the detection of radiation from one or more laser devices. Similar optimized families of variants exist for several other common generic photoemissive surfaces.

The spectral response of most NEA emitters with thresholds above 1 μm are shaped similarly to one another regardless of the bandgap of the material. A typical curve is that for (cooled) NEA InGaAsP in Fig. 5.18, where a sharp threshold is followed by a flat quantum efficiency plateau. For longer wavelength threshold devices and in particular for InGaAs, the response curves show inflection points which do vary with bandgap (see Fig. 5.16). At present

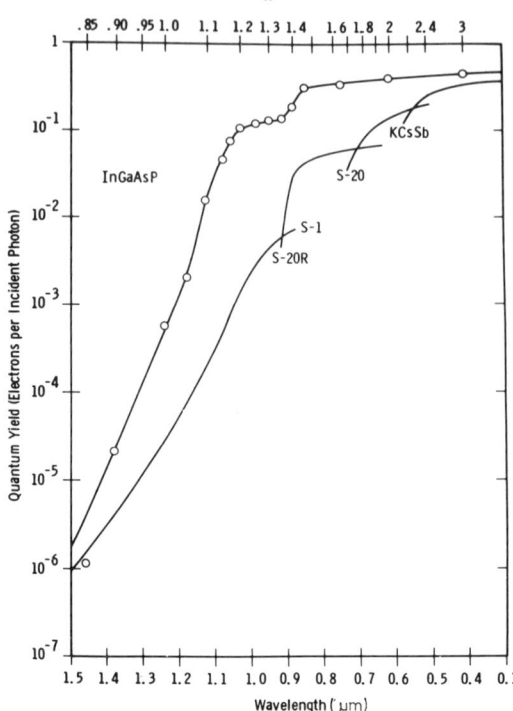

Fig. 5.22. Comparison of response at visible and near IR wavelengths of best classical and NEA InGaAsP photoemitters [5.124]

three InGaAs devices are offered commercially [5.137], each with a different In/Ga ratio to control bandgap and thus threshold. Each material has a unique response curve, which differs also from that of the one InGaAsP offering [5.138]. The latter seems to outperform InGaAs near 1.1 µm.

Generally, as stated earlier, NEA devices outperform their classical counterparts across the entire available spectral range. NEA devices provide equal or smaller dark current for a given threshold, along with higher quantum efficiency and flatter response. Sacrifices with NEA entail low maximum recommended emission, cathode size limitations, availability only in the opaque form, and, at least for the long-wavelength threshold devices, a long-term decrease in emission with time [5.120, 168; cf. 5.169]. A summary of response of NEA InGaAsP and the best classical surfaces in each wavelength range is shown in Fig. 5.22 [5.124].

NEA tubes are about a factor of five more expensive than their classical counterparts, but GaAs devices are dropping in price and amorphous GaAsP (comparable to S–20 in response) is less expensive and suffers from no lifetime problems [5.18, 125]. With increasing production quantities, both limitations (price and lifetime) will tend to disappear, although the very critical NEA surface required for 1 µm devices may make increasing operational life more problematical.

Although much discussion of the TM NEA imaging surface is given in the text, NEA imaging devices remain special-order items. Classification of performance data limits promulgation of results, and these devices will not replace S–20 and S–1 image intensifiers for several years. Any limitations on MTF and resolution, response time characteristics, and surface blemish quality all seem only technological; the remaining structure of the imaging tube (multipliers and readout mechanism) limits these parameters more than the NEA photocathode itself [5.14, 15, 116, 170]. When results for imaging devices below 1.2 eV become more openly published, new IR imaging devices will be found to outperform classical surfaces by the same orders of magnitude by which opaque devices outperform the S–1 surface. The very low dark current of room temperature 1 μm photocathodes will make high quality IR imaging a reality.

5.5.5 Specialized NEA Analysis Tools

In addition to the advancement of III–V semiconductor technology in the past ten years, other scientific advances have combined to make possible the organized development of NEA photoemission [5.21]. Of prime importance are recent developments in ultrahigh vacuum (UHV) equipment, where advances in vacuum pumps and surface-analysis equipment have transformed photocathode research from an art into a science. The most important analytical device currently in use in the laboratory is the Auger-electron analyzer, in which both surface layer and deeper, sputter-etched layers of a material are probed with a beam of focussed electrons of energy up to 5 kV [e.g., 5.98]. "Secondary" electrons emitted from the surface are collected and separated as a function of energy. With this tool in ultrahigh vacuum one can determine the chemical (element) composition of the surface and, with sufficient additional data, attempt to measure such quantities as surface-ion thickness below a monolayer. A similar device is the photoelectron energy analyzer, in which the energy distribution of electrons emitted from a monochromatically illuminated surface is measured [e.g., 5.171]. This information, along with work function data, is invaluable in attempting to model the electron barriers present at the NEA surface [e.g., 5.13, 31, 87]. Such experimental data are often cited in the literature, particularly in defense of theoretical models. Further breakthroughs in the field of photoemissive surfaces will be aided by this type of detailed investigation. Low electron energy diffraction (LEED) analysis, in which the surface crystal structure is probed (similar to bulk analysis with X-rays) is another tool. It has made possible the orderly fabrication and analysis of NEA Si [5.30], and it may provide assistance in reducing to acceptable levels the room-temperature dark emission from Si or Si/Ge. All of these tools are used, together with the quadrapole residual gas analyzer, in examining contamination in sealed-off photoemissive tubes, an effort which is required if the operational life of NEA devices is to be extended.

References

5.1 A.H.Sommer: *Photoemissive Materials* (John Wiley and Sons, New York, London, Sydney, Toronto 1968) p. 222 ff.
5.2 *RCA Photomultiplier Manual* (RCA, Electronic Components, Harrison, New Jersey 1970) p. 3 ff.
5.3 *RCA Electro-Optics Handbook* (RCA, Commercial Engineering, Harrison, New Jersey 1974), pp. 145 ff. and pp. 173 ff.
5.4 H.Sonnenberg: IEEE Solid State Circ. **SC-5**, 272 (1970)
5.5 *RCA Electro-Optics Handbook* (RCA, Commercial Engineering, Harrison, New Jersey 1974) p. 145
5.6 L.M.Biberman, S.Nudelman (eds.): *Photoelectronic Imaging Devices*, Vols. 1 and 2 (Plenum Press, New York, London 1971)
5.7 *RCA Photomultiplier Manual* (see charts within covers)
5.8 A.H.Sommer: *Photoemissive Materials* (John Wiley and Sons, New York, London, Sydney, Toronto 1968) p. 176
5.9 L.VanSpeybroeck, E.Kellogg, S.Murray: IEEE Trans. Nucl. Sci. **NS-21**, 408 (1974)
5.10 J.A.R.Samson: *Techniques of Vacuum Ultraviolet Spectroscopy* (John Wiley and Sons, New York, London, Sydney 1967) p. 212
5.11 A.H.Sommer: J. de Phys. **34**, C6–51 (1973)
5.12 D.G.Fisher, R.E.Enstrom, J.S.Escher, B.F.Williams: J. Appl. Phys. **43**, 3815 (1972)
5.13 R.L.Bell, W.E.Spicer: Proc. IEEE **58**, 1788 (1970)
5.14 D.G.Fisher, R.U.Martinelli: *Advances in Image Pickup and Display*, Vol. 1 (Academic Press, New York 1974) p. 71
5.15 R.U.Martinelli, D.G.Fisher: Proc. IEEE **62**, 1339 (1974)
5.16 N.N.Petrov: Soviet Phys.—Tech. Phys. **16**, 1965 (1972)
5.17 R.E.Simon: IEEE Spectrum **9**, 74 (1972)
5.18 A.H.Sommer: RCA Rev. **34**, 95 (1973); also published as
A.H.Sommer: *Gallium Arsenide and Related Compounds, 1972 Proceedings of the Fourth International Conference, Conference Series Number 17* (Institute of Physics, London 1973) p. 143
5.19 W.E.Spicer, R.L.Bell: Publication of the Astron. Soc. of the Pacific **84**, 110 (1972)
5.20 B.F.Williams, J.J.Tietjen: Proc. IEEE **59**, 1489 (1971)
5.21 R.L.Bell: *Negative Electron Affinity Devices* (Clarendon Press, Oxford 1973)
5.22 R.L.Bell, L.W.James, R.L.Moon: Appl. Phys. Lett. **25**, 645 (1974)
5.23 V.L.Dalal: J. Appl. Phys. **43**, 1160 (1972)
5.24 J.R.Howorth, A.L.Harmer, E.W.L.Trawny, R.Holtom, C.J.R.Sheppard: Appl. Phys. Lett. **23**, 123 (1973)
5.25 A.G.Milnes, D.L.Feucht: Appl. Phys. Lett. **19**, 383 (1971)
5.26 R.L.Bell: *Negative Electron Affinity Devices* (Clarendon Press, Oxford 1973) p. 61
5.27 A.H.Sommer: *Photoemissive Materials* (John Wiley and Sons, New York, London, Sydney, Toronto 1968) p. 54
5.28 A.H.Sommer: *Photoemissive Materials* (John Wiley and Sons, New York, London, Sydney, Toronto 1968) p. 175
5.29 A.H.Sommer: *Photoemissive Materials* (John Wiley and Sons, New York, London, Sydney, Toronto 1968) p. 53
5.30 B.Goldstein: Surface Science **35**, 227 (1973)
5.31 J.D.Levine: Surface Science **34**, 90 (1973)
5.32 C.Kittel: *Introduction to Solid State Physics* (John Wiley and Sons, New York, London 1976)
5.33 J.P.McKelvey: *Solid State and Semiconductor Physics* (Harper and Row, New York, Evanston, London 1966)
5.34 R.A.Smith: *Semiconductors* (Cambridge University Press, London, New York 1959)
5.35 S.Wang: *Solid State Electronics* (McGraw Hill, New York, San Francisco, St. Louis, Toronto, Sydney 1966)

5.36 A. H. Sommer: *Photoemissive Materials* (John Wiley and Sons, New York, London, Sydney, Toronto 1968) p. 23
5.37 A. H. Sommer: *Photoemissive Materials* (John Wiley and Sons, New York, London, Sydney, Toronto 1968) p. 91
5.38 R. L. Bell: *Negative Electron Affinity Devices* (Clarendon Press, Oxford 1973) p. 62
5.39 A. H. Sommer: *Photoemissive Materials* (John Wiley and Sons, New York, London, Sydney, Toronto 1968) p. 56
5.40 A. H. Sommer: *Photoemissive Materials* (John Wiley and Sons, New York, London, Sydney, Toronto 1968) p. 124
5.41 A. H. Sommer: *Photoemissive Materials* (John Wiley and Sons, New York, London, Sydney, Toronto 1968) p. 33
5.42 R. L. Bell: *Negative Electron Affinity Devices* (Clarendon Press, Oxford 1973) p. 20
5.43 W. E. Spicer: Phys. Rev. **112**, 114 (1958)
5.44 R. L. Bell: *Negative Electron Affinity Devices* (Clarendon Press, Oxford 1973) p. 4
5.45 E. M. Conwell: *High Field Transport in Semiconductors, Solid State Physics Supplement 9* (Academic Press, New York, London 1967)
5.46 L. W. James, G. A. Antypas, J. Edgecumbe, R. L. Moon, R. L. Bell: J. Appl. Phys. **42**, 4976 (1971)
5.47 L. W. James, J. L. Moll, W. E. Spicer: *Gallium Arsenide: 1968 Symposium Proceedings: Proceedings of the Second International Symposium, Conference Proceedings No. 7* (Institute of Physics and Physical Society, London 1969) p. 230
5.48 H. E. Ives, H. B. Briggs: J. Opt. Soc. Am. **28**, 330 (1938)
5.49 J. J. Uebbing, L. W. James: J. Appl. Phys. **41**, 4505 (1970)
5.50 R. U. Martinelli: Appl. Phys. Lett. **16**, 261 (1970)
5.51 W. E. Spicer: J. Appl. Phys. **31**, 2077 (1960)
5.52 *RCA Electro-Optics Handbook* (RCA, Commercial Engineering, Harrison, New Jersey 1974) p. 152
5.53 A. H. Sommer: *Photoemissive Materials* (John Wiley and Sons, New York, London, Sydney, Toronto 1968) p. 62 ff.
5.54 G. K. Bhide, L. M. Rangarian, B. M. Bhat: J. Phys. D, **4**, 568 (1971)
5.55 K. Deutscher, K. Hirschberg: Phys. Stat. Sol. **27**, 145 (1968)
5.56 B. R. C. Garfield, R. F. Thumwood: Brit. J. Appl. Phys. **17**, 1005 (1966)
5.57 A. H. Sommer: *Photoemissive Materials* (John Wiley and Sons, New York, London, Sydney, Toronto 1968) p. 89, 93
5.58 F. Wooten, W. E. Spicer: Surface Science **1**, 367 (1964)
5.59 A. H. Sommer: *Photoemissive Materials* (John Wiley and Sons, New York, London, Sydney, Toronto 1968) p. 73
5.60 A. H. Sommer: *Photoemissive Materials* (John Wiley and Sons, New York, London, Sydney, Toronto 1968) p. 132
5.61 A. H. Sommer: J. Appl. Phys. **42**, 567 (1971)
5.62 A. H. Sommer: *Photoemissive Materials* (John Wiley and Sons, New York, London, Sydney, Toronto 1968) p. 155 ff.
5.63 A. H. Sommer: *Photoemissive Materials* (John Wiley and Sons, New York, London, Sydney, Toronto 1968) p. 159
5.64 A. H. Sommer: *Photoemissive Materials* (John Wiley and Sons, New York, London, Sydney, Toronto 1968) p. 152
5.65 A. F. Milton, A. D. Baer: J. Appl. Phys. **42**, 5095 (1971)
5.66 J. J. Scheer, J. van Laar: Solid State Commun. **3**, 189 (1965)
5.67 J. van Laar: Acta Electronica **6**, 215 (1973)
5.68 J. J. Scheer, J. van Laar: Solid State Commun. **5**, 303 (1967)
5.69 R. L. Bell, L. W. James, G. A. Antypas, J. Edgecumbe, R. L. Moon: Appl. Phys. Lett. **19**, 513 (1971)
5.70 J. J. Scheer, J. van Laar: Surface Science **18**, 130 (1969)
5.71 S. Garbe, G. Frank: *Gallium Arsenide and Related Compounds, 1970: Proceedings of the Third International Symposium, Conference Series Number 9* (Institute of Physics, London 1971) p. 208

5.72 R.U. Martinelli: J. Appl. Phys. **44**, 2566 (1973)
5.73 H. Sonnenberg: Appl. Phys. Lett. **14**, 289 (1969)
5.74 J.J. Uebbing, R.L. Bell: Proc. IEEE **56**, 1624 (1968)
5.75 A.A. Turnbull, G.B. Evans: Brit. J. Appl. Phys. (J. Phys. D) Series 2 **1**, 155 (1968)
5.76 A.H. Sommer, H.H. Whitaker, B.F. Williams: Appl. Phys. Lett. **17**, 273 (1970)
5.77 A.H. Sommer: J. Appl. Phys. **42**, 2158 (1971)
5.78 H. Sonnenberg: Appl. Phys. Lett. **21**, 278 (1972)
5.79 H. Sonnenberg: Appl. Phys. Lett. **19**, 431 (1971)
5.80 H. Sonnenberg: Appl. Phys. Lett. **21**, 103 (1972)
5.81 L.W. James, J.J. Uebbing: Appl. Phys. Lett. **16**, 370 (1970)
5.82 L.W. James, G.A. Antypas, J.J. Uebbing, J. Edgecumbe, R.L. Bell: *Gallium Arsenide and Related Compounds, 1970: Proceedings of the Third International Symposium, Conference Series Number 9* (Institute of Physics, London 1971) p. 195
5.83 L.W. James, G.A. Antypas, J.J. Uebbing, T.O. Yep, R.L. Bell: J. Appl. Phys. **42**, 580 (1971)
5.84 H. Sonnenberg: Appl. Phys. Lett. **16**, 245 (1970)
5.85 J.M. Chen: Surface Science **25**, 457 (1971)
5.86 J.J. Uebbing, R.L. Bell: Appl. Phys. Lett. **11**, 357 (1967)
5.87 J.S. Escher, H. Schade: J. Appl. Phys. **44**, 5309 (1973)
5.88 D.G. Fisher, R.E. Enstrom, B.F. Williams: Appl. Phys. Lett. **18**, 371 (1971)
5.89 G.A. Allen: Acta Electronica **16**, 229 (1973)
5.90 S. Garbe: Solid-State Electron. **12**, 893 (1969)
5.91 R.L. Bell: *Negative Electron Affinity Devices* (Clarendon Press, Oxford 1973) p. 65
5.92 G.A. Allen: J. Phys. D.: Appl. Phys. **4**, 308 (1971)
5.93 G.A. Antypas, L.W. James, J.J. Uebbing: J. Appl. Phys. **41**, 2888 (1970)
5.94 D.G. Fisher, R.E. Enstrom, J.S. Escher, H.F. Gossenberger, J.R. Appert: IEEE Trans. Electron Devices **ED-21**, 641 (1974)
5.95 L.W. James: J. Appl. Phys. **45**, 1326 (1974)
5.96 Y.Z. Liu, J.L. Moll, W.E. Spicer: Appl. Phys. Lett. **17**, 60 (1970)
5.97 R.L. Bell: *Negative Electron Affinity Devices* (Clarendon Press, Oxford 1973) p. 75
5.98 J.J. Uebbing: J. Appl. Phys. **41**, 802 (1970)
5.99 H. Schade, H. Nelson, H. Kressel: Appl. Phys. Lett. **18**, 121 (1971)
5.100 G.A. Antypas, J. Edgecumbe: Appl. Phys. Lett. **26**, 371 (1975)
5.101 D. Andrew, J.P. Gowers, J.A. Henderson, M.J. Plummer, B.J. Stocker, A.A. Turnbull: J. Phys. D.: Appl. Phys. **3**, 320 (1970)
5.102 S.B. Hyder: J. Vacuum Science Tech. **8**, 228 (1971)
5.103 R.E. Enstrom, D.G. Fisher: J. Appl. Phys. **46**, 1976 (1975)
5.104 Y.Z. Liu, C.D. Hollish, W.W. Stein, D.E. Bolger, P.D. Greene: J. Appl. Phys. **44**, 5619 (1973)
5.105 G.A. Antypas, L.W. James: J. Appl. Phys. **41**, 2165 (1970)
5.106 G.A. Antypas, T.O. Yep: J. Appl. Phys. **42**, 3201 (1971)
5.107 G.A. Antypas, R.L. Moon, L.W. James, J. Edgecumbe, R.L. Bell: *Gallium Arsenide and Related Compounds, 1972 Proceedings of the Fourth International Conference, Conference Series Number 17* (Institute of Physics, London 1973) p. 48
5.108 G.A. Antypas, R.L. Moon: J. Electrochem. Soc. **120**, 1574 (1973)
5.109 R.L. Moon, G.A. Antypas, L.W. James: J. Electron. Mater. **3**, 635 (1974)
5.110 M.B. Allenson, P.G.R. King, M.C. Rowland, G.J. Steward, C.H.A. Syms: J. Phys. D.: Appl. Phys. **5**, L89 (1972)
5.111 G. Frank, S. Garbe: Acta Electronica **16**, 237 (1973)
5.112 W.A. Gutierrez, H.L. Wilson, E.M. Yee: Appl. Phys. Lett. **25**, 482 (1974)
5.113 T.G.J. van Oirschot: Appl. Phys. Lett. **24**, 211 (1974)
5.114 T.G.J. van Oirschot, G.A. Acket, W.J. Bartels: J. Appl. Phys. **46**, 1893 (1975)
5.115 W.A. Gutierrez, H.D. Pommerrenig: Appl. Phys. Lett. **22**, 292 (1973)
5.116 F.R. Hughes, E.D. Savoye, D.L. Thoman: J. Electron. Mater. **3**, 9 (1974)
5.117 R.L. Bell, J.J. Uebbing: Appl. Phys. Lett. **12**, 76 (1968)
5.118 D.L. Schaefer: J. Appl. Phys. **40**, 445 (1969)

5.119 R.E.Enstrom, D.Richman, M.S.Abrahams, J.R.Appert, D.G.Fisher, A.H.Sommer, B.F.Williams: *Gallium Arsenide and Related Compounds, 1970: Proceedings of the Third International Symposium, Conference Series Number 9* (Institute of Physics, London 1971) p. 30
5.120 D.G.Fisher: IEEE Trans. Electron Devices **ED-21**, 541 (1974)
5.121 D.A.Jackson, E.M.Yee: Proc. IEEE **59**, 90 (1971)
5.122 W.Klein: J. Appl. Phys. **40**, 4384 (1969)
5.123 J.S.Escher, G.A.Antypas, J.Edgecumbe: Appl. Phys. Lett. **29**, 153 (1976)
5.124 L.W.James, G.A.Antypas, R.L.Moon, J.Edgecumbe, R.L.Bell: Appl. Phys. Lett. **22**, 270 (1973)
5.125 R.E.Simon, A.H.Sommer, J.J.Tietjen, B.F.Williams: Appl. Phys. Lett. **15**, 43 (1969)
5.126 R.U.Martinelli: J. Appl. Phys. **45**, 1183 (1974)
5.127 B.F.Williams, R.U.Martinelli, E.S.Kohn: Advan. Electron Electron Phys. **33A**, 447 (1972)
5.128 R.U.Martinelli: Appl. Phys. Lett. **17**, 313 (1970)
5.129 E.S.Kohn: IEEE Trans. Electron Devices **ED-20**, 321 (1973)
5.130 H.Schade, H.Nelson, H.Kressel: Appl. Phys. Lett. **18**, 413 (1971)
5.131 H.Schade, H.Nelson, H. Kressel: Appl. Phys. Lett. **20**, 385 (1972)
5.132 B.F.Williams, R.E.Simon: Appl. Phys. Lett. **14**, 214 (1969)
5.133 *RCA Photomultiplier Manual* (RCA, Electronic Components, Harrison, New Jersey 1970) p. 30
5.134 W.E.Spicer, F.Wooten: Proc. IEEE **51**, 1119 (1963)
5.135 *EMI Photomultiplier Tubes*, Containing *An Introduction to the Photomultiplier*, available from local representative (Gencom Division, Varian EMI, Plainview, New York, in USA)
5.136 *EMR Photoelectric Multiplier Phototubes*, from EMR Photoelectric, Princeton, New Jersey
5.137 *RCA Photomultiplier Tubes*, from local representative or RCA Electronic Components, Harrison, New Jersey
5.138 Varian LSE "Excellence in Photo Detection", from Varian LSE, Palo Alto, CA
5.139 R.L.Bell: Solid-State Electron. **12**, 475 (1969)
5.140 R.L.Bell: Solid-State Electron. **13**, 397 (1970)
5.141 *RCA Photomultiplier Manual* (RCA, Electronic Components, Harrison, New Jersey 1970) p. 40
5.142 A.H.Sommer: *Photoemissive Materials* (John Wiley and Sons, New York, London, Sydney, Toronto 1968) pp. 68 and 229
5.143 M.Cole, D.Ryder: Electro Optical System Design **4**, 16 (1972)
5.144 R.B.Murray, J.J.Manning: IRE Trans. Nuc. Sci. **NS-7**, 80 (1960)
5.145 *RCA Photomultiplier Manual* (RCA, Electronic Components, Harrison, New Jersey 1970) p. 31
5.146 A.H.Sommer: *Photoemissive Materials* (John Wiley and Sons, New York, London, Sydney, Toronto 1968) p. 126
5.147 A.H.Sommer: *Photoemissive Materials* (John Wiley and Sons, New York, London, Sydney, Toronto 1968) p. 68
5.148 B.F.Williams: IEEE Trans. Nuc. Sci. **NS-19**, 39 (1972)
5.149 *RCA Photomultiplier Manual* (RCA, Electronic Components, Harrison, New Jersey 1970) p. 13
5.150 R.E.Simon, B.F.Williams: IEEE Trans. Nucl. Sci. **NS-15**, 167 (1968)
5.151 *RCA Photomultiplier Manual* (RCA, Electronic Components, Harrison, New Jersey 1970) pp. 25 and 48
5.152 R.E.Simon, A.H.Sommer, J.J.Tietjen, B.F.Williams: Appl. Phys. Lett. **13**, 355 (1968)
5.153 *RCA Phototubes and Photocells* (RCA, Lancaster, PA 1963; out of print) p. 38
5.154 A.H.Sommer: J. Appl. Phys. **43**, 2479 (1972)
5.155 *RCA Photomultiplier Manual* (RCA, Electronic Components, Harrison, New Jersey 1970) p. 19
5.156 G.W.Goodrich, W.C.Wiley: Rev. Sci. Instr. **33**, 761 (1962)
5.157 W.C.Wiley, C.F.Hendee: IEEE Trans. Nucl. Sci. **NS-9**, 103 (1962)
5.158 R.U.Martinelli, M.L.Schultz, H.F.Gossenberger: J. Appl. Phys. **43**, 4803 (1972)
5.159 R.U.Martinelli: J. Appl. Phys. **45**, 3203 (1974)

5.160 R.U.Martinelli, M.Ettenberg: J. Appl. Phys. **45**, 3896 (1974)
5.161 H.R.Krall, F.A.Helvy, D.E.Persyk: IEEE Trans. Nucl. Sci. **NS-17**, 71 (1970)
5.162 D.E.Persyk, D.D.Crawshaw: RCA Rev. **34**, 344 (1973)
5.163 H.R.Krall, D.E.Persyk: IEEE Trans. Nuc. Sci. **NS-19**, 45 (1972)
5.164 B.Leskovar, C.C.Lo: IEEE Trans. Nuc. Sci. **NS-19**, 50 (1972)
5.165 G.A.Morton: Appl. Opt. **7**, 1 (1968)
5.166 *RCA Photomultiplier Manual* (RCA, Electronic Components, Harrison, New Jersey 1970) p. 61
5.167 *RCA "ERMA" Photodetector, Application Note AN–4637* (RCA Electronic Components, Harrison, New Jersey 1971)
5.168 E.M.Yee, D.A.Jackson,Jr.: Solid-State Electron. **15**, 245 (1972)
5.169 A.H.Sommer: Appl. Opt. **12**, 90 (1973)
5.170 R.L.Bell: *Negative Electron Affinity Devices* (Clarendon Press, Oxford 1973) p. 111
5.171 T.E.Fischer: Surface Science **13**, 30 (1969)

6. Charge Transfer Devices for Infrared Imaging

A. F. Milton

With 18 Figures

6.1 Historical

A wide variety of infrared photodetectors has been developed for use in thermal imaging systems operating in either the 3–5 µm or 8–12 µm atmospheric windows. The detectors of most interest include photovoltaic InSb, photoconductive $Hg_{1-x}Cd_xTe$, photovoltaic $Pb_xSn_{1-x}Te$ and various extrinsic photoconductors fabricated by incorporating deep dopants into Ge and Si. Quantum efficiencies for the intrinsic photodetectors are typically above 50%. With sufficient cooling and terrestrial backgrounds, most of these detectors combined with low noise preamplifiers can provide background limited performance (BLIP) for the signal frequencies of interest to thermal imaging systems which use a linear array of detectors in a parallel scan mode (see Fig. 6.1a). Some of them (notably photoconductive $Hg_{1-x}Cd_xTe$) are also suitable for 8–12 µm serial scan systems which use time delay and integration (TDI) (see Fig. 6.1b).

If BLIP operation is obtained in the chosen spectral region, the theoretical limit to the sensitivity of an infrared focal plane becomes proportional to the square root of the product of the quantum efficiency of the detectors η and the focal plane filling efficiency, defined as the ratio of the total photoactive area of the detector array to the total area in the focal plane of the image being scanned. With a mechanically scanned imaging system using N detectors to scan a total of N_{TOT} resolution elements, the focal plane filling efficiency is approximately N/N_{TOT}. This efficiency stands near 10^{-3} for present technology infrared imaging systems which use up to several hundred detectors.

As mentioned previously, modern photodetectors provide quantum efficiency η approaching unity so that for terrestrial applications the only way to significantly increase focal plane sensitivity without sacrificing resolution is to increase the number of IR detectors used by the imaging system. At present the number of detectors used in IR imaging systems is limited by the requirement to connect each detector to its own preamplifier. It is hoped that the use of charge transfer devices (principally charge coupled devices, CCDs, and charge injection devices, CIDs) at or near the focal plane for multiplexing and signal processing will make practical IR focal planes with thousands of IR detectors and focal plane filling efficiencies which approach 0.1. This would result in a tenfold decrease in the apparent minimum resolvable temperature difference over that obtainable with current technology. For terrestrial thermal imaging applications

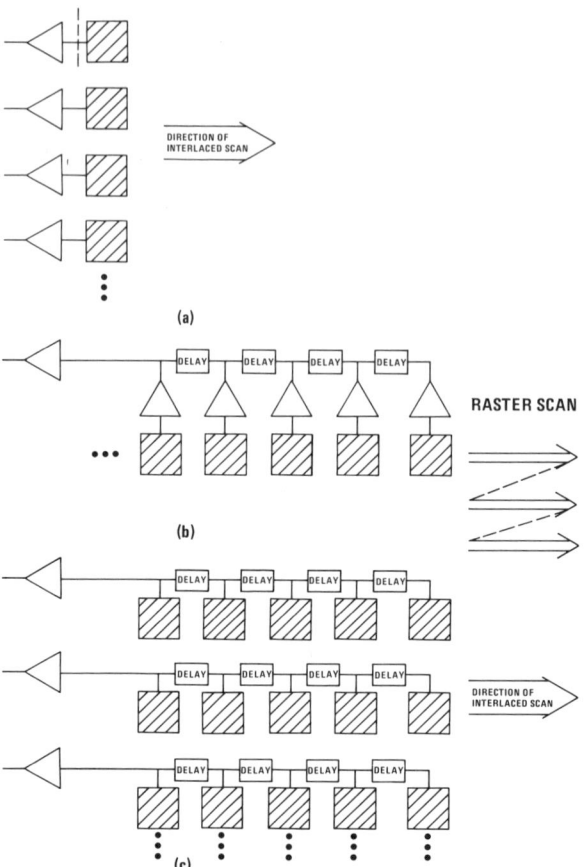

Fig. 6.1a–c. Different approaches to image scanning with an array of detectors. (a) Parallel scan with a linear array. (b) Serial scan with a linear array using time delay and integration (TDI). (c) Series/parallel scan with a two-dimensional array using TDI

the signal consists of small temperature or emissivity variations. Equivalent temperature differences at the target are often as small as 1 K which leads to a 4% contrast in the 3–5 µm window and a 1.6% contrast in the 8–12 µm window. These contrasts at the target are reduced by atmospheric extinction between the target and the imager.

CCD and CID technology in silicon has already been used to form two-dimensional area arrays for the visible and very near infrared; however, the low contrasts typical of thermal imagery preclude the straightforward adaptation of these electronically scanned dc-coupled staring arrays to the IR, even though they have the advantage that their focal plane filling efficiency can be over 50%. In a low contrast situation pattern noise becomes dominant. Any dc-coupled staring sensor has difficulty distinguishing between variations in detector responsivity and true variations in the image. If the detectors are dc-coupled to the image and measure the absolute irradiance at each detector site, then responsivity variations acting on the large IR uniform background will mask the true image. On the other hand, if the image is mechanically scanned over the focal plane and the detectors are ac coupled to the image, the uniform background will not produce any pattern noise even if detector responsivity variations exist. This is also true for an ac-coupled staring focal plane which can only serve as a moving target indicator.

Net detector to detector response uniformities which would be necessary to provide a competitive dc-coupled staring imager in the IR are beyond the state of the art. The responsivity uniformity requirement for performance competitive with a system using mechanical scanning is, of course, more severe for a staring system designed for the 8–12 μm window since thermal contrasts are lower in that spectral region. However even in the 3–5 μm region, net response uniformities of a few tenths of a percent would still be needed to make it practical to do away with mechanical scanning in high performance imaging systems. Even if detector material uniformity could be achieved, 1 μm sized errors in the photolithographic definition of the active areas will keep detector cell response nonuniformities above 1 %.

Frame to frame subtraction could be used with a staring sensor to improve net response uniformity; however the main thrust so far for thermal imaging has been to develop two-dimensional IR detector modules for use in a series-parallel scan, time delay and integration mode (see Fig. 6.1c). As we shall see, CCDs are particularly well suited to perform the TDI function. The principal advantage of this approach is that the equivalent of ac coupling and responsivity averaging can be provided. This relaxes responsivity uniformity requirements. As an additional bonus the TDI function provides built-in redundancy. Another advantage of the series-parallel scan approach is that an evolutionary growth in the number of detectors used for a particular application is possible since the number of detectors used to cover a given number of resolution elements can be varied according to focal plane design.

The 8–12 μm window has proven itself superior to the 3–5 μm window for thermal imaging with present technology; nevertheless for advanced systems with increased sensitivity there is interest in both IR windows. With increased imager sensitivity, operating ranges will be longer so that atmospheric transmission and the diffraction limited modulation transfer function (MTF) will be more important in comparison to the amount of signal generated by a temperature difference at the target. Achieving background limited performance in the 3–5 μm region for scan frequencies of interest does, however, represent a significant challenge.

There exists a wide variety of approaches to the use of charge transfer devices in infrared focal planes. We shall discuss five high packing density, high quantum efficiency, approaches appropriate for series-parallel scan: 1) IR sensitive CCD, 2) direct injection: hybrid, 3) direct injection: extrinsic silicon, 4) accumulation mode: extrinsic silicon, and 5) infrared sensitive CID with silicon CCD signal processing. The reader is referred to a review article by *Steckl* et al. for a comprehensive discussion of a number of other approaches not discussed here which include indirect injection pyroelectric detectors and Schottky barrier photoemissive injection [6.1]. Three approaches in our list of five do not require charge transfer devices in any material other than silicon; however the IR sensitive CCD and CID require the development of MIS technology in an infrared sensitive material. Although the technique used to input photosignal into the CCD register varies, all the approaches we shall discuss use CCD action to perform time delay and integration. It is appropriate at this point to discuss basic charge transfer device operation with a view to understanding how MIS system parameters affect performance.

6.2 Charge Coupled Devices

6.2.1 Basic Operating Principles

Since the invention of charge coupled devices by *Boyle* and *Smith* [6.2] in 1970, a remarkable variety of different structures has been investigated in silicon using SiO_2 as the insulating layer. A number of excellent reviews of silicon CCD technology exists [6.3, 4] and only a summary of the basic operating principles will be given here. A charge coupled device operating in the inversion mode is a metal-oxide-semiconductor (MOS) or a metal-insulator-semiconductor (MIS) structure which can collect, store and transfer localized packets of minority carrier charge along the semiconductor-insulator interface.

Voltages applied to the gate electrodes are used to control the surface potential ϕ_s under the oxide and to form localized potential wells which can store minority charge. Operation of the basic three-phase device is shown in Fig. 6.2. The relationship between ϕ_s and the applied gate voltage V_G can be understood for surface channel devices by reference to Fig. 6.3. The potential wells are operated in a nonequilibrium condition, i.e., charge is removed from the well before the condition depicted by Figs. 6.3d or 6.3h is reached.

An elementary electrostatic analysis using the depletion region approximation leads to a relationship between ϕ_s, the applied gate voltage V_G and the surface density of mobile minority carriers N (for an n-channel device):

$$\phi_s = V_G - V_{FB} - \frac{WN_A e}{C_{ox}} - \frac{eN}{C_{ox}} \qquad (6.1)$$

where the C_{ox} is the oxide capacitance per unit area, W is the depletion layer width, e is the electronic charge, N_A is the acceptor density, and

$$W = \left(\frac{2\varepsilon_s \phi_s}{eN_A}\right)^{1/2} \qquad (6.2)$$

where ε_s is the dielectric constant of the semiconductor. V_{FB} is the flat band voltage:

$$V_{FB} = \phi_{ms} - Q_{ss}/C_{ox}. \qquad (6.3)$$

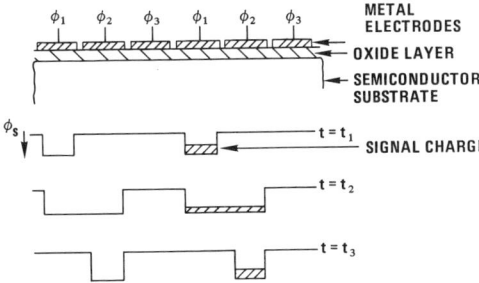

Fig. 6.2. The basic MIS CCD structure illustrating charge transfer with three-phase clock voltage operation. In silicon devices overlapping polysilicon electrode gates are often used

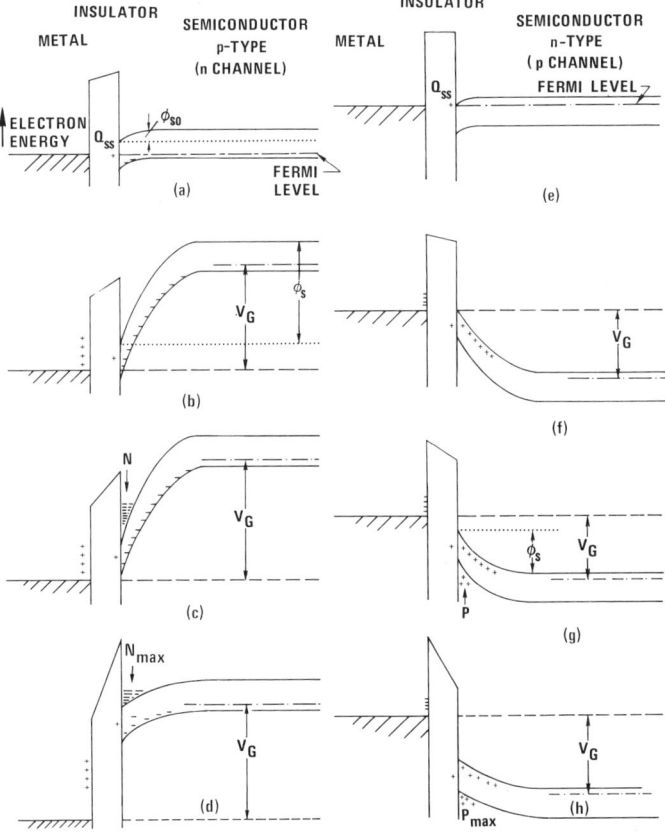

Fig. 6.3a–h. Electron energy as a function of distance from the surface for an n channel MIS structure (a, b, c, d) and for a p channel MIS structure (e, f, g, h). These graphs for surface channel structures illustrate how the band bending changes after a gate voltage, V_G is applied and charge builds up at the surface until the equilibrium condition depicted by (d) and (h) is approached

METAL INSULATOR SEMICONDUCTOR

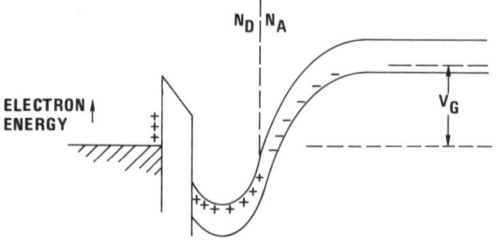

Fig. 6.4. Electron energy as function of distance from the surface for a buried channel MIS structure. The $p-n$ junction is formed by ion implantation. Notice that the potential minimum is not at the surface

Here ϕ_{ms} is the metal semiconductor work function difference and Q_{ss} is the charge density trapped in interface states.

The capacitance (per unit area) from gate electrode to substrate C_G is the series combination of the oxide capacitance C_{ox} and the depletion capacitance and is given by

$$C_G = C_{ox} \frac{1}{1 + C_{ox} W/\varepsilon_s}. \tag{6.4}$$

C_G thus increases as the well fills up and W decreases. Before the well fills up, if $(2e\varepsilon_s N_A)^{1/2}/C_{ox} < (V_G)$, then the oxide capacitance will be much larger than the depletion layer capacitance and ϕ_s will track V_G. For this reason lightly doped material and thin gate oxide layers (~ 1000 Å) are needed. Examination of (6.1) also shows that barriers can be permanently built into the device by varying N_A and/or by changing the oxide thickness (which changes C_{ox}). These techniques are used to form channel stops to define the CCD channel and can also be used to provide the built-in barriers necessary for two-phase CCD operation.

p-type substrates can be used to make n-channel devices and n-type substrates can be used to make p-channel devices. n-channel devices are usually preferred due to higher electron mobilities; however they have not been fabricated in any material other than silicon. For the usual CCD device structures it is necessary that the semiconductor surface not be inverted when $V_G = 0$. Since Q_{ss} is always positive, a low surface state density is required to achieve this condition for an n-channel device. With an n-channel device, to avoid inversion with $V_G = 0$ we must have

$$Q_{ss} < 2C_{ox}\phi_F - (4eN_A\phi_F\varepsilon_s)^{1/2} + \phi_{ms}C_{ox} \tag{6.5}$$

where $2\phi_F = (2kT/e)\ln N_A/n_i$; n_i is the intrinsic carrier density. Usually the last two terms can be neglected so that for an n-channel device we need an interface charge density less than 3×10^{-8} coulombs/cm^2 assuming an SiO$_2$ gate oxide thickness d of 1000 Å, which means a density of elementary positive charges less than 2×10^{11} cm^{-2}. With a p-channel device a higher positive interface charge density only drives the surface into accumulation when no bias is applied.

Buried channel CCDs are fabricated using ion implantation to form a $p-n$ junction several thousand Ångströms below the surface. With a structure such as the one shown in Fig. 6.4, the potential minimum is not quite at the surface so that the minority carriers do not come into contact with surface states. This has a number of advantages associated with more complete charge transfer and lower noise. A relationship similar to (6.1) can be developed for buried channel CCDs [6.3].

6.2.2 Limitations

Signal Handling Capabilities

When a well is full, the surface potential is such that the surface would be inverted under equilibrium conditions, i.e., for a full well $\phi_s = 2\phi_F = (2kT/e) \ln N_A/n_i$. Using (6.1) and the fact that V_{FB}, $2\phi_F$ and $(4e\varepsilon_s N_A \phi_F)^{1/2}/C_{ox}$ are usually much smaller than V_G, the full well surface density N_{max} is just

$$N_{max} = C_{ox} V_G / e = \frac{\varepsilon_{ox} V_G}{ed}. \tag{6.6}$$

ε_{ox} is the dielectric constant of the insulating layer. Thus the amount of charge that can be stored on the surface is related to the maximum electric field for the oxide (V_G/d), which for SiO_2 ($\varepsilon_{ox}/\varepsilon_0 = 3.9$) is in the range of 1×10^6 V/cm. With this value for V_G/d

$$N_{max} = 2 \times 10^{12} \text{ electrons/cm}^2. \tag{6.7}$$

A typical CCD well of dimension $10\,\mu m \times 20\,\mu m$ thus has a storage capacity of 4×10^6 electrons. Buried channel CCDs have a storage capacity a factor of two to three smaller than this [6.3]. For $d = 1000\,\text{Å}$ the oxide capacitance of the $10\,\mu m \times 20\,\mu m$ well would be 0.07 pF. In materials other than silicon where the impurity dopant density is larger (leading to smaller depletion widths) and the bandgap is smaller the limit on V_G can be set by avalanche breakdown in the semiconductor itself or tunneling.

Charge-Up Time (Dark Current)

To avoid saturation effects, charge must be removed from the potential wells before $N \to N_{max}$ and $\phi_s \to 2\phi_F$. In the absence of charge being transferred into the cell by CCD action there will still be a gradual accumulation of minority carriers at the surface due to thermal generation and background generation (if the CCD cells are being used as sensor elements).

$$\frac{dN}{dt} = \frac{n_i W}{2\tau} + \frac{n_i S_0}{2} + \frac{n_i^2 L_D}{N_A \tau} + \eta J_b \tag{6.8}$$

where the first term on the right represents depletion region generation, the second term interface state generation, the third term bulk generation, and the last term background generation.

The depletion region generation will decrease as W decreases so that (6.8) can be written as

$$\frac{dN}{dt} = \frac{n_i}{2\tau}\left(\frac{2\varepsilon_s\phi_s}{eN_A}\right)^{1/2} + J_x \qquad (6.9)$$

where J_x is constant as a function of time.

If we assume that $\phi_s = \phi_{s0}$ at $t=0$ and $\phi_s = 2\phi_F$ at the charge up time $t=\tau_c$ then (6.9) leads to

$$\tau_c = \frac{C_{ox}(\phi_{s0} - 2\phi_F)}{eJ_x} \qquad (6.10)$$

for

$$J_x \gg \frac{n_i}{2\tau}\left(\frac{2\varepsilon_s\phi_{s0}}{eN_A}\right)^{1/2}. \qquad (6.11)$$

Cooling can reduce n_i and ensure that background generation (ηJ_b) dominates J_x, which is the desired operating condition for an infrared sensitive CCD or CID. If the CCD or CID operates as the infrared detector this is a necessary but not sufficient condition for BLIP operation. In any case, excess thermal dark current will cause unnecessary shot noise and can also cause pattern noise if the thermal generation is nonuniform. The necessary operating temperatures to avoid these problems for an IR sensitive CCD or CID are very similar to those that would be necessary to ensure background limited operation with a back-biased photodiode made of the same IR sensitive material. The generation terms are similar except for the influence of the Fermi level position on surface state generation.

Transfer Inefficiency

If CCD action is used to transfer signal charge originating from an image, it is essential that high transfer efficiency be maintained. If any charge in an individual cell is removed or left behind after a clocking sequence, a distortion of the image will occur upon passing the signal through the CCD register. With significant transfer inefficiency, point images will be broadened and have their position shifted; sinusoidal images will experience a phase shift and have their amplitude reduced compared to the uniform background. In other words, the passage of the signal information through an imperfect CCD register can result in a degradation of the modulation transfer function (MTF) of the system.

The effects of transfer inefficiency can be fairly well described in terms of a transfer inefficiency parameter ε defined by reference to an isolated charge packet. Every time a signal packet containing minority charge Q_s is moved over one CCD cell, charge εQ_s is left behind. For a dc-coupled staring sensor where the image is read out by transferring charge along a CCD register, the MTF

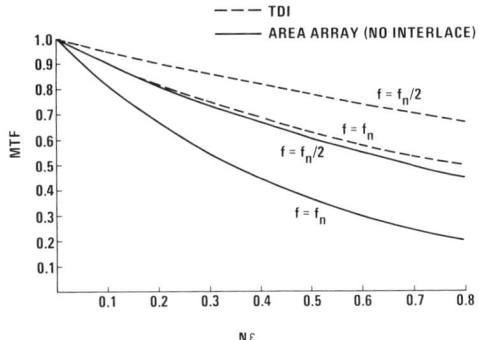

Fig. 6.5. Modulation transfer function (MTF) as a function of $N\varepsilon$ where N is the total number of elementary transfers and ε is the transfer inefficiency per elementary transfer. The curves are drawn for different spatial frequencies [f_n is the Nyquist frequency ($f_c/2$)] and different array configurations. The solid curves are for a staring array and the dashed curves for an array which uses series-parallel scan with TDI

degradation along the direction of transfer due to a finite value of ε_s can be calculated to be

$$\mathrm{MTF} = \exp\left[-n\varepsilon\left(1 - \cos\pi\frac{f}{f_n}\right)\right] \tag{6.12}$$

where n is the number of elementary transfers in the readout process and f/f_n is the ratio of the spatial frequency in the image under consideration to the Nyquist frequency of the CCD transfer register itself (in the time domain $f_n = f_c/2$ where f_c is the clock frequency) [6.5]. This expression is plotted for $f/f_n = 1/2$ and $f = f_n$ in Fig. 6.5. The expression must be modified for an imager which uses interlace since in that case the spatial frequency carried by the CCD register is not always equivalent to the spatial frequency in the image.

Since generally an imager which uses a CCD readout must operate with a value of $n\varepsilon \leq 0.1$, transfer inefficiencies on the order of 10^{-4} are required for the conventional staring CCD imager which has TV resolution. CID operation, on the other hand, operates with only a single transfer so such low values of ε are not required. With both CCD and CID operation MTF in all directions is of course influenced by the integration and sampling functions of the matrix of detection cells, and for some focal plane designs is also degraded by diffusion in the light sensitive material.

At low clocking frequencies CCD transfer inefficiency is independent of clock frequency and is caused by trapping at interface states (for a surface channel device) whereas at higher clocking frequencies, the inefficiency will increase with clocking frequency due to the limitations of free charge transfer. The effect of interface state trapping on pulses in a CCD register is very complex due to the fact that the interface states fill nearly instantaneously upon being exposed to minority carrier charge and emit more slowly. If we assume an energy independent interface state density N_{ss} (in states per unit area per unit energy) and a constant capture cross section σ, then the mean number of carriers emitted in time t per unit area from a set of traps initially full at time $t=0$ can be written as [6.6]

$$N(t) = kTN_{ss}\ln(\sigma V_{th}N_c t) \tag{6.13}$$

where V_{th} is the mean thermal velocity of the charge carrier and N_c is the density of states in the band under consideration. This expression is valid for $(\sigma V_{th} N_c)^{-1} \ll t \ll \tau_{mb}$ where τ_{mb} is the emission time constant of the interface states ($\sim 10^{-3}$ s for silicon). The amount of charge carriers per unit area lost to the first one following an isolated zero for a single transfer considering reemission into the one during a time $1/(pf_c)$ is just

$$N_e = N\left(\frac{2}{f_c} - \frac{1}{pf_c}\right) - N\left(\frac{1}{pf_c}\right) = kTN_{ss}\ln(2p-1) \tag{6.14}$$

where p is the number of phases in the clocking process. For small losses, the transfer inefficiency can thus approximately be expressed as

$$\varepsilon = \frac{eN_e A}{Q_s} = \frac{eA}{Q_s} kTN_{ss}\ln(2p-1) \tag{6.15}$$

where A is the surface area which can reemit into a single well.

Clearly the effect of interface states is more significant when the amount of charge being transferred is small; however, even with a large signal (1×10^{12} el/cm^2), N_{ss} would have to be less than 1×10^{10} eV cm^2 for $\varepsilon = 10^{-4}$. The use of a constant background charge or "fat zero" can help, however. For one thing the fat zero ensures that the wells are never close to being empty [ensuring a reasonably sized Q_s in (6.15)]. A fat zero charge density Q_{s0} greater than the interface density ($Q_{s0}/e > N_{ss}kT$) is required to reduce the influence of interface states by keeping them filled. Unfortunately since the area occupied by the charge in a well is somewhat a function of the amount of charge in the well, the fat zero cannot fill all the states that will be exposed to the signal charge and a certain fraction of the cell interface area will still contribute to a trapping inefficiency [6.7]. Wider CCD channels will evidence less loss due to this "edge effect".

The use of a fat zero will help transfer efficiency; however the electrical introduction of the fat zero can introduce an additional source of temporal and pattern noise depending upon the input circuit. With infrared focal planes the fat zero is often automatically provided by the uniform background illumination. The low contrast situation thus automatically reduces the influence of surface interface states on transfer inefficiency. The transfer efficiency of buried channel CCDs is not influenced by trapping at interface states. Trapping at bulk defects plays the same role although the density of bulk traps is usually smaller than typical interface state densities [6.3].

At higher clock frequencies (typically over several MHz with silicon devices), other processes limit the transfer efficiencies. Overlapping electrodes are often used to eliminate barriers and pockets between wells; however even with the elimination of these spurious effects, at high clock frequencies there is not enough time for all the charge to transfer under the influence of self-induced drift, thermal diffusion and fringing fields. For large charge packets the transfer is dominated by self-induced drift caused by the electrostatic repulsion of the carriers. For times during which the shape of the carrier distribution remains stationary the decay can be described by

$$\frac{Q(t)}{Q_0} = \frac{\tau_0}{t+\tau_0} \tag{6.16}$$

where Q_0 is the original amount of charge in the packet and

$$\tau_0 = (2/\pi)\frac{L^2}{\mu}\left(\frac{LWC_{ox}}{Q_0}\right). \tag{6.17}$$

Here L is the cell length, W is the cell width and μ is the surface mobility [6.8]. Note that τ_0 decreases with increasing Q_0 and is equal to $(4/\pi)L^2/(\mu V_G)$ for $N=1/2 \cdot N_{max}$. With a half full well, $L=10\,\mu m$ and $V_G=10\,V$, τ_0 will be approximately $3 \times 10^{-10}\,s$ in silicon which means that from electrostatic repulsion alone we could not expect $\varepsilon < 10^{-4}$ unless $t \gg 3\,\mu s$. Fortunately when N drops below $kTC_{ox}/e^2 = N_{max}(kT/eV_G)$, the diffusion current $J = -kT\mu\,\mathrm{grad}\,N$ will exceed the current caused by self-induced drift [6.9]. This current is

$$J = (\mu e^2 N/C_{ox})\,\mathrm{grad}\,N. \tag{6.18}$$

By this point thermal diffusion and fringing fields will control the transfer and the remaining charge will decay exponentially, i.e.,

$$Q(t) = C_2 e^{-t/\tau_f} \tag{6.19}$$

with a final decay constant of

$$\frac{1}{\tau_f} = C_1^2 \frac{\pi^2 kT\mu}{e4L^2} + \frac{\mu e E_{my}^2}{4kT} \tag{6.20}$$

where C_1 is a parameter that varies from one to two as the normalized fringing field varies from zero to ∞ [6.10]. The derivation of this expression for the final time constant assumed that $N=0$ at the edge of the well so that it will not apply as the thermal diffusion constant approaches zero. E_{my} is the minimum fringing field and is given by

$$E_{my} = \frac{2\pi}{3}\frac{\varepsilon_s}{\varepsilon_{ox}}\frac{d_{ox}(\Delta V)}{L^2}\left[\frac{5W/L}{(5W/L)+1}\right]^4 \tag{6.21}$$

for a surface channel device when ΔV is the potential difference between different electrodes and W is the depletion layer width [6.11]. The fringing field thus can play a key role in the transfer of the last small portion of signal charges if

$$e\Delta V \frac{d_{ox}}{L}\left(\frac{5W/L}{5W/L+1}\right)^4 \frac{\varepsilon_s}{\varepsilon_{ox}} > 3kT. \tag{6.22}$$

Matching of the hyperbolic decay of (6.16) with the final exponential (6.19), following *Steckl* et al. [6.1], leads to an expression for C_2

$$C_2 = Q_0 \frac{\tau_0}{\tau_f} \exp\frac{\tau_f - \tau_0}{\tau_f} \tag{6.23}$$

so that if only a time t is allowed for transfer

$$\varepsilon = \frac{\tau_0}{\tau_f} \exp\frac{\tau_f - \tau_0 - t}{\tau_f}. \tag{6.24}$$

Thus for $\varepsilon \leq 10^{-4}$ the minimum time for transfer will be

$$t_4 = \tau_f \ln\{10^4(\tau_0/\tau_f)\exp[(\tau_f - \tau_0)/\tau_f]\}. \tag{6.25}$$

With a nearly full well $t_0 \ll t_f$. For that case

$$t_4 = \tau_f[1 + \ln 10^4(\tau_0/\tau_f)] = 5 \times 10^{-7} \text{ s} \tag{6.26}$$

using $\tau_0 = 3 \times 10^{-10}$ s and $\tau_f = 1.2 \times 10^{-7}$ s. This value for τ_f is for SiO_2 on Si with $\Delta V = 5$ V, $L = 10\,\mu\text{m}$, $W/L = 0.3$, $d/L = 0.01$, $T = 300$ K. Free charge transfer processes can therefore limit transfer efficiency for clock frequencies over a few MHz. Note the strong dependence of τ_f on cell length L. Only devices with small cell lengths and high surface mobility can operate efficiently at high clock frequencies. Fringing fields with a buried channel device tend to be larger than with a surface channel device since the charge is held further from the electrodes. The highest speed devices therefore use a buried channel [6.12].

Noise

There are a number of sources of noise which must be considered in the design of an IR imager which uses a CCD readout. Apart from the pattern noise discussed in Section 6.1 which may result from responsivity variations, thermal dark current variations or clock noise feed through, there can be noise associated with the uniform background flux, noise associated with uniform thermal generation, noise associated with the input circuit used to put signal into the CCD, noise associated with signal transfer in the CCD and noise associated with readout from the CCD.

In the IR with terrestrial backgrounds and cooled infrared sensitive devices, the ultimate limit in performance is set by the shot noise arising from fluctuations in the arrival rate of the background photons. For the wavelengths of interest the background photons follow Poisson statistics, and the standard deviation of the number of collected photoelectrons from a detector in a sampling time will be the square root of the number. For an individual detection cell of area A with quantum efficiency η, the noise spectrum of the photocurrent will be $S_i = e^2 J_b \eta A$ where J_b is the background photon flux at the detection cell and η is the quantum efficiency. Deterministic gain processes operating on the photosignal after collection will not alter the ratio of the standard deviation to the mean but random processes will. We shall defer discussion of noise from input circuits until a description of specific focal plane designs and shall concentrate here on noise sources arising from transfer in the CCD register itself.

In the absence of trapping there should be no noise added by the charge transfer process, even if there is variation in the clock voltages, as long as the CCD is operated in a complete charge transfer mode and there is no transfer inefficiency. With transfer inefficiency there is an added noise since the amount of charge left behind is a random variable. Since charge added to one packet is taken away from adjacent packets, the fluctuations are strongly correlated. For reconstructed analogue signals which are properly band limited to frequencies below the Nyquist limit $f_c/2$, the correlation has the effect of enhancing the noise at high frequencies and suppressing it at low frequencies. The transfer noise spectrum of the current after n transfers will be

$$S(f) = 4nf_c \langle \Delta Q^2 \rangle (1 - \cos 2\pi f/f_c) \tag{6.27}$$

where $\langle \Delta Q^2 \rangle$ is the mean square fluctuation for each transfer process [6.13]. When transfer inefficiency is limited by free charge transfer processes, a shot noise point of view can be used [6.14] such that

$$\langle \Delta Q^2 \rangle = e\varepsilon Q_s. \tag{6.28}$$

When trapping dominates ε then the dominant noise source will be fluctuations in the charge retained in traps. Trapping states having reemission time constants on the order of the transfer time will contribute most to the noise. For surface trapping states, assuming that the capture cross section σ and the surface state density N_{ss} are independent of energy, it can be shown [6.4] that

$$\langle \Delta Q^2 \rangle = e^2 kTAN_{ss} \ln(2) \tag{6.29}$$

which is approximately the result obtained by substituting (6.15) into (6.28). This is nearly equal to the shot noise in the carriers emitted from an energy band of interface states of width kT. At room temperature with $A = 200\,\mu m^2$ and $N_{ss} = 2 \times 10^{10}\,cm^{-2}\,eV^{-1}$ the mean square fluctuation is about 36 electrons so that this can be an important source of noise after a large number of transfers. Due to the smaller number of trapping states which can interact with the signal, the transfer noise will be significantly lower for buried channel devices [6.3, 15].

With small signals in buried channel devices cooled sufficiently to reduce thermal generation, the limiting noise is often associated with the readout process. Besides dumping the charge into the substrate there are two basic techniques for measuring the size of a packet of minority charge in a CCD register. The first is to transfer it into a sensing $p-n$ diode. The current can then be sensed by a current mode preamplifier as shown in Fig. 6.6a. The second is to

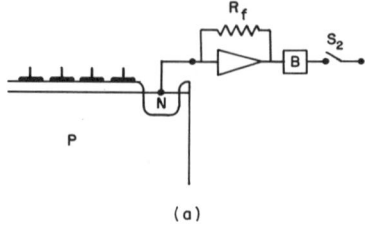

(a)

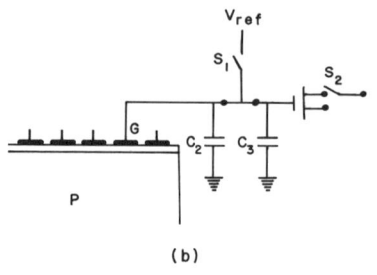

(b)

Fig. 6.6a and b. CCD readout circuits: (a) Current mode readout from a $p-n$ diode. R_f is the feedback resistance and B a low pass filter. (b) Voltage mode readout with a floating gate preamplifier. The voltage on G is reset to V_{ref} by closing and opening switch S_1 before the signal charge is transferred under gate G

dump the charge under a floating gate whose voltage has been set to a reference voltage and then measure the voltage difference caused by this charge transfer with a high input impedance preamplifier as shown in Fig. 6.6b. With both these techniques several parts of the same chip can be read out simultaneously.

The use of the sensing $p-n$ diode with the current mode preamplifier has the advantage of superior dynamic range since a small diode can handle a large amount of charge and kTC noise will not be present since no reset operation is involved. This approach is hard to integrate on the chip since a high impedance feedback resistor must be attached to the detection node. A back-biased $p-n$ sensing diode can also be used with a voltage sensitive preamplifier. This approach or the approach with the floating gate preamplifier is much easier to integrate on the chip (reducing the capacitance at the input to the preamplifier). In this case the area of the floating gate or the $p-n$ junction must be comparable to the CCD well size to handle the largest signals. Both of these voltage sensing techniques suffer from kTC noise introduced by resetting the voltage between samples through a resistance with Johnson noise. kTC noise, however, can be suppressed with correlated double sampling [6.16]. The floating gate preamplifier approach is a nondestructive readout process so that the same charge packet can be sensed again and again as it moves down a CCD register as in the sophisticated distributed floating gate amplifier (DFGA) [6.17]. The repeated sensing in the DFGA leads to a coherent addition of the signal and an incoherent addition of the noise for a signal-to-noise ratio improvement proportional to the square root of the number of stages.

With the floating gate preamplifier, the potential of the floating gate is first clamped to V_{ref} by closing the switch S_1 shown in Fig. 6.6b. The switch is then opened, which leaves the gate with a standard deviation about V_{ref} of $\sqrt{(2/3)kT/C}$ where C is the total capacitance between the gate and ground. The

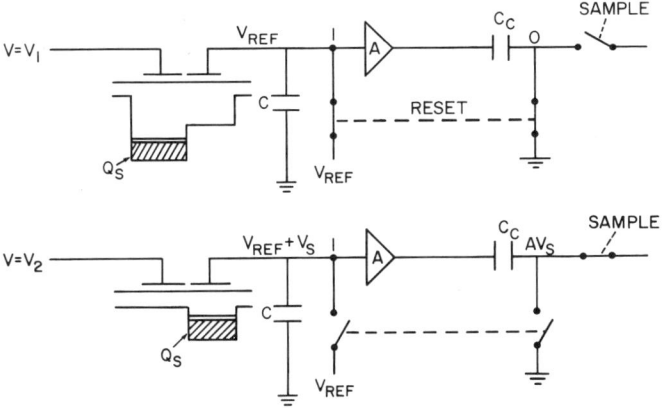

CID READOUT WITH
CORRELATED DOUBLE SAMPLING

Fig. 6.7. Circuit for correlated double sampling. The preamplifier input voltage is reset to V_{ref} and the output voltage to zero by closing and then opening the reset switch. The charge is then transferred and the output voltage is sampled. With this approach the output measures only the change in voltage caused by the charge transfer operation. kTC voltage noise caused by the reset operation is not transmitted to the output

correlation time of this noise is $(1/2)RC$ where R is the impedance to ground with the switch open. This time is usually much longer than the time used to read out a charge packet. Transfer of minority charge Q_s under the floating gate results in a change in its potential of

$$\Delta V = \frac{Q_s}{C_D + (C_2 + C_3)(1 + C_D/C_{\text{ox}})} \tag{6.30}$$

The depletion layer capacitance C_D is usually smaller than both the gate oxide capacitance C_{ox} and the external capacitance so that approximately

$$\Delta V = Q_s/(C_2 + C_3). \tag{6.31}$$

$C_2 + C_3$ encompasses both stray capacitance and preamplifier capacitance. With a one-stage preamplifier (not DFGA) which experiences kTC noise, the minimum detectable signal will therefore be

$$Q_{S\min} = \sqrt{(2/3)kT(C_2 + C_3)} \tag{6.32}$$

which for $C_2 + C_3 = 0.3$ pF and $T = 77°$ comes out to a minimum detectable packet size of 91 electrons. To detect signals less than this without the DFGA, correlated double sampling must be used to suppress kTC noise. With correlated double sampling implemented as shown in Fig. 6.7, the difference in voltage on the floating gate between before and after charge transfer is measured. Due to its long correlation time kTC noise will be cancelled out.

With correlated double sampling, the limiting readout noise at megahertz frequencies will be the equivalent input noise voltage of the first FET of the preamplifier. Neglecting $1/f$ noise the rms noise voltage can be expressed as $V_n B^{1/2}$ where B is the electrical bandwidth at the output of the preamplifier before any sample and hold function. B must be large enough to allow the signal to pass through unattenuated in the time allowed, considering the various functions which must be performed within a clock period with correlated double sampling. With correlated double sampling the minimum detectable signal will be

$$Q_{Smin} \approx \sqrt{2} V_n B^{1/2} (C_2 + C_3) \tag{6.33}$$

where the $\sqrt{2}$ comes from the fact that two samples are taken to suppress kTC noise. This expression also approximately applies for a sensing diffusion used with a current mode preamplifier if we were to incorporate the diffusion capacitance in C_2 and make use of a large enough feedback resistor.

In the megahertz frequency range V_n is caused by thermal noise in the FET channel and can be expressed by

$$V_n = \sqrt{(8/3)kT/g_m} \tag{6.34}$$

where g_m is the FET transconductance.

For a MOSFET with gate width W and gate length L in the current saturation region

$$g_m = \frac{W}{L} \mu C_{ox} (V_G - V_{T'}) \tag{6.35}$$

where under the usual operating conditions the MOSFET gate voltage $V_G > V_{T'}$ [6.18]. $V_{T'}$ is approximately the threshold voltage. Most of the preamplifier capacitance C_3 is just WLC_{ox} so that we may write

$$g_m = \frac{\mu C_3}{L^2} (V_G - V_{T'}). \tag{6.36}$$

By examination of (6.33) it becomes clear that in the absence of device area constraints if $1/f$ noise can be avoided, it is desirable to increase the MOSFET gate width W until $C_3 = C_2$. In this case C_2 includes the capacitance introduced by the switch S_1. With $C_3 = C_2$ the minimum detectable charge will be

$$Q_{Smin} = 2L \left(\frac{16}{3} \frac{C_2 BkT}{\mu V_G} \right)^{1/2}. \tag{6.37}$$

For small values of C_2 (below 1 pF) V_G will be set by the oxide breakdown voltage whereas for large values of C_2 (the conventional detector case) V_G will be limited by power dissipation constraints. For $V_G = 8$ V, MOSFET gate length $L = 8$ μm, channel mobility $\mu = 400$ cm^2/Vs, $C_2 = 0.3$ pF, $T = 77$ K and $B = 1$ MHz, the minimum detectable signal from (6.37) will be very near 1 electron. $1/f$ noise has been neglected; however the correlated double sampling process itself helps to suppress $(1/f)$ noise in the preamplifier since that noise which does not fluctuate between the two samples will not be recorded at the output [6.16]. Such low values of minimum detectable signal have not been observed, probably due to clock pickup; however these predictions point to the very low noise performance which is possible if kTC noise is suppressed.

In all cases a low value of capacitance at the detection node is desirable. This points up the advantages of a CCD with preamplifier electronics integrated on the same chip. Here C_2 can be on the order of 0.1–0.2 pF. As will be discussed later, use of a CID tends to lead to more capacitance at the sensing electrode. With more capacitance at the detection node, power dissipation constraints become important for the preamplifier especially if a large number are to be used at cryogenic temperatures (as is often the case with IR imagers).

6.3 Time Delay and Integration (TDI) and IR Sensitive CCD

As was discussed in Section 6.1, due to the low contrasts in the IR there is strong motivation to use some form of mechanical scanning in thermal imaging systems. The most straightforward approach to the use of a large number of detectors with mechanical scanning is to use a number of time delay and integration channels with series-parallel scan as shown in Fig. 6.1c. The TDI function essentially converts each row of detectors (TDI channel) into a single superdetector. If the delay between each detector matches the scanning rate charge generated by illumination from the same point on the image will be added together. The delay between detectors must therefore be adjusted to the time it takes to scan one center to center detector spacing. At the output of an ideal M element TDI channel the signal will be increased by a factor of M due to the TDI process, whereas the noise from each detector will be added incoherently so that its rms value will be increased by a factor of $\sqrt{M}$. If the TDI process itself does not introduce additional noise, the signal-to-noise ratio at the output of the TDI register will thereby be improved by a factor of $\sqrt{M}$ over that which would be obtainable with a single detector.

With mechanical scanning of the image across the focal plane, the detectors will experience a time varying signal due to structure in the image. The maximum signal frequency of interest for a TDI series-parallel scan system will be the same as for a parallel scan focal plane if both use the same number of interlace steps and do not segment the focal plane. For 1/30s frame times and TV compatible

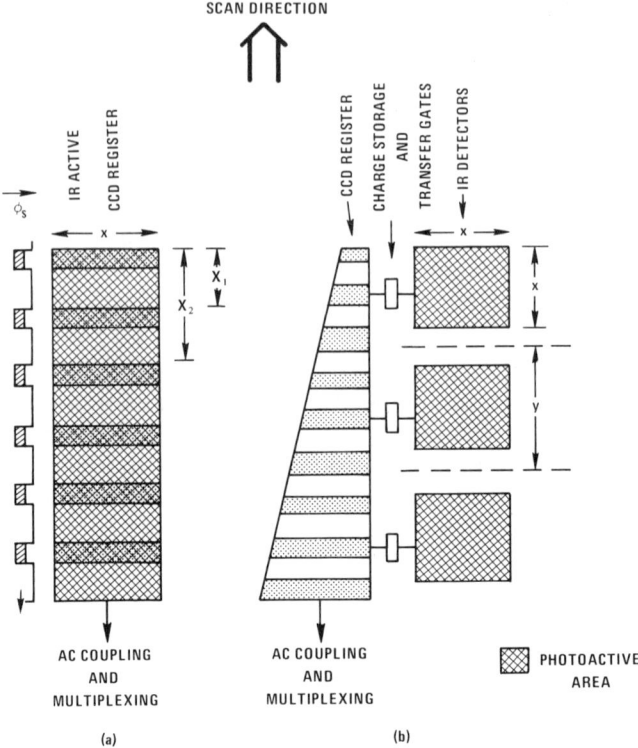

Fig. 6.8a and b. The use of CCD registers for TDI. (a) IR active CCD register (monolithic focal plane array with photogeneration in the register). (b) Direct injection with Si parallel in series out CCD register; hybrid with IR photodiodes or monolithic with silicon IR extrinsic photoconductors

resolution, the maximum signal frequency of interest will lie in the 20 to 30 kHz range for a 2:1 interlace. Clock frequencies will need to be about 50 kHz. For many applications it is important to obtain the highest signal-to-noise ratios at the maximum signal frequency. Since for a given implementation of TDI the signal-to-noise ratio will only increase as the square root of the number of detectors, it is generally essential to maintain BLIP operation as the number of detectors is increased.

Since a CCD register is essentially a discrete analog delay line, TDI can be easily implemented using CCDs. Coherent addition of the signal will be obtained if the clock frequency is synchronized with the scan rate. There are two basic configurations as shown in Fig. 6.8. With the first approach the CCD material is IR sensitive and detection takes place within the CCD register itself. A high pass filter which suppresses the uniform (DC) background signal is inserted after the TDI register. Either backside illumination or transparent electrodes are required. With diffusion along the surface the individual CCD wells can collect photogenerated minority charge from an area larger than the well size itself so that the effective detector dimension becomes the center-to-center spacings (X_1 in Fig. 6.8).

There will be MTF degradations due to the finite detector size, the discrete nature of the charge transfer, and due to any lack of synchronism between the clock frequency and the scan rate [6.19]. These MTF degradations can be particularly troublesome since they act on the signal without filtering the shot noise due to the uniform background. Transfer inefficiency in the TDI register itself can cause additional MTF problems, especially since MIS technology is not as advanced in IR sensitive materials [6.20] (see Fig. 6.5). Transfer inefficiency will decrease the gain of a TDI register, reducing the value of longer registers. CCDs have been demonstrated in InSb but the transfer efficiencies were not high [6.21].

With 3–5 μm operation the most serious obstacle to BLIP operation with an IR sensitive CCD TDI register is likely to be caused by interface state trapping noise [see (6.27) and (6.29)] which can easily dominate the background shot noise at the clock frequencies of interest to imagers. In the 8–12 μm region the background shot noise can more easily dominate the interface state trapping noise but the larger background means that saturation will be a serious problem towards the end of a TDI register. Saturation will occur whenever

$$\eta J_b \geq \frac{f_c C_{ox} V_G}{eM} \frac{A}{A_{tot}} \tag{6.38}$$

where J_b is the background on the focal plane in photons/cm²/s and A/A_{tot} represents the ratio of cell area to collection area. For the $C_{ox}V_G/e$ characteristic of SiO$_2$ on Si, $\eta = 0.5$, $f_c = 50$ kHz and $A/A_{tot} = 0.25$, saturation would occur for

$$J_b \geq \frac{5 \times 10^{16}}{M} \text{ photons/cm}^2 \text{ s}. \tag{6.39}$$

Just for the sake of comparison let us consider the more conventional staring system which does not use mechanical scanning. Such a system would saturate for

$$\eta J_b \geq \frac{C_{ox} V_G}{\tau_F e} (A/A_{tot}) \tag{6.40}$$

where τ_F is the frame time. For 1/30 s frame times, saturation would be encountered for backgrounds over 3×10^{12} photon/cm² s. One approach with a staring system to avoid saturation problems would be to use shorter frame times and scan faster. If BLIP performance could be maintained, no penalty would be caused by the shorter frame time since in a real time imaging system the final filter is controlled by the integration time of the eye on the display. In general however saturation problems will be more serious with a staring system. Maximum values for $V_G C_{ox}$ will be somewhat reduced for IR material systems.

6.4 Direct Injection: Hybrid and Extrinsic Silicon

The second approach to using CCDs for TDI with a system which does use mechanical scanning is to input signals from IR detectors (such as InSb photodiodes which can operate at 77 K) into an ordinary silicon CCD register as shown on the right in Fig. 6.8. This approach has the advantage that CCD action is not required in an IR sensitive material, so that it can be used with different types of infrared detectors. Usually a hybrid sandwich structure is used with a connection between each individual detector and the silicon CCD TDI register [6.22], but the approach can also be implemented with extrinsic silicon photoconductors in a monolithic form with the CCD material grown epitaxially on the IR detector material (see Fig. 6.9) [6.23–25]. It is easiest to provide dc coupling between the detectors and the CCD register and to suppress the low frequency signal components (ac couple) after exiting the TDI channel. However, in principle, ac coupling could be introduced between the detector and the TDI register thereby greatly enhancing dynamic range and avoiding the saturation problems which can be expected with operation in the 8–12 μm region if all the dc photocurrent caused by the uniform background is injected into the CCD register.

A fairly simple input circuit (direct injection) which fills the CCD wells with photoinduced current (signal and background) can be used with high impedance detectors such as InSb photodiodes or extrinsic photoconductors; however buffer preamplifiers are needed for low impedance intrinsic photoconductors. With a low impedance intrinsic photoconductor detector resistance is not controlled by background illumination and most of the bias current is not photocurrent, so that to avoid saturation most of the bias current must be shunted off through a load resistor and a voltage preamplifier introduced between the detector and the CCD input. Buffer preamplifiers use up chip area and cause power dissipations usually over 1 mW per preamp, so that the higher packing density, largest number of detector systems will tend to use high impedance detectors.

Without preamplification surface state trapping noise will still be a problem for 3–5 μm systems for TDI clocking frequencies over 10 kHz; however since the CCD can be fabricated in silicon a buried channel device could be used to circumvent this noise problem. In any case a tapered register is usually used for TDI as shown in Fig. 6.8 to provide a constant fat zero from the background. With this hybrid approach the CCD well width is not directly related to the photoactive area so that saturation problems can be relieved by using a wide register near the end, but this increases the separation requirement between TDI channels, complicating focal plane layout since no more than a 2:1 interlace is desirable.

The use of a hybrid approach increases flexibility as to detector choice at the expense of packing density and problems associated with inputting the photosignal into the CCD. The photocurrent is first integrated and then periodically dumped into the CCD, which causes a sampling of the signal in the

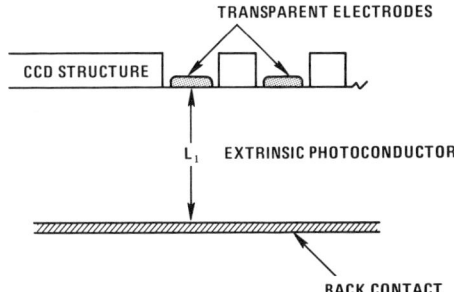

Fig. 6.9. Monolithic IR focal plane using extrinsic photoconductivity in silicon. The CCD material can be grown epitaxially on the IR detector material. A longitudinal detector bias is used with L_1 being the interelectrode spacing for the photoconductor

direction of scan. The MTF of the overall system in the direction of scan is degraded first by the finite aperture of the detector and then secondly by this sampling function. This sampling problem with hybrid systems is analogous to the discrete motion problem with the detection in the register systems, which has already been discussed. If the sampling rate is inadequate, background limited performance will be degraded since the phase averaged MTF effect of limited sampling acts only on the signal and will not affect the noise. High sampling rates require more CCD bits per detector.

The input technique called direct injection is shown in Fig. 6.10. The detector anode is connected to a p-type diffusion. V_G is held at a fixed bias to deplete the vicinity of the input diffusion while V_D is biased more negatively to create a potential well where photocurrent can be collected. After an integration period t_s, this stored charge is transferred into the CCD TDI register by changing ϕ. The charge is then transferred down the TDI register and the cycle is repeated an integration period later.

The input circuit has been analyzed in [6.22, 26, 27]. The ac equivalent circuit is shown in Fig. 6.10. The direct injection input structure is conveniently analyzed as a grounded gate MOSFET where the input diffusion acts at a source, V_G as a gate and the potential well under V_D as a virtual drain. The gate to drain voltage is fixed and the current is a function of the voltage of the source relative to the gate. For the circuit shown in Fig. 6.10 the dc source to drain current will arise from the background illumination falling on the detector.

The principal difficulty with this input circuit is that since g_m is finite, a fairly large effective resistance is in series between the detector and the CCD register itself. At high frequencies the capacitance C will short out the photocurrent, and the channel thermal noise in the equivalent MOSFET can come to dominate the shot noise caused by the uniform background. Without special precautions, devices with this input circuit have severe problems in obtaining background limited operation in the 3–5 μm region for scan frequencies of interest. If the resistance between the point S and ground is large enough, for the circuit shown in Fig. 6.10 the maximum signal frequency for which the device will be background limited will be

$$f^* = \frac{\sqrt{2e^2 \dot{N}_B g_m}}{2\pi C \sqrt{(8/3)kT}} \tag{6.41}$$

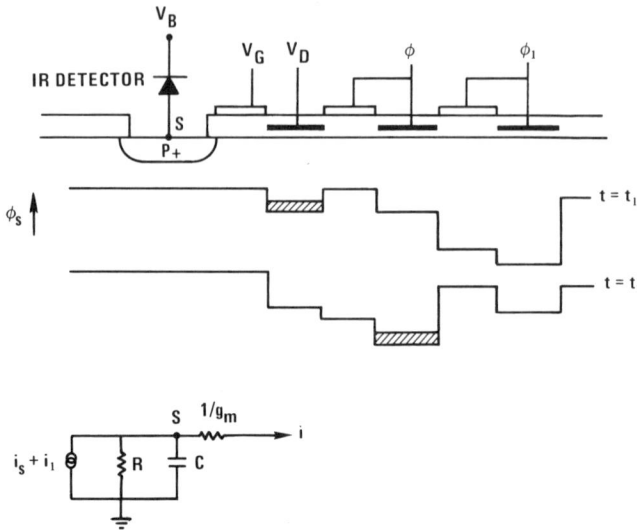

Fig. 6.10. Direct injection CCD input structure showing the IR detector connected to an input diffusion. Charge is collected under V_D and then transferred to the CCD register (which runs perpendicular to the plane of the figure) by changing the voltage on the ϕ electrodes. The lower portion of the figure shows the ac equivalent circuit where the current i' flows into the CCD register

where $\dot{N}_B$ is the number of photoelectrons collected by the photodiode per second. Since g_m depends upon the amount of dc current flowing through the grounded gate MOSFET, for low backgrounds it is advantageous to leak extra low noise dc current I through the device to enhance g_m and provide a larger f^*. For currents less than about 4×10^{-10} amps the MOSFET will be in the weak inversion regime and $g_m = (1/2)eI/kT$ [6.22, 28, 29]. No f^* problem would exist for 8–12 μm detectors unless the detector capacitance were unusually large as with $Pb_{1-x}Sn_xTe$ diodes.

The use of an extrinsic photoconductor with direct injection has the advantage that dc gain can enhance I and therefore g_m, but gain saturation due to sweepout will limit ac gain to 1/2 at frequencies near f^* [6.30]. The capacitance C of an extrinsic photoconductor can be an order of magnitude lower than for a photodiode, which will lead to a higher f^*. With an extrinsic photodetector, crosstalk problems must be considered for the detector thicknesses necessary to provide reasonable quantum efficiency, and even for the 3–5 μm window operating temperatures will tend to be below 50 K.

Even though there are sometimes problems with input to the CCD register all approaches which introduce the TDI function before preamplification have an advantage in that the gain of the TDI register ($\sqrt{M}$ on the background noise) makes it easier for the post TDI preamplifier to deliver BLIP performance.

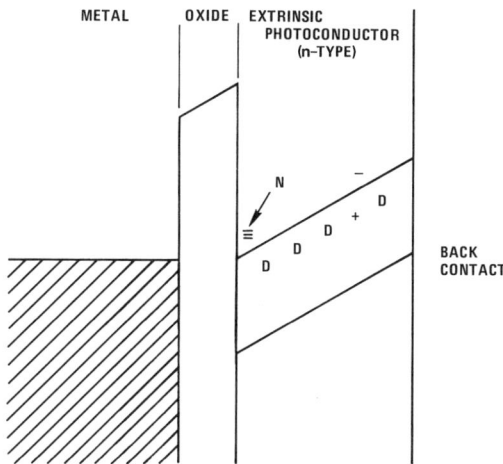

Fig. 6.11. Electron potential as a function of distance from the surface for an accumulation mode MIS CCD structure which uses n-type extrinsic photoconductivity. Note that the electric field extends all the way to the back contact

6.5 Accumulation Mode: Extrinsic Silicon

Since extrinsic silicon photoconductor material has high resistivity at cryogenic temperatures it can be used to form the substrate of an accumulation mode CCD as shown in Fig. 6.11. With an accumulation mode MIS structure the gates are biased so that majority carriers are stored and transferred down the insulator semiconductor interface. Local potential wells are formed under the gates; however the dynamics of the charge transfer process will be very different from those for an inversion mode device since with an accumulation mode device the transverse electric fields will extend all the way to the back contact instead of being confined to the depletion region of an inversion mode structure.

In general the accumulation mode structure forms an RC circuit. At cryogenic temperature the resistance of the extrinsic silicon material is controlled by photogeneration. The current flowing from the back contact to the insulator semiconductor interface is therefore controlled by the amount of IR radiation on that portion of the device. If the back contact is ohmic the accumulation layer charge-up time is just $\varrho t C_{ox}$ where ϱ is the resistivity of the silicon substrate, t its thickness and C_{ox} the capacitance per unit area of the insulator. Since charge which accumulates at the interface must be removed before all the voltage appears across the insulator, saturation will occur with a TDI register whenever

$$\varrho < \frac{M}{f_c t C_{ox}} \frac{A}{A_{tot}}. \tag{6.42}$$

At low temperatures if the substrate has a thickness such that most of the IR radiation is absorbed we can estimate the resistivity by

$$\varrho = (e\mu\tau J_b \eta / t)^{-1} \tag{6.43}$$

where τ is the recombination time of the majority carriers and μ their mobility. In that case saturation will occur whenever

$$\eta J_b > \frac{t^2 f_c C_{ox}}{Me\mu\tau} \frac{A}{A_{tot}}. \tag{6.44}$$

Although an IR accumulation mode device seems particularly straightforward, the reported transfer inefficiencies have been only $\varepsilon = 0.01$ [6.31]. In general this is not good enough for device operation, which suggests that accumulation mode devices will only be used in configurations similar to a CID where good transfer efficiency is not required. Focal plane temperature well below 77 K will of course be needed to use extrinsic silicon in the IR.

6.6 IR CID

The easiest kind of MIS detector to fabricate in an IR sensitive material is a MIS capacitor with a transparent electrode or charge injection device (CID). With a CID operating in the inversion mode, minority carriers are photogenerated near the surface and collected at the semiconductor insulator interface in isolated potential wells formed by voltages applied to electrodes on the surface. After a collection time the amount of charge at a particular location is then sensed as the well is collapsed and the minority charge is injected into the bulk to combine with majority carriers. After recombination the wells are reformed and the collection process starts all over again. With a CID the minority charge is not clocked along the surface but the charge can be sensed by manipulating the voltages on the xy readout lines. If low resistance readout lines are used, CID readout speeds are limited by the requirement for minority carrier recombination after injection before the well can be reestablished.

The collected charge can be sensed by intergrating the substrate current during injection into the substrate or by monitoring the voltages on the electrodes. To read out a two-dimensional matrix of detector sites, a two well unit cell is used with one electrode of the cell connected to a row line and the other electrode connected to a column line. By manipulating the voltages on these x and y readout lines as shown in Fig. 6.12 charge can be transferred back and forth between the two electrode wells and injected into the substrate. The signal can be read out by comparing the voltage on the column line before and after a transfer sequence (a nondestructive readout) or by comparing the voltage on the column line before and after a capacitively coupled injection pulse is applied to the column line [6.32].

An electrode organization with the two well unit cell as shown in Fig. 6.13 allows horizontal and vertical scan registers external to the CID chip to read out the unit cells. Thus an $m \times n$ CID array would require $m+n$ connections to the IR sensitive material. Since the CID is xy addressed, random access to any

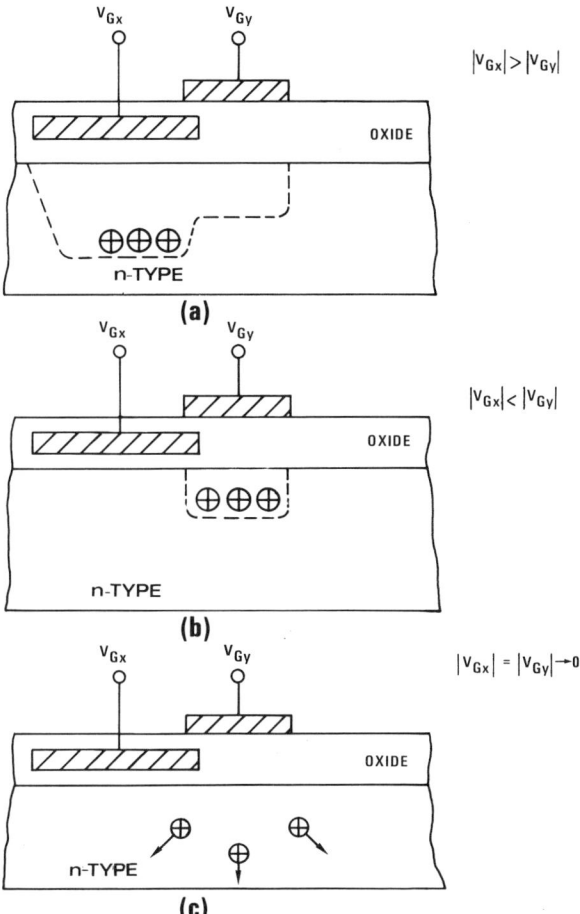

Fig. 6.12a–c. Two well unit cell of a two-dimensional CID array demonstrating the readout and injection sequence. The electrode marked V_{Gx} is attached to an x row line and the electrode marked V_{Gy} is attached to y column line. The column line will be attached to a preamplifier for readout. Note that the area under V_{Gy} can be less than half the area of the unit cell

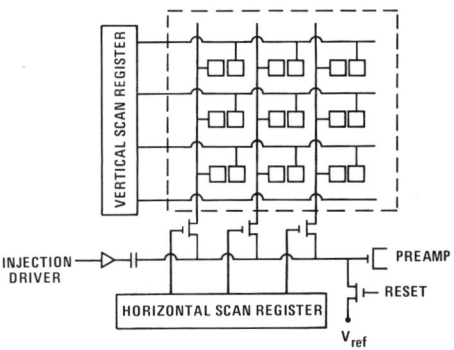

Fig. 6.13. Two-dimensional CID array using the two well unit cell. The horizontal scan register controls column select switches and the voltage to provide charge injection is capacitively coupled in from the injection driver. By manipulating the voltages provided by the scan registers the CID array can be read out in any desired sequence

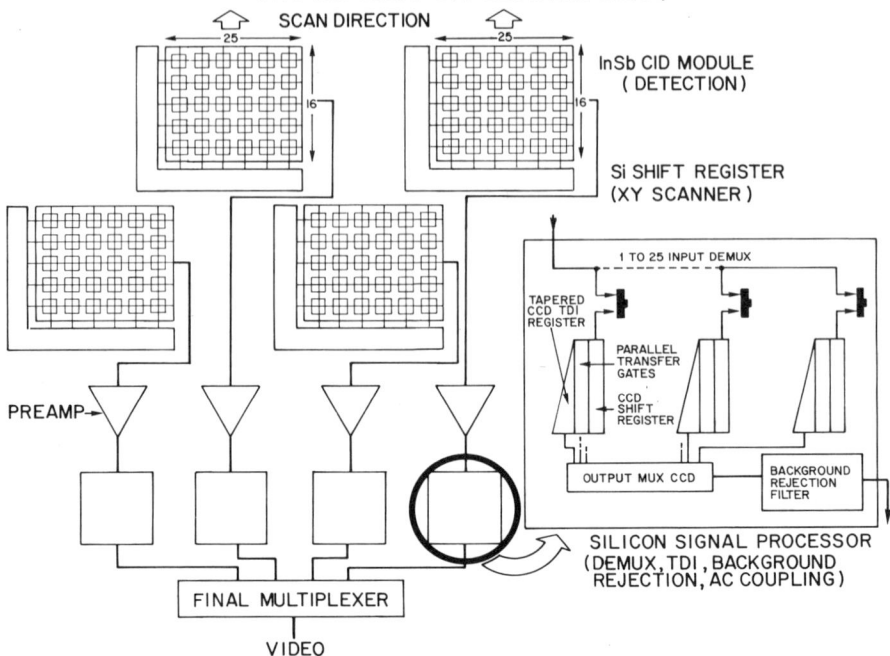

Fig. 6.14. Series-parallel scan IR CID. Infrared sensitive IR CID modules are completely read out more than once a dwell time through a preamplifier into a silicon CCD signal processor. In the signal processor TDI is performed and then the individual lines of imagery are ac coupled and multiplexed. Only the CID module itself is fabricated from IR sensitive material

particular unit cell is possible. This would not be possible with a CCD which requires a sequential access.

More than one preamplifier can be used per CID chip. With multiple preamplifiers attached to the column lines, different columns on the same chip can be read out simultaneously. In all cases with a two-dimensional matrix of CID detection cells the readout line is connected to a whole column of detectors. Because of this the input capacitance to the preamplifier is necessarily larger than with a CCD where only a single output well is attached to the preamplifier. For staring systems with high backgrounds and long collection times a symmetric cell is used to avoid saturation. (Saturation occurs when the transfer process leads to overflow of the readout cell.) On the other hand with a scanned CID an asymmetric unit cell can be used (as shown in Fig. 6.12) to reduce column capacitance so that the electrode area under the readout line can be less than half the photoactive area. With modest numbers of detection cells using an asymmetric design, stray capacitance not associated with the photoactive area becomes dominant so that careful control of interline crossover capacitance and interconnection capacitance is necessary for low noise operation. Even with

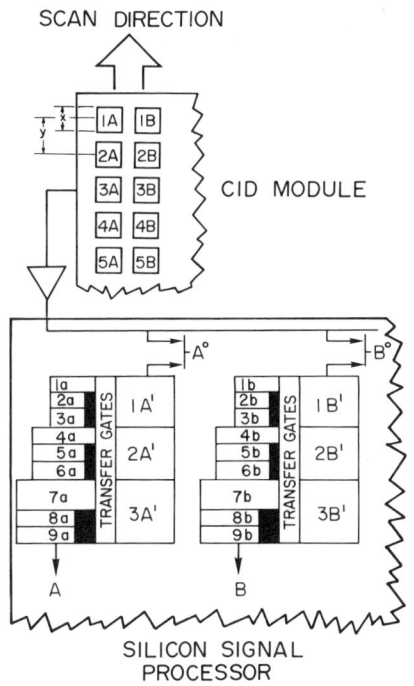

Fig. 6.15. Detail of part of the CID and signal processor modules showing how the charge pattern in the CID is recreated in the signal processor before TDI is accomplished with a parallel in series out CCD register (in this case with a 3:1 interleave so that three samples can be taken in the time the image moves the distance y)

much care it is hard to achieve column capacitance below several pF which is higher than the readout capacitance for a CCD; however, since only a single transfer is needed during the readout process, high transfer efficiency is not required. Two-dimensional CID arrays which can operate at temperatures greater than 77 K have been developed in InSb for the 3–5 μm region [6.33].

It is of course very straightforward to use a large xy scanned two-dimensional CID array in a staring mode. In that case the charge collection or integration time for each cell is just the frame time of the imager. The problem of course is that since cell to cell responsivity nonuniformities are likely to be over 1 %, a staring device will have pattern noise which will interfere with the detection of low contrast IR imagery. For low contrast applications it is preferable to use smaller two-dimensional CID arrays in a series-parallel scan mode, in which case the entire array is read out through a preamplifier more than once per dwell time into a CCD TDI signal processor (which can be fabricated in silicon) [6.27]. In the signal processor the charges are integrated together to provide the TDI function and then the lines of imagery are passed through a high pass background rejection filter to provide the ac coupling function. Although a certain amount of dc clipping can be provided at the preamplifier to avoid saturation in the processor, the required ac coupling cannot be performed before the TDI function, since at the preamplifier adjacent data samples do not represent data from the same detector.

One implementation of the series-parallel scanned CID is shown in Fig. 6.14. In this case one preamplifier is used per CID chip. The CID xy readout is used in

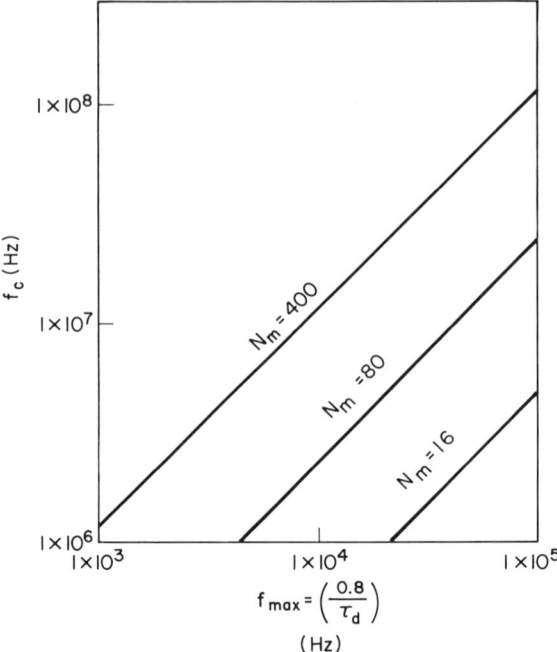

Fig. 6.16. Relationship between the elemental CID readout rate f_c and the maximum signal frequency of interest in a line of imagery f_{max} for different numbers of CID detectors per preamplifier $N_m \cdot f_{max} = 0.8/\tau_d$ and the use of 2.4 samples per dwell time τ_d is assumed

a premultiplexing mode to pass columns of CID readout signals through a preamplifier into the signal processor. The streams of IR signals are then demultiplexed into the CCD TDI processor such that the charge pattern observed in the IR sensitive CCD is recreated in the TDI signal processor. As shown in Fig. 6.15 the charge pattern from CID cells 1 A, 2 A, etc., will be recreated in CCD cells 1 A′, 2 A′, etc. At this point the TDI function is performed with a parallel in series out CCD register just as was the case for the direct injection hybrid focal plane array discussed in Section 6.4. If this register has a three-to-one interleave as shown in Fig. 6.15 then three samples of the signal can be obtained in the time the image moves one interdetector spacing (the distance y in Fig. 6.15). Fewer samples would lead to degradation in BLIP limited signal to noise performance.

With readout on the electrodes more than one preamplifier can be used per CID chip so that the numbers of detectors per preamplifier N_m is a design variable. Fewer detectors per preamplifier means a lower readout rate and lower required bandwidths after the preamplifier. This bandwidth must be large enough not to distort the signal but not too large since sampling occurs after the preamp and noise aliasing can occur at that point. If we assume the maximum signal frequency of interest f_{max} to be $(0.8/\tau_d)$ and that 2.4 samples are taken per dwell time τ_d (the time it takes the image to move the distance x in Fig. 6.15), then we have the following relationship for the required readout rate f_c

$$f_c = 3 N_m f_{max} \tag{6.45}$$

which is plotted in Fig. 6.16.

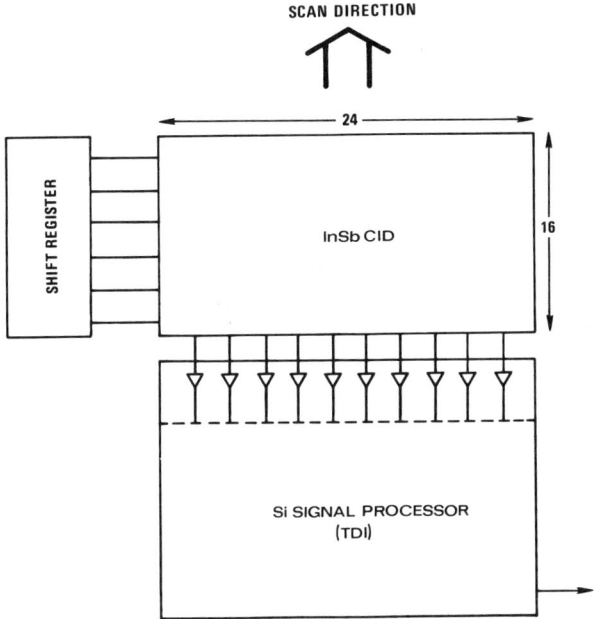

Fig. 6.17. Series-parallel scan IR CID with one preamplifier per column. All the columns are read out simultaneously. This lowers the readout speed and avoids the column select switches shown in Fig. 6.13.

If more than one column is read out through one preamplifier then a column select switch must be in series between the preamplifier and the CID cell. This switch need not consume any power but it must be low resistance in the on state for low noise operation. If we use one preamplifier per column then this column selection switch is not required and we will then have the rather simple configuration shown in Fig. 6.17.

If correlated double sampling is not used, besides background noise the most important source of noise for the series-parallel scan CID is likely to be kTC noise. If this is the case then the maximum signal frequency for which we can design the device to obtain BLIP operation will be

$$f^* = \frac{\dot{N}_B e^2}{2kTC} \tag{6.46}$$

where C is the full capacitance of the CID column and preamplifier input. Note that this value does not depend directly upon the number of detectors per preamplifier. If correlated double sampling is used then f^* can be higher as the device will be limited by preamplifier noise. If we assume that the preamplifier bandwidth B is adjusted such that $B = 3f_c$ then

$$f^* = \left[\frac{\dot{N}_B e^2}{54 V_n^2 C^2 N_m}\right]^{1/2} \tag{6.47}$$

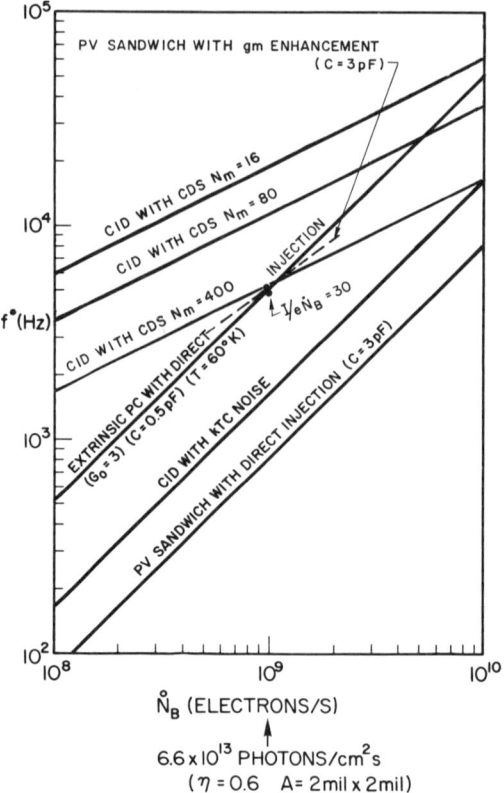

Fig. 6.18. Maximum signal frequency for background limited operation f^* as a function of the background on the detector $\dot{N}_B$ for different approaches to focal plane arrays. The following assumptions were used:

CID with CDS $N_m = 16$: $C = 5\,\text{pF}$, $V_n = 1.7 \times 10^{-9}$ V Hz$^{-1/2}$
CID with CDS $N_m = 80$: $C = 6\,\text{pF}$, $V_n = 7.5 \times 10^{-10}$ V Hz$^{-1/2}$, $R_s = 400\,\Omega$
CID with CDS $N_m = 400$: $C = 6\,\text{pF}$, $V_n = 7.5 \times 10^{-10}$ V Hz$^{-1/2}$, $R_s = 400\,\Omega$
CID with kTC noise: $C = 7\,\text{pF}$.

C includes all the capacitance at the input to the preamplifier. (The calculation assumes that the capacitance of a single CID cell is small compared to C.) The operating temperature was taken to be 77 K unless otherwise noted

where N_m is the number of detectors per preamp and $\dot{N}_B$ is the background in collected photons/detector s. If we neglect $1/f$ noise then for a MOSFET preamplifier the equivalent input voltage noise spectral density V_n is just

$$V_n = \sqrt{(8/3)kT/g_m} \,. \tag{6.48}$$

Comparative plots of f^* vs $\dot{N}_B$ are given in Fig. 6.18 for representative values of detector and preamplifier capacitance. As can be seen from Fig. 6.18 it is not easy to provide an f^* high enough for a TV compatible imager with 3–5 µm backgrounds.

From (6.47) it would appear that higher f^* could be obtained by decreasing N_m and providing more preamplifiers. However in terms of chip area and power dissipation it does not matter very much if we build one large very low noise preamplifier or a number of smaller ones with higher V_n. In any case preamplifiers with very large W/L ratios will be needed to control power dissipation. The tradeoff between many columns per preamplifier (as shown in Fig. 6.14) and one column per preamplifier (as shown in Fig. 6.17) depends upon preamplifier location. If the preamps are to be located in the dewar at the focal plane temperature, then one column per preamp is preferable since the column select switches are avoided. However if the preamps were located at a higher temperature, a lower number of output lines from the CID would be desirable.

In general the premultiplexing function of the CID readout is very attractive in that it physically separates the detection and signal processing functions and allows a preamplifier to be introduced before input to the CCD. With the CID, the signal processing function has no chip area constraint and noise problems associated with direct injection are avoided. However the CID approach requires more preamplifier gain and power dissipation than those approaches which introduce TDI before preamplification.

6.7 Conclusions

There are a number of promising approaches to IR focal plane arrays which use CCD and/or CID technology for multiplexing and TDI. The development of these series-parallel scan technologies should offer order of magnitude improvement in IR sensor performance. Besides obtaining satisfactory production yields with these complex structures, there exist a number of challenging problems; namely obtaining BLIP performances in the 3–5 µm region for scan frequencies of interest and avoiding saturation for focal plane arrays operating in the 8–12 µm region.

For high packing density there are two principal series-parallel scan approaches, one involving direct injection and an electrical connection between each IR detector and the CCD TDI register and the other which uses CID for premultiplexing. With backgrounds typical for the 3–5 µm region obtaining a high enough f^* is not easy in either case. The comparison goes roughly as

$$f^*_{\text{CID}}/f^*_{\text{DI}} = \frac{C_{\text{DI}}}{C_{\text{CID}}} \sqrt{(g_{\text{mCID}}/N_m)/g_{\text{mDI}}} \tag{6.49}$$

with the capacitance ratio in favor of the direct injection approach and the ratio of $(g_{\text{mCID}}/N_m/g_{\text{mDI}})$ usually in favor of the CID approach. As far as saturation is concerned the CID series-parallel scan approach should do well since charge is stored in the IR material for only a fraction of a dwell time. The CID structure also makes high packing density easy to achieve; however preamplifier power dissipation will be a problem. An IR CCD with generation in the register has the advantage of simplicity and TDI gain; however the development of this approach is further away since high transfer efficiency in an IR material will be needed and for 3–5 μm operation it will suffer from surface state trapping noise problems.

References

6.1 A.J.Steckl, R.D.Nelson, B.T.French, R.A.Gudmundsen, D.Schechter: Proc. IEEE **63**, 67 (1975)
6.2 W.S.Boyle, G.E.Smith: Bell System. Tech. J. **49**, 587 (1970)
6.3 D.F.Barbe: Proc. IEEE **63**, 38 (1975)
6.4 C.H.Sequin, M.F.Tompsett: *Charge Transfer Devices* (Academic Press, New York, San Francisco, London 1975)
6.5 W.B.Joyce, W.J.Bertram: Bell System. Tech. J. **50**, 1741 (1971)
6.6 R.J.Strain: IEEE Trans. Electron Devices **ED-19**, 1119 (1972)
6.7 M.F.Tompsett: J. Vac. Sci. Technol. **9**, 1166 (1972)
6.8 J.E.Carnes, W.F.Kosonocky, E.G.Ramberg: IEEE Trans. Electron Devices **ED-19**, 798 (1972)
6.9 W.E.Engeler, J.J.Tiemann, R.D.Baertsch: Appl. Phys. Lett. **17**, 469 (1970)
6.10 Y.Daimon, A.M.Mohsen, T.C.McGill: IEEE Trans. Electron Devices **ED-21**, 226 (1974)
6.11 J.E.Carnes, W.F.Kosonocky, E.G.Ramberg: IEEE J. Solid-State Circuits **SC-6**, 322 (1971)
6.12 L.J.M.Esser, M.G.Collet, J.G.vanSanten: IEDM, Washington, D.C. Tech. Digest (1973)
6.13 K.K.Thornber, M.F.Tompsett: IEEE Trans. Electron Devices **ED-20**, 456 (1973)
6.14 D.F.Barbe: Electronics Lett. **8**, 207 (1972)
6.15 A.M.Mohsen, M.F.Tompsett: IEEE Trans. on Electron Devices **ED-21**, 701 (1974)
6.16 M.H.White, D.R.Lampe, F.C.Blaha, I.A.Mack: IEEE J. Solid-State Circuits **SC-9**, 1 (1974)
6.17 D.D.Wen, J.M.Early, C.K.Kim, G.F.Amelio: ISSCC, Philadelphia, Digest of Tech. Papers, 24 (1975)
6.18 S.M.Sze: *Physics of Semiconductor Devices* (John Wiley and Sons, New York 1969) p. 522
6.19 H.V.Soule: *Electro-Optical Photography at Low Illumination Levels*, (John Wiley and Sons, New York 1968) pp. 332–333
6.20 *Moving Target Sensors*, Texas Instruments, Navy Contract N00039-73-C-0070, Final Report (1973)
6.21 R.D.Thom, R.E.Eck, J.D.Phillips, J.B.Scroso: Proc of the Int. Conf. on the Application of CCDs, p. 31, San Diego (1975)
6.22 M.R.Hess, J.C.Fraser, K.Nummedal, B.J.Tilley, R.D.Thom: The MOSART (Monolithic Signal Processor and Detector Array Integration Technology) Program, Proceedings of the 22nd National IRIS (1974)
6.23. M.Y.Pines, R.Baron: Proceedings of the International Electron Devices Meeting, p. 446, Washington (1974)
6.24 M.Y.Pines, D.Murphy, D.Alexander, R.Baron, M.Young: Proceedings of the International Electron Devices Meeting, p. 582, Washington (1975)
6.25 K.Nummedal, J.C.Fraser, S.C.Su, R.Baron, R.M.Finnila: Proceedings of the International Conference on the Applications of CCDs, p. 19, San Diego (1975)
6.26 D.M.Erb, K.Nummedal: Proceedings 1973 CCD Applications Conference, p. 157, San Diego (1973)
6.27 A.F.Milton, M.Hess: Proceedings of the International Conference on the Applications of CCDs, p. 71, San Diego (1975)
6.28 E.O.Johnson: RCA Rev. **34**, 80 (1973)
6.29 R.M.Swanson, J.D.Meindl: IEEE J. Solid-State Circuits **SC-7**, 146 (1972)
6.30 A.F.Milton, M.M.Blouke: Phys. Rev. **3**, 4312 (1971)
6.31 R.D.Nelson: Appl. Phys. Lett. **25**, 568 (1974)
6.32 G.J.Michon, H.K.Burke, D.M.Brown, M.Ghezzo: Proceedings of the International Conference on the Application of CCDs, p. 93, San Diego (1975)
6.33 J.C.Kim: Proceedings of the International Conference on the Application of CCDs. p. 1, San Diego (1975)

7. Nonlinear Heterodyne Detection

M. C. Teich

With 24 Figures

7.1 Two-Frequency Single-Photon Heterodyne Detection

Conventional heterodyne detection is useful in a number of configurations, including the detection of scattered or reflected radiation from a moving target (Doppler radar), communications, spectroscopy, and radiometry. Its use has been demonstrated in many regions of the electromagnetic spectrum including the radiowave, microwave, infrared, and optical. Its advantages as a detection technique are well known: high sensitivity, frequency selectivity, and strong directivity. For radar applications, it provides a major method of recovering desired signals and removing clutter. The significant improvement in sensitivity that it provides over direct detection arises from knowledge of the Doppler frequency (also called the heterodyne frequency or the intermediate frequency (IF)) which permits a narrow receiver bandwidth centered about the IF. In such applications, obtaining a reasonably high signal-to-noise ratio (SNR) requires 1) a good knowledge of the velocity of the source or target, 2) a stable yet tunable local oscillator, 3) a target or source which presents a minimum of frequency broadening and 4) at least several photons per measurement interval. These conditions are frequently not adhered to by actual systems, particularly in the infrared and optical, giving rise to detection capabilities which are well below optimum. In this chapter, we study the performance and requirements of a number of alternative heterodyne receiver configurations. In particular, we consider two basic systems which are intrinsically nonlinear, the first by virtue of the multiple-quantum detection process itself, and the second by virtue of the mixing configuration and the electronics following the detector.

After briefly reviewing conventional optical and infrared heterodyne detection, we examine the behavior of a multiphoton absorption heterodyne receiver. Expressions are obtained for the detector response, signal-to-noise ratio, and minimum detectable power for a number of cases of interest. Receiver performance is found to depend on the higher-order correlation functions of the radiation field and on the local oscillator irradiance. This technique may be useful in regions of the spectrum where high quantum efficiency detectors are not available since performance similar to that of the conventional unity quantum efficiency heterodyne receiver can theoretically be achieved. Practical problems which may make this difficult are discussed. A physical interpretation of the process in terms of the absorption of monochromatic and nonmonochromatic photons is given. The double-quantum case is treated in particular detail; the results of a preliminary experiment are presented and suggestions for future experiments to ascertain the usefulness of the technique are provided.

We then investigate the operation and performance of a three-frequency nonlinear heterodyne system which eliminates some of the stringent conditions required for conventional heterodyne detection while maintaining its near-ideal SNR. The technique is similar to heterodyne radiometry, but carefully takes into consideration the effects of Doppler shift and signal statistics. It makes use of a two-frequency transmitter and a nonlinear second detector, and is particularly useful for signal acquisition; for signals of unknown Doppler shift, in fact, performance is generally superior to that of the conventional system because of a reduction in the effective noise bandwidth. While primary emphasis is on the infrared and optical because of the large Doppler shifts encountered there, application of the principle in the microwave and radiowave is also discussed. For cw radar and analog communications, the signal-to-noise, power spectral density, and minimum detectable power are obtained and compared with the standard configuration. Both sinewave and Gaussian input signals are treated. A variety of specific cases is discussed including the optimum performance case, the typical radar case, and the AM and FM communications case. The technique is shown to have similar advantages for pulsed radar and digital communications applications, both in the absence and in the presence of the lognormal atmospheric channel. Computer-generated error probability curves as a function of the input signal-to-noise ratio are presented for a variety of binary receiver parameters and configurations, and for various levels of atmospheric turbulence. Orthogonal and nonorthogonal signaling schemes, as well as dependent and independent fading, are considered.

In the last part of the chapter, we extend three-frequency nonlinear heterodyne mixing to n frequencies and examine the performance of a Doppler-insensitive radiometer that detects the radiation from known species moving with unknown velocities. Expressions for the signal-to-noise ratio and the minimum detectable total power are obtained for sinusoidal signals and for Gaussian signals with both Gaussian and Lorentzian spectra. In distinction to conventional heterodyning, knowledge of absolute line rest frequencies and a stable, tunable local oscillator are not required. This configuration may find use in the detection of certain remote species such as interstellar molecules and pollutants. A number of potential applications are examined. Finally, attention is drawn to a recently proposed variation of the technique, called heterodyne correlation radiometry, that incorporates a radiating sample of the species to be detected as part of the laboratory receiver. This configuration should be useful for the sensitive detection of species whose radiated energy is distributed over a large number of lines, with frequencies that are not necessarily known, when the Doppler shift is given.

Conventional photomixing in the infrared and optical is a useful detection technique for applications such as optical communications, spectroscopy, and radiometry, and it has been studied in great detail. The effect was first observed by *Forrester* et al. [7.1] in a classic experiment using two Zeeman components of a visible (incoherent) spectral line. With the development of the laser, photomixing became considerably easier to observe and was studied by *Javan* et al. [7.2] at 1.15 μm using a He–Ne laser, and by *Siegman* et al. [7.3] at 6943 Å with a ruby laser. Extending this work into the middle infrared, *Teich* et al. used a CO_2 laser at 10.6 μm in conjunction with a copper-doped germanium photoconductive detector operated at 4 K [7.4], and subsequently with a lead-tin selenide photovoltaic detector operated at 77 K [7.5].

The observed signal-to-noise power ratio for these experiments was found to behave in accordance with the theoretical expression obtained for parallel, plane-polarized beams incident on a quantum-noise-limited detector under ideal conditions [7.4–7], i.e.,

$$\text{SNR}^{(1)} = \eta_1 P_1 / h v \Delta f. \tag{7.1}$$

Here, η_1 is the detector quantum efficiency (electrons/photon), P_1 is the received signal radiation power, hv is the photon energy, and Δf is the receiver bandwidth. Heterodyne detection which is Johnson-noise rather than quantum-noise limited will be considered in Section 7.4.3. For radiation beams which are not parallel to within an angle $\theta = \lambda/d$, with d the detector aperture and λ the radiation wavelength, the SNR is reduced below the value given in (7.1) by spatial averaging of the mixing signal over the detector aperture. This effect was studied in detail by *Siegman* [7.7], and is often referred to as "washboarding". Similar calculations have been effected for focused radiation beams, first considered by *Read* and *Fried* [7.8]. Other factors must also be accounted for in a real system [7.8a]. Furthermore, it is clear that the signal-to-noise ratio is useful as a criterion only under certain conditions. *Jakeman* et al. [7.8b] have recently examined homodyne detection for signal detection and estimation experiments in which performance is more naturally linked to other measures such as error probability or various estimators.

Other experimental and theoretical studies focused on the statistical nature of the heterodyne signal resulting from the beating of a coherent wave with a Gaussian (scattered) wave [7.9, 10]. Although the stochastic nature of this signal was found to depend in detail on the irradiance statistics, the SNR turned out to be essentially independent of the higher-order correlation functions of the field [7.9, 10]. Furthermore, using the first-order coherent field results of *Titulaer* and *Glauber* [7.11] for absorption detectors, an explicit calculation for the case of two-beam photomixing showed that sum- and double-frequency components did not appear in the detected current, and that the heterodyne process could be interpreted in terms of the annihilation of a single (nonmonochromatic) photon

[7.12, 13] as was qualitatively appreciated by *Forrester* et al. [7.1]. A concise review of the basic theoretical and experimental aspects of heterodyne detection in the infrared and optical, as well as a partial review of the literature, was prepared by *Teich* in 1970 [7.14]. Recently, *Mandel* and *Wolf* [7.15] used geometrical and statistical arguments to show that an optimum receiver area exists for conventional heterodyne detection; this result complements *Siegman*'s antenna theorem [7.7], according to which the product of the receiver area and the angular field of view is of the order of the wavelength λ squared. A number of authors have also examined the photon counting statistics of the superposition of coherent and chaotic signal components, with the same or different mean frequencies [7.15a–15f]. In short, conventional optical heterodyne detection is well understood both theoretically and experimentally, and is a highly useful technique from a practical point of view.

7.2 Two-Frequency Multiphoton Heterodyne Detection

Since multiple-quantum optical direct detection has also been studied in great detail, it seems natural to investigate the behavior of such a detector in the presence of more than one frequency [7.16]. In this section, we obtain the response and the signal-to-noise power ratio (SNR) for a multiple-quantum absorption heterodyne receiver, with particular attention devoted to the simplest case, i.e., the mixing of two waves in a double-quantum infrared or optical device.

After briefly considering the relevant results pertinent to multiple-quantum direct detection (Sec. 7.2.1), we derive the combination device response for the general multiple-quantum photomixing process (including the important two-quantum case) in Section 7.2.2. In Section 7.2.3, we obtain the SNR for a receiver using a multiphoton optical heterodyne device, and compare it with the SNR for conventional optical heterodyne detection. The results of a two-photon experiment are presented in Section 7.2.4, while a suggested setup for future experiments, as well as the applicability of the scheme in general, is reserved for Section 7.2.5.

7.2.1 Multiple-Quantum Direct Detection

The ordinary photoeffect was discovered by *Hertz* in 1887 and explained in terms of the absorption of a single quantum of light by *Einstein* in his now famous work published in 1905 [7.17]. It was not until 1959, however, that the relationship between the statistics of an arbitrary incident radiation field and the emitted photoelectrons was firmly established by *Mandel* [7.18]. Consideration of the general photodetection process in terms of quantum-electrodynamic coherent states of the radiation field was undertaken by *Glauber* [7.19] in 1963, and by *Kelley* and *Kleiner* [7.20] in 1964, and provides a convenient starting point for calculations involving multiple-photon as well as single-photon absorptions.

Multiple-quantum photoemission, being a higher-order effect, is most easily observed in the absence of ordinary (first-order) photoemission. For the two-quantum case, it becomes important when

$$\tfrac{1}{2}H < hv < H, \tag{7.2}$$

where hv is the photon energy of the incident radiation, and H is the work function of the material under consideration. (Even when (7.2) is satisfied, however, it should be kept in mind that small amounts of single-quantum photoemission can arise from excited electrons in the Fermi tail [7.21].) The two-quantum photoeffect was first experimentally observed in 1964. Using a GaAs laser, *Teich* et al. [7.22] observed the effect in sodium metal, while *Sonnenberg* et al. [7.23] induced it in Cs_3Sb with a Nd-doped glass laser. Since that time, there have been a number of experimental measurements of second- and higher-order photoelectric yields in a variety of materials [7.24–28].

Theoretical work has focused on two aspects of the problem: perturbation theory and other calculations of the transition probabilities in the material, and the effect on the transition probability of the statistical nature of the radiation field. *Makinson* and *Buckingham* [7.29] were the first to predict the second-order effect and calculate its magnitude based on a surface model of photoemission; this work was expanded by *Smith* [7.30], *Bowers* [7.31], and *Adawi* [7.32]. The analogous volume calculation was performed by *Bloch* [7.33] and later corrected by *Teich* and *Wolga* [7.24, 25].

All of the models predict a two-quantum dc photocurrent $W_{dc}^{(2)}$ (expressed in amperes) proportional to the square of the incident radiation power P and inversely proportional to the irradiated area A. Using the results of a number of authors [7.24, 25, 30–33], we can therefore write the double-quantum dc photocurrent as

$$W_{dc}^{(2)} = \Lambda^{(2)}(\lambda, T) P \propto IP. \tag{7.3}$$

Here $\Lambda^{(2)}$ is the two-quantum yield expressed in amperes/watt [7.8], λ is the radiation wavelength, T is the sample temperature, P is the radiation power expressed in watts, and I is the irradiance at the detector expressed in watts/cm^2. The two-quantum efficiency (electrons/photon) is denoted by η_2, and is related to the two-quantum yield by the relationship

$$\Lambda^{(2)} = (e/hv)\eta_2 \propto I. \tag{7.4}$$

Here, the quantity (hv/e) is the incident photon energy expressed in eV and is of order unity. For the k-photon process, defining $W_{dc}^{(k)}$, $\Lambda^{(k)}$, and η_k as the k-photon analogs of the quantities defined above, the following generalized results are obtained:

$$W_{dc}^{(k)} = \Lambda^{(k)}(\lambda, T) P \propto I^{k-1} P \tag{7.5}$$

and

$$\Lambda^{(k)} = (e/hv)\eta_k \propto I^{k-1}. \tag{7.6}$$

Typical numerical values for the two-quantum yield are [7.25] $\Lambda_{Na}^{(2)}$ (8450 Å, 300 K) $\sim 8 \times 10^{-16} I$ and $\Lambda_{Cs_3Sb}^{(2)}$ (10600 Å, 300 K) $\sim 5 \times 10^{-11} I$ amperes/watt. Again I represents the irradiance at the detector in watts/cm^2. These values, even when precisely measured, can vary by a factor (usually ≤ 2) depending on the coherence properties of the inducing radiation, as we now consider.

Theoretical work relating to multiple-quantum statistical effects began in 1966 with an examination of the higher-order field correlation functions by *Teich* and *Wolga* [7.34] and by *Lambropoulos* et al. [7.35]. This was followed by more detailed calculations by *Mollow* [7.36] and by *Agarwal* [7.37]. All of these studies predicted a factor of $k!$ enhancement for the magnitude of certain k-quantum processes induced by chaotic (rather than coherent) sources. This enhancement was later observed in the two-quantum photoeffect by *Shiga* and *Imamura* [7.26], and in second harmonic generation (SHG) by *Teich* et al. [7.38]. The theoretical relationship between two-quantum photocurrent spectra and the incident radiation statistics was then obtained by *Diament* and *Teich* [7.39], and compared with the analogous single-quantum results previously given by *Freed* and *Haus* [7.40]. In 1969, two-quantum photocounting distributions were calculated for amplitude-stabilized, chaotic, and generalized laser sources by *Teich* and *Diament* [7.41]. This work was extended to higher-order photocounting distributions by *Barashev* in 1970 [7.42], who also wrote a comprehensive review article on multiple-quantum photoemission and photostatistics in 1972 [7.43]. Detailed calculations of the generalized higher-order photocounting statistics have also been reported by *Peřina* et al. [7.15e].

7.2.2 Theory of Multiphoton-Photomixing

We begin this section by considering a two-quantum absorption detector initially in the ground state. The detector response $W^{(2)}$ at the space-time point $x_a = r_a, t_a$ may be written in terms of the second-order correlation function $G^{(2)}$ [7.19, 34, 36, 41], and is given by

$$W^{(2)} \propto \mathrm{tr}\{\varrho E^-(x_a)E^-(x_a)E^+(x_a)E^+(x_a)\} \equiv G^{(2)}(x_a x_a x_a x_a). \tag{7.7}$$

Here, ϱ is the density operator for the field, and E^- and E^+ represent the negative- and positive-frequency portions of the electric field operator E, respectively. We assume that the final state of the detector is much broader than the bandwidth of the incident radiation, and that a broad band of final states is accessible [7.36, 37].

If we specifically consider the mixing of two single-mode, amplitude-stabilized, first-order coherent waves, both of which are well collimated, parallel, plane polarized along a common unit vector, and normally incident onto a photosensitive material, we may write the positive portion of the electric field operator E^+ as the superposition of two scalar fields

$$E^+ = \varepsilon_1^0 e^{-i\omega_1 t} + \varepsilon_2^0 e^{-i\omega_2 t} \tag{7.8}$$

with angular frequencies ω_1 and ω_2. This is equivalent to assuming a semiclassical approach which makes use of the analytic signal [7.44]. The complex wave amplitude ε_i^0 can be expressed in terms of its absolute magnitude

$|\varepsilon_i^0|$ and a phase factor $\exp(i\alpha_i)$ such that

$$\varepsilon_1^0 = |\varepsilon_1^0| e^{i\alpha} \tag{7.9}$$
$$\varepsilon_2^0 = |\varepsilon_2^0| e^{i\beta}.$$

Under these conditions, the quantum-statistical detector responses can be written in terms of the fields as

$$\text{tr}\{\varrho E^- E^+\} \Rightarrow |\varepsilon_1^0|^2 + |\varepsilon_2^0|^2 + 2|\varepsilon_1^0||\varepsilon_2^0|\cos[(\omega_1-\omega_2)t + (\beta-\alpha)] \tag{7.10}$$

and

$$\text{tr}\{\varrho E^- E^- E^+ E^+\}$$
$$\Rightarrow \{|\varepsilon_1^0|^2 + |\varepsilon_2^0|^2 + 2|\varepsilon_1^0||\varepsilon_2^0|\cos[(\omega_1-\omega_2)t + (\beta-\alpha)]\}^2. \tag{7.11}$$

These expressions are scalar quantities and contain no spatial dependence because of the assumptions of plane polarization, parallel beams, and normal incidence.

Generalizing these results to sinusoidal beam photomixing in which the k-photon detector response is the normally ordered product [7.19]

$$W^{(k)} \propto \text{tr}\{\varrho[E^-(x_a)]^k[E^+(x_a)]^k\}, \tag{7.12}$$

and using the binomial theorem leads to a heterodyne signal given by

$$W^{(k)} = \zeta_k(\{|\varepsilon_1^0|^2 + |\varepsilon_2^0|^2\}^k$$
$$+ \binom{k}{1}\{|\varepsilon_1^0|^2 + |\varepsilon_2^0|^2\}^{k-1}\{2|\varepsilon_1^0||\varepsilon_2^0|\cos[(\omega_1-\omega_2)t + (\beta-\alpha)]\}$$
$$+ \binom{k}{2}\{|\varepsilon_1^0|^2 + |\varepsilon_2^0|^2\}^{k-2}\{2|\varepsilon_1^0||\varepsilon_2^0|\cos[(\omega_1-\omega_2)t + (\beta-\alpha)]\}^2 + \dots +$$
$$+ \binom{k}{r}\{|\varepsilon_1^0|^2 + |\varepsilon_2^0|^2\}^{k-r}\{2|\varepsilon_1^0||\varepsilon_2^0|\cos[(\omega_1-\omega_2)t + (\beta-\alpha)]\}^r + \dots +$$
$$+ \{2|\varepsilon_1^0||\varepsilon_2^0|\cos[(\omega_1-\omega_2)t + (\beta-\alpha)]\}^k). \tag{7.13}$$

Here ζ_k represents a proportionality constant for the k-photon process. The leading dc terms are proportional to $|\varepsilon_1^0|^{2k}$ and $|\varepsilon_2^0|^{2k}$, and may be associated with the absorption of k monochromatic photons, each of which arises from a given beam (1 and 2, respectively). The highest frequency current component is proportional to $|\varepsilon_1^0|^k|\varepsilon_2^0|^k \cos[k(\omega_1-\omega_2)t + \phi]$, and corresponds to the absorption of k nonmonochromatic photons, each of which must be associated with both of the beams. It is evident from the above that multiple- and sum-frequency terms do not appear in the k-photon absorption heterodyne detector output, in analogy with the result for the one-quantum case [7.12–14].

Inserting the constants ζ for the one- and two-quantum cases in (7.13) above, the detector responses for coherent signal mixing are, respectively,

$$W^{(1)} = \zeta_1\{|\varepsilon_1^0|^2 + |\varepsilon_2^0|^2 + 2|\varepsilon_1^0||\varepsilon_2^0|\cos[(\omega_1-\omega_2)t + (\beta-\alpha)]\} \tag{7.14}$$

and

$$W^{(2)} = \zeta_2\{|\varepsilon_1^0|^4 + |\varepsilon_2^0|^4 + 2|\varepsilon_1^0|^2|\varepsilon_2^0|^2 + 4|\varepsilon_1^0|^3|\varepsilon_2^0|\cos[(\omega_1-\omega_2)t + (\beta-\alpha)]$$
$$+ 4|\varepsilon_1^0||\varepsilon_2^0|^3 \cos[(\omega_1-\omega_2)t + (\beta-\alpha)]$$
$$+ 4|\varepsilon_1^0|^2|\varepsilon_2^0|^2 \cos^2[(\omega_1-\omega_2)t + (\beta-\alpha)]\}. \tag{7.15}$$

Using the double-angle formula for the last term in (7.15), $W^{(2)}$ may also be written as

$$W^{(2)} = \zeta_2 \{|\varepsilon_1^0|^4 + |\varepsilon_2^0|^4 + 4|\varepsilon_1^0|^2 |\varepsilon_2^0|^2 + 4|\varepsilon_1^0|^3 |\varepsilon_2^0| \cos[(\omega_1 - \omega_2)t + (\beta - \alpha)]$$
$$+ 4|\varepsilon_1^0| |\varepsilon_2^0|^3 \cos[(\omega_1 - \omega_2)t + (\beta - \alpha)]$$
$$+ 2|\varepsilon_1^0|^2 |\varepsilon_2^0|^2 \cos[2(\omega_1 - \omega_2)t + 2(\beta - \alpha)]\}, \qquad (7.16)$$

when this $\cos^2$ term is present. As noted previously, double- and sum-frequency terms are absent.

It is not difficult to associate various second-order correlation functions $G^{(2)}(x_a x_b x_c x_d) \equiv [abcd]$ with (7.15). (When two beams are present, we must consider a space-time point for each of the beams so that the index in $G^{(2)}$ takes on two values [7.34].) Thus, the first term, $|\varepsilon_1^0|^4$, may be associated with [1111], the second with [2222], the third with [1221] and [2112], the fourth with the four permutations of [1112], the fifth with the four permutations of [2221], and the sixth with the four permutations of [1212], with $b \neq c$. The coefficient of each term in (7.15) is therefore equal to the number of permutations in the appropriate form of the correlation function for that term. The physical interpretation follows immediately: the first two dc terms in (7.15) arise from the absorption of two monochromatic photons, both from the same beam. The third dc term, which exists in two permutations with $b = c$, arises from the two ways in which two single monochromatic photons can be absorbed, one from each beam. The fourth and fifth terms correspond to the absorption of a single monochromatic photon from one of the beams plus a single nonmonochromatic photon which must be associated with both beams. These terms therefore contribute currents at the difference frequency $(\omega_1 - \omega_2)$, in analogy with the single-quantum heterodyne interference term [7.12–14]. The final term corresponds to the absorption of two nonmonochromatic photons, and therefore varies at double the difference frequency, i.e., at $2(\omega_1 - \omega_2)$; clearly there is no analogous process possible in the one-quantum case.

We note that the absorption of two nonmonochromatic photons imparts an additional dc value to the double-difference-frequency term, as may be seen by comparing (7.15) and (7.16). This additional term, of magnitude $2|\varepsilon_1^0|^2 |\varepsilon_2^0|^2$, appears in the presence of double-quantum photomixing; in the absence of such photomixing, we must obtain $W^{(2)}$ from (7.15) and *not* from (7.16). In this latter case, the detector response reduces to the previously obtained result [7.34]

$$W^{(2)}(\text{mixing absent}) = \zeta_2(|\varepsilon_1^0|^4 + |\varepsilon_2^0|^4 + 2|\varepsilon_1^0|^2 |\varepsilon_2^0|^2) = \zeta_2'(I_1 + I_2)^2, \qquad (7.17)$$

where I_i represents the intensity of the ith beam and ζ_2' is a new proportionality constant.

The results presented above can be expanded to modulated, noncoherent, and nonparallel beam mixing. As an example, we consider two ideal amplitude-stabilized nonparallel $(\theta > \lambda/d)$ plane traveling waves impinging on a two-quantum detector, so that washboarding can occur. In contrast to the one-quantum case, the detector responds to the *square* of this spatiotemporal intensity variation, resulting in a factor of 2 enhancement in the dc cross term, as obtained with pure temporal mixing. Thus, the two-quantum dc photocurrent will in general be enhanced due to spatiotemporal intensity variations (interference fringes); the magnitude of this enhancement depends on the system configuration. Experimental evidence for two-quantum enhancement due to spatial variations has, in fact, been provided by *Shiga* and *Imamura* [7.26] and by *Teich* et al. [7.38].

As a final example, we consider mixing due to radiation which is nonsinusoidal (i.e., not coherent to all orders). We consider two parallel, plane-polarized normally incident superimposed beams of radiation from the same chaotic source, one of which is a time-delayed version of the other (delay τ_δ) entering one double-quantum detector. This was previously shown to be equivalent to a self-integrating Hanbury-Brown–Twiss device [7.34]. For a thermal source in the absence of a beat signal, we find

$$W^{(2)}(\text{mixing absent}) = 2\zeta_2''(I_1^2 + 2I_1 I_2 + I_2^2),\ \tau_\delta < \tau_c \tag{7.18}$$

and

$$W^{(2)}(\text{mixing absent}) = 2\zeta_2''(I_1^2 + I_1 I_2 + I_2^2),\ \tau_\delta > \tau_c, \tag{7.19}$$

where τ_c is the coherence time of the source. For $\tau_\delta < \tau_c$, (7.18) represents the enhancement of both the single-beam and the mixed-beam counting rates, arising from the tendency of these photons to arrive in correlated pairs (assuming that the detector intermediate state lifetime $\tau_1 \ll \tau_c$). For $\tau_\delta > \tau_c$, however, there is no correlation between the arrival time of a photon from one beam and the arrival time of a photon from the other. Thus, the absorption of two photons from a single beam is enhanced by a factor of 2 relative to the absorption of one photon from each beam, leading to a cross term of 1. This can also be qualitatively understood from the point of view of additive Gaussian fields; the sum of two fully correlated Gaussian random processes ($\tau_\delta < \tau_c$) has a greater variance than that of two independent Gaussian random processes ($\tau_\delta > \tau_c$) leading to an enhanced value for the cross term when $\tau_\delta < \tau_c$. An arrangement to observe spatial effects of a similar type has also been proposed [7.44a].

From the foregoing, it is clear that the double-quantum current can be calculated for a variety of configurations involving different relative time scales, angular separations, polarization properties, and statistical characteristics. Some additional examples are treated in [7.34]. Clearly, the second-order correlation functions of the field play an important role in determining the magnitude of the signal, in distinction to the one-quantum case.

7.2.3 Signal-to-Noise Ratio and Minimum Detectable Number of Photons

We now follow the usual procedure used for the single-quantum case [7.4–6, 10] to calculate the approximate SNR for k-photon sinusoidal heterodyne detection. We begin with two-quantum photomixing, neglecting the double-difference-frequency component and assuming that the ac signal is at the fundamental-difference-frequency (IF) between the two waves. Thus, considering the mixing of two parallel, coherent waves as described earlier, (7.16) yields

$$W_{IF}^{(2)} = 4\zeta_2(|\varepsilon_1^0|^3|\varepsilon_2^0| + |\varepsilon_1^0||\varepsilon_2^0|^3)\cos[(\omega_1 - \omega_2)t + (\beta - \alpha)] \quad (7.20)$$

and

$$W_{dc}^{(2)} = \zeta_2(|\varepsilon_1^0|^4 + |\varepsilon_2^0|^4 + 4|\varepsilon_1^0|^2|\varepsilon_2^0|^2), \quad (7.21)$$

so that

$$W^{(2)} = \left\{1 + \frac{4(|\varepsilon_1^0|^3|\varepsilon_2^0| + |\varepsilon_1^0||\varepsilon_2^0|^3)\cos[(\omega_1 - \omega_2)t + (\beta - \alpha)]}{|\varepsilon_1^0|^4 + |\varepsilon_2^0|^4 + 4|\varepsilon_1^0|^2|\varepsilon_2^0|^2}\right\} W_{dc}^{(2)}. \quad (7.22)$$

We now assume that one of the waves (which we call the local oscillator or LO) is strong, i.e., $E_2 \gg E_1$, in which case

$$W_{IF}^{(2)} \simeq 4(|\varepsilon_1^0|/|\varepsilon_2^0|)W_{dc}^{(2)}\cos[(\omega_1 - \omega_2)t + (\beta - \alpha)] \quad (7.23)$$

and

$$\langle [W_{IF}^{(2)}]^2 \rangle \simeq 8(|\varepsilon_1^0|^2/|\varepsilon_2^0|^2)[W_{dc}^{(2)}]^2. \quad (7.24)$$

The noise power can be obtained from the two-quantum photocurrent spectrum [7.39] which, in turn, is related to the stochastic nature of the radiation source. For a coherent and strong LO, however, the k-quantum counting statistics will be Poisson [7.41, 42], and the two-quantum (shot) noise power is then

$$\langle [W_n^{(2)}]^2 \rangle = 2e[W_{dc}^{(2)}]\Delta f. \quad (7.25)$$

Thus, using (7.3), (7.24), and (7.25), the two-quantum SNR can be written as

$$\text{SNR}^{(2)} \simeq \frac{4P_1}{e\Delta f}\left[\frac{W_{dc}^{(2)}}{P_2}\right] = \frac{4P_1}{e\Delta f}\Lambda^{(2)}. \quad (7.26)$$

Using the relationship between the two-quantum yield $\Lambda^{(2)}$ and the two-quantum efficiency η_2 given in (7.4), we finally obtain

$$\text{SNR}^{(2)} \simeq 4\eta_2 P_1/h\nu\Delta f. \quad (7.27)$$

We recall from (7.4) that η_2 is itself proportional to the irradiance of the LO, and we must have $4\eta_2 < 1$. The result is therefore similar to that for the single-quantum heterodyne detector given in (7.1); in that case, however, η_1 is independent of the LO. The two-quantum minimum detectable power (MDP) [7.4, 5] therefore becomes

$$\text{MDP}^{(2)} \simeq h\nu\Delta f/4\eta_2. \quad (7.28)$$

This corresponds to a minimum number of photons $\mathcal{N}_{\min}^{(2)}$, detectable in the resolution time of the receiver $[\tau_r \sim (\Delta f)^{-1}]$, given by

$$\mathcal{N}_{\min}^{(2)} \simeq (4\eta_2)^{-1}. \quad (7.29)$$

In contradistinction to the single-quantum case, performance is not limited by the detector (single) quantum efficiency since η_2 may be increased by increasing the LO irradiance. This technique may therefore be useful in regions of the electromagnetic spectrum where detectors with high (single) quantum efficiency are not available.

We note that the SNR at the double-difference-frequency, corresponding to the absorption of two nonmonochromatic photons, is reduced by the factor (P_1/P_2). Clearly, using methods similar to those presented above and in [7.34] and [7.10], we can obtain analogous SNR expressions for photomixing with nonsinusoidal beams.

The SNR at the fundamental-difference-frequency $(\omega_1 - \omega_2)$ may also be obtained for coherent beam mixing in the k-photon absorption heterodyne detector. Following a series of steps similar to those given above, we find

$$\text{SNR}^{(k)} \simeq c_k \eta_k P_1 / h\nu \Delta f , \qquad (7.30)$$

$$\text{MDP}^{(k)} \simeq h\nu \Delta f / c_k \eta_k , \qquad (7.31)$$

and

$$\mathcal{N}^{(k)}_{\min} \simeq (c_k \eta_k)^{-1} , \qquad (7.32)$$

where c_k is a constant (dependent on k), and η_k is proportional to I^{k-1}, where I is the LO irradiance [see (7.6)]. Here $c_k \eta_k < 1$.

7.2.4 Experiment

In this section we describe a preliminary set of experiments in which double-quantum photoemission was observed from a sodium surface simultaneously illuminated by two superimposed beams of laser radiation. While ac photomixing terms were not observed in these experiments, the measurements are consistent with the theoretical calculations given in Section 7.2.2.

The apparatus used for the experimental measurement of two-quantum photomixing is shown in the block diagram of Fig. 7.1. The radiation source was a pulsed GaAs multimode semiconductor on laser operated at 77 K and emitting a peak radiation power of 400 mW at about 8450 Å. Mode shifts due to laser heating occurred during the pulse duration, which was about 35 µs. The radiation was collimated by a 10-cm focal length lens, passed through an iris and then through a configuration of dielectric beam splitters and antireflection coated prisms resembling a Mach-Zender interferometer. The beam splitters were approximately 2/3 transmitting and were flat only to about 1 wavelength; the optical phase across the beam could therefore be considered to vary. The purpose of the interferometer configuration was to allow the irradiance of each beam (denoted as 1 and 2) to be independently controlled by means of calibrated attenuating filters. Beam 1 could also be time delayed with respect to beam 2 by means of a sliding prism (see Fig. 7.1), but this capability was not important in these experiments where τ_δ was always greater than τ_c due to the very small value of τ_c. After passing through a second iris and a (6-cm focal length) focusing lens, the radiation was allowed to impinge on a specially constructed Na-surface photomultiplier tube, which has been described previously [7.22, 24, 25]. A Polaroid type HN-7 sheet polarizer was almost always placed at the front face of the photomultiplier as shown in Fig. 7.1 (the one exception will be noted later). The electron-multiplied current was passed through a 1-MΩ load resistor which fed a Princeton Applied Research (PAR) low-noise preamplifier followed by a PAR lock-in amplifier. Phase-sensitive detection was performed as 2.2 kHz, which is the fundamental repetition frequency of the pulsed

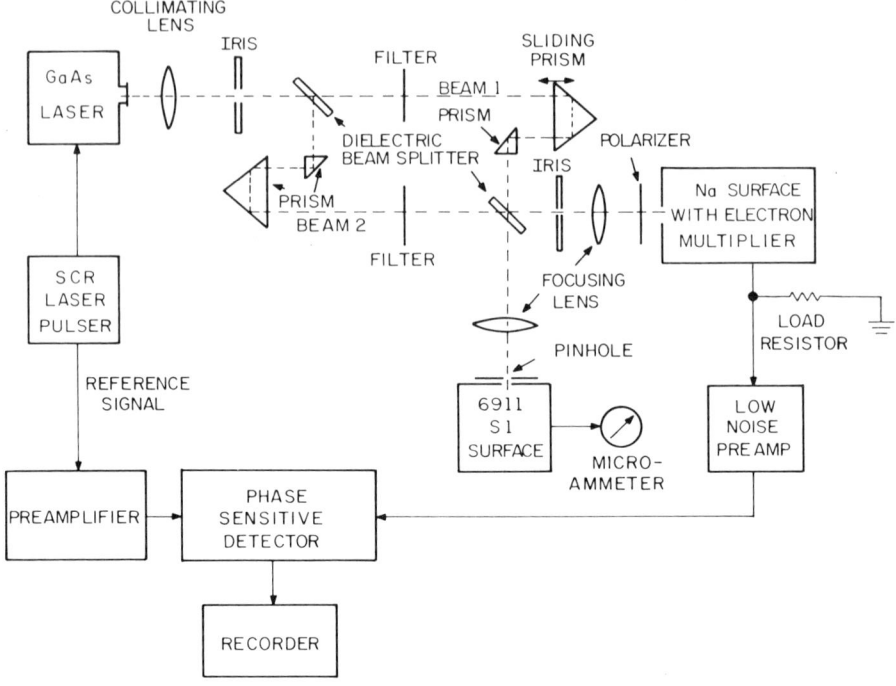

Fig. 7.1. Block diagram of the double-quantum photomixing experimental arrangement

laser output. Large integration times were used so that only the dc or average value of this current component was measured. The reference signal for the lock-in amplifier was obtained directly from the silicon-controlled-rectifier power supply [7.45] used to drive the laser.

Radiation from the other leg of the interferometer was focused onto a 25-μm diameter pinhole which acted as an aperture stop at the face of a standard Dumont 6911 type S–1 photomultiplier. This provided a relatively accurate method for superimposing the two beams [7.24]. This is critical since the double-quantum response is inversely proportional to the illuminated area A. The beams were adjusted to achieve maximum output from the 6911 photomultiplier tube, a procedure which was often difficult and required a great deal of care.

The following procedure was used in making a measurement: 1) The beams were aligned to provide maximum current from the 6911 photomultiplier. 2) Beam 1 was blocked and the double-quantum current $\overline{W_2^{(2)}}$ from beam 2 was maximized by imaging the laser junction on the sodium surface, and then recorded. Using a calibrated attenuating filter, it was ascertained that pure two-quantum emission was occurring, i.e., that $\overline{W_2^{(2)}} \propto I_2^2$, where I_2 represents the irradiance of beam 2. 3) Beam 2 was blocked and the double-quantum current from beam 1 was recorded, after verifying that it was $\propto I_1^2$. (The constant of proportionality was taken to be the same in both cases.) 4) Both beams were then unblocked and, after once again verifying that pure double-quantum emission was occurring, the total average double-quantum current $\overline{W^{(2)}}$ (at the fundamental repetition frequency of 2.2 kHz) was recorded.

Experiments were performed with different values of I_1/I_2, obtained by attenuating one of the beams relative to the other by means of thin gelatin (Kodak Wratten) filters. Ordinary glass filters could not be used to provide the decrements of light intensity because refraction in the glass caused the imaged spot size and position to change thus altering the two-quantum current in an unpredictable way.

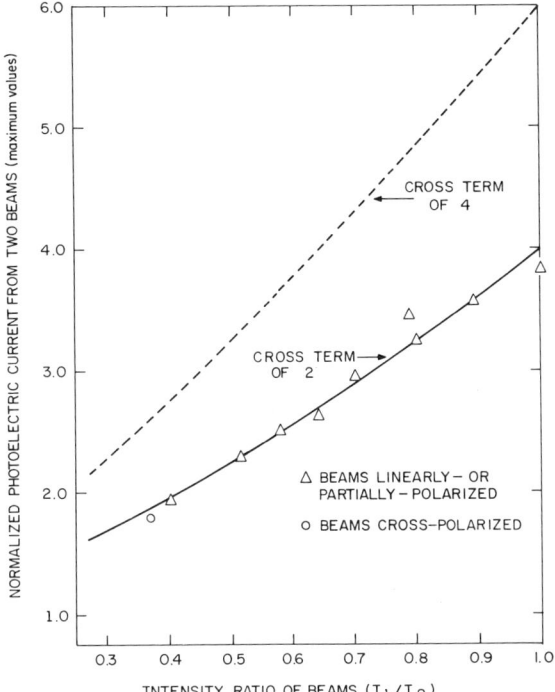

Fig. 7.2. The average double-quantum photocurrent from two beams (maximum values) vs the intensity ratio of the beams

The total average fundamental-repetition-frequency two-quantum photoelectric current $\overline{W^{(2)}}$, for different intensity ratios of the two constituent beams (I_1/I_2), is presented in Fig. 7.2. The solid curve represents the equation $W^{(2)} \propto (I_1 + I_2)^2$ which is simply the parabola $(1 + I_1/I_2)^2$ when the intensity I_2 is normalized to unity. This represents a cross term of $2I_1 I_2$, and is so labeled. The cross term of $4I_1 I_2$, on the other hand, is shown by the dashed line in Fig. 7.2. Only the highest observed values of $\overline{W^{(2)}}$ are plotted in Fig. 7.2, many more points having been found to lie below the curves. This has been attributed to the difficulty in obtaining precise alignment of the two radiation beams, and therefore superposition of the focused spots on the sodium surface.

The triangles in Fig. 7.2 represent data for linearly or partially polarized radiation, while the circle is for cross-polarized radiation (this is the one exception mentioned previously). The experimental measurements are consistent with the following interpretation. The laser output consists of a number of more-or-less independent Fabry-Perot modes which are changing during the pulse width due to heating of the laser junction [7.24]. The radiation may therefore be considered to behave as a Gaussian source with a coherence time $\tau_c \sim (\Delta v)^{-1} \sim 10^{-13}$ s. Since the intermediate-state lifetime for the double-quantum sodium photodetector is much shorter than the radiation coherence time, the irradiance fluctuations result in a factor of 2 enhancement of the single-beam photocurrents. As far as the irradiance cross-term is concerned, the two-quantum current will also be enhanced by a factor of 2 owing to the random spatial irradiance fluctuations across the detector for the superimposed beams. Thus, we obtain a relative cross term of 2, i.e., $\overline{W^{(2)}} \propto I_1^2 + 2I_1 I_2 + I_2^2$, in agreement with the data.

7.2.5 Discussion

From the foregoing, it is clear that multiple-quantum heterodyne detection is somewhat more complex than the analogous single-quantum process. In particular, the average detector response and the SNR are found to depend on the higher-order correlation functions of the radiation field and on the LO irradiance. From a physical point of view, it has been possible to associate various terms in the detected current with specific kinds of photon absorptions. Calculations of the SNR and MDP for a number of cases have been carried out. The results of a preliminary two-quantum photomixing experiment are in agreement with the theory.

Although it appears that the k-photon heterodyne detector can be made to perform as well as or better than the single-photon heterodyne detector by simply increasing the LO intensity, a number of practical problems would likely make this difficult. In as much as the transition probabilities decrease rapidly as k is increased, and are furthermore proportional to A^{1-k}, it appears that very high LO intensities would be required to place η_k anywhere in the vicinity of 0.1 for $k > 2$. Aside from alignment problems, these high intensities could result in thermionic emission from cathode heating, or possibly cathode damage.

The two-quantum case is therefore likely to be the most interesting, and also the easiest to examine experimentally. A possible arrangement for studying the effect in a more detailed and controlled fashion is the following. The radiation from a 0.5 mW He-Ne laser, operating at a wavelength of 1.15 µm, is passed through an acoustooptic modulator (which splits it into two frequencies) and a focusing lens. A 5 µm focused spot size, corresponding to an area of 2.5×10^{-7} cm^2, would then provide an incident irradiance $I \sim 2 \times 10^3$ W/cm^2. Using a Cs$_3$Sb photocathode with a work function ~ 2.05 eV and a yield [7.23] $\sim 5 \times 10^{-11} I$ amperes/watt, a two-quantum current $\sim 5 \times 10^{-11}$ amperes may then be obtained. An experiment of this nature would allow the validity of (7.15), (7.16), and (7.27) to be examined. A YAG:Nd laser could be substituted for the He-Ne laser for an even simpler experimental configuration, since focusing would not then be required.

Although the emphasis in this section has been on linearly polarized incident radiation, considerable enhancement of the k-quantum photocurrent may occur for circularly (or elliptically) polarized radiation, as recently discussed by a number of authors [7.46]. We note that information relating to the intermediate-state lifetime of the detector (τ_1) can be obtained by measuring the two-quantum detector output for various values of τ_c.

Finally, we observe that the use of a two-quantum photomixer in a three-frequency nonlinear heterodyne detection receiver would result in a reduction of the SNR by the factor (P_1/P_2), corresponding to the absorption of 2 nonmonochromatic photons as discussed earlier. It therefore does not appear to be suitable for this application. The next section is devoted to a discussion of the three-frequency technique, but using a single-photon detector, in which case the nonlinearity is derived from a circuit element rather than from a multiphoton process and the (undesirable) reduction factor does not appear.

7.3 Three-Frequency Single-Photon Heterodyne Detection Using a Nonlinear Device

The extension of conventional microwave heterodyne techniques into the infrared and visible, and the attendant increase in the Doppler shift by many orders of magnitude for a target of a given velocity, has provided improved target resolution capabilities [7.47], but there have been attendant difficulties as indicated previously. If the radial velocity of a target or the frequency difference between the transmitter and LO in a communications system has not been established, for example, then the heterodyne frequency is not known and it may be very difficult indeed to acquire a weak signal using the standard single- or multiphoton heterodyne technique. An unknown IF necessitates the use of broad bandwidth detection and electronics, resulting in a degraded signal-to-noise ratio, or perhaps the use of frequency scanning of the receiver or of the LO. The rate of such scanning is of course limited by the time response of the system. In order that the heterodyne signal remain within the electrical passband of the system, furthermore, the LO frequency must be relatively stable with reference to the signal frequency, and yet tunable so that it can track the Doppler shift, which varies in time. These various difficulties are more acute in the infrared and optical where large values of the IF are encountered (Doppler shift is proportional to the radiation frequency).

In this section we discuss the operation of a three-frequency single-photon nonlinear heterodyne detection scheme useful for cw and pulsed radar, and for analog and digital communications. The concept, which is similar to heterodyne radiometry, was first proposed in 1969 by *Teich* [7.48], and experimentally examined in 1972 by *Abrams* and *White* [7.49]. Applications for the system have since been studied in detail from a theoretical point of view [7.50–52]. The system appears to provide the near-ideal SNR offered by conventional heterodyne detection, while eliminating some of the difficulties discussed above. It often obviates the need for high-frequency electronics, improving impedance matching, system noise figure, and range of operation over the conventional case.

The usual frequency scanning is eliminated as is the necessity for a stable LO. And, it allows targets to be continuously observed with Doppler shifts of considerably greater magnitude and range than previously possible. This is particularly important in the infrared and optical, where Doppler shifts are generally large [7.47], and the importance of this technique is expected to be emphasized in these regions. Furthermore, it is possible for the three-frequency system to have a higher output SNR than the conventional system by providing a reduced noise bandwidth, as we will show later.

As defined by our usage in this section, the designation "nonlinear" refers to the electronics following the detector, and not to the process itself, which could be referred to as nonlinear in any case since it involves mixing or multiplication. This is again a different kind of nonlinearity than that discussed previously in Section 7.2. In this system, two signals of a small, but well-known difference frequency Δv are transmitted. With the LO frequency f_L, we then have a three-frequency mixing system. Aside from the heterodyne mixer, a nonlinear element such as a square-law device is included to provide an output signal at a frequency very close to Δv regardless of the Doppler shift of the transmitted signals.

The over-all system configuration is presented in Section 7.3.1. In Sections 7.3.2, 7.3.3, and 7.3.4, we consider applications of the system to a cw radar with sinewave, Gaussian/Gaussian, and Gaussian/Lorentzian input signals, respectively. Section 7.3.5 deals with its use in an analog communications system, whereas Section 7.3.6 is concerned with low-frequency applications of the technique. A numerical example in Section 7.3.7 is followed by evaluations of system performance for binary communications and pulsed radar in the vacuum channel (Sec. 7.3.8) and in the lognormal atmospheric channel (Sec. 7.3.9). A discussion is presented in Section 7.3.10. The main results are expressed as the output SNR for the system in terms of the input SNR.

7.3.1 System Configuration

In Fig. 7.3, we present a block diagram for a radar version of the system. A transmitter emits two waves of frequencies f_1 and f_2 whose difference $f_c = |f_1 - f_2| = \Delta v$ is known to high accuracy. (This is particularly easy to accomplish if the transmitter is a two-mode laser, since the modes tend to drift together keeping f_c constant, or if it is a single-frequency laser modulated into two frequency components.) The waves are Doppler shifted by the moving target, the nature of which is unimportant. Thus, a wave of frequency f will return with a frequency f' given by the standard nonrelativistic Doppler-shift formula

$$f' = f(1 \pm 2v_{\parallel}/c), \tag{7.33}$$

where $v_{\parallel}$ is the radial velocity of the target and c is the speed of light. Therefore, after scattering from the target, and choosing $f_1 > f_2$, the new frequency difference between the two waves f'_c is given by

$$f'_c = |f'_1 - f'_2| = f_c \pm (2v_{\parallel}/c) f_c . \tag{7.34}$$

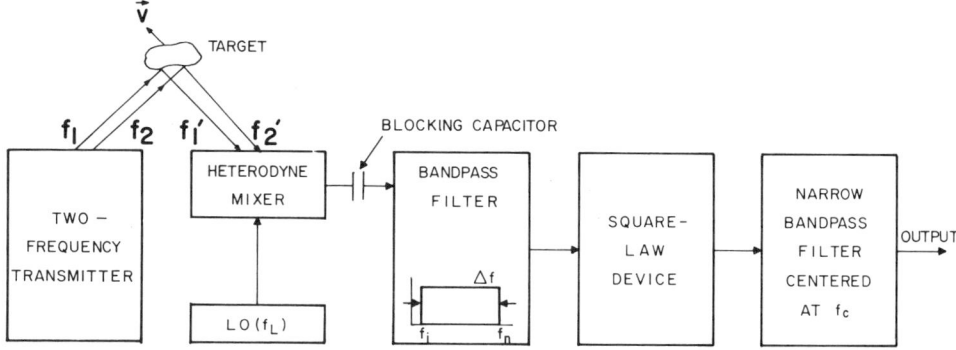

Fig. 7.3. Block diagram of the three-frequency nonlinear heterodyne system for radar application

Aside from the frequency shift which results from the Doppler effect, there may be a frequency broadening of each wave associated with the scattering by a moving target in a typical radar configuration. For a rotating target, this broadening is of the order $4R\omega_\perp/\lambda$, where R is the "radius" of the target, $\omega_\perp$ is its component of angular velocity perpendicular to the beam direction, and λ is the wavelength of the transmitted signal [7.9]. For practical systems, as will be shown later, the difference $f'_c - f_c$ may be made much smaller than the broadening effects and thereby neglected. We can therefore choose $f'_c = f_c$ with high accuracy.

The receiver includes a heterodyne mixer with an LO followed by a blocking capacitor and a bandpass filter with bandwidth $\Delta f = f_n - f_i$. Here f_n and f_i are the upper and lower cutoffs, respectively, of the bandpass filter. When Doppler information is poor (in which case the three-frequency system is particularly useful), Δf will be large so that it will cover a wide frequency range. In the following, we therefore pay particular attention to the case where $f_i \to 0$ so that $\Delta f = f_n$. The noise arising from the strong LO is assumed to be shot noise which, in the high current limit, becomes Gaussian as illustrated by *Davenport* and *Root* [7.53]. To good approximation, the spectrum may be taken to be white. The latter part of the receiver is a square-law (or other nonlinear) device and a narrow bandpass filter centered at frequency $f_c = |f_1 - f_2|$. The details of the system are given below.

The mixer consists of a photodetector and a local oscillator. We are interested in determining the signal-to-noise ratio at the output of the photodetector. The input electric field consists of three plane, parallel, coincident electromagnetic waves, which are assumed to be polarized and to impinge normally on the photodetector. Spatial first-order coherence is assumed over the detector aperture. The total incident electric field E_t may therefore be written as

$$E_t = A_1 \cos(\omega_1 t + \phi_1) + A_2 \cos(\omega_2 t + \phi_2) + A_L \cos(\omega_L t + \phi_L). \tag{7.35}$$

Here ω_1 and ω_2 are the angular frequencies of the two incoming signals (we have omitted the primes for simplicity), ϕ_1 and ϕ_2 are their phases, and ω_L is the angular frequency of the LO beam. The quantities A_1, A_2, and A_L are the amplitudes of the three waves, all of which are assumed to have the same plane polarization. In the infrared and optical, the output of a photodetector or mixer is proportional to the total intensity of the incoming waves. Taking into account the quantum electrodynamics of photon absorption by optical and infrared detectors [7.10, 12–14], the output signal r consists only of difference-frequency terms and dc terms. Thus, for $f_L < f_1', f_2'$ or $f_L > f_1', f_2'$,

$$r = \beta\{A_1^2 + A_2^2 + A_L^2 + 2A_1 A_L \cos[(\omega_1 - \omega_L)t + (\phi_1 - \phi_L)]$$
$$+ 2A_2 A_L \cos[(\omega_2 - \omega_L)t + (\phi_2 - \phi_L)]$$
$$+ 2A_1 A_2 \cos[(\omega_1 - \omega_2)t + (\phi_1 - \phi_2)]\}, \qquad (7.36)$$

where β is a proportionality constant containing the detector quantum efficiency. If the incident waves are not spatially first-order coherent and/or polarized in the same direction, the usual decrease in r will occur [7.5–8, 15].

Since the LO beam may be made much stronger than the two signal beams, i.e., $A_L \gg A_1, A_2$, we can write

$$r \simeq \beta A_L^2 \left\{ 1 + \frac{2A_1}{A_L} \cos[(\omega_1 - \omega_L)t + (\phi_1 - \phi_L)] \right.$$
$$\left. + \frac{2A_2}{A_L} \cos[(\omega_2 - \omega_L)t + (\phi_2 - \phi_L)] \right\}. \qquad (7.37)$$

The term containing $\omega_1 - \omega_2$ has been neglected because of its relatively small amplitude. We now define

$$r_{dc} \equiv \beta(A_1^2 + A_2^2 + A_L^2) \simeq \beta A_L^2, \qquad (7.38a)$$

and

$$r_{IF} \equiv 2\beta A_1 A_L \cos[(\omega_1 - \omega_L)t + (\phi_1 - \phi_L)]$$
$$+ 2\beta A_2 A_L \cos[(\omega_2 - \omega_L)t + (\phi_2 - \phi_L)]$$
$$= r_{dc} \left\{ \frac{2A_1}{A_L} \cos[(\omega_1 - \omega_L)t + (\phi_1 - \phi_L)] \right.$$
$$\left. + \frac{2A_2}{A_L} \cos[(\omega_2 - \omega_L)t + (\phi_2 - \phi_L)] \right\}. \qquad (7.38b)$$

The mean-square photodetector response is then given by

$$\langle r_{IF}^2 \rangle = \left(\frac{2A_1^2}{A_L^2} + \frac{2A_2^2}{A_L^2} \right) r_{dc}^2 = 2r_{dc}^2 \frac{P_1 + P_2}{P_L}, \qquad (7.39)$$

where P_1, P_2, and P_L are the radiation powers in the two signal beams and in the LO beam, respectively.

If we consider the noise response r_n of the detector as arising from shot noise, which is the case for the photoemitter and the ideal reverse-biased photodiode [7.5, 10, 14], the mean-square noise response is given by the well-known shot-noise formula [7.5, 54]

$$\langle r_n^2 \rangle = 2er_{dc}\Delta f, \tag{7.40}$$

in which Δf is the noise response bandwidth and is determined by the Doppler uncertainty, and e is the electronic charge. For a comparatively strong LO, we have

$$r_{dc} = \frac{\eta e}{h f_L} P_L, \tag{7.41}$$

where $\eta \equiv \eta_1$ is the quantum efficiency and h is Planck's constant.

From (7.39), (7.40), and (7.41), the signal-to-noise power ratio $(SNR)_{power} \equiv SNR^{(1)}$ is given by

$$(SNR)_{power} = \frac{\langle r_{IF}^2 \rangle}{\langle r_n^2 \rangle} = \frac{\eta(P_1 + P_2)}{h f_L \Delta f}, \tag{7.42a}$$

which is seen to be independent of P_L. If we define $P_r \equiv P_1 + P_2$, and let $v = f_L \approx f_1 \approx f_2$, and $(SNR)_i = (SNR)_{power}$, we obtain

$$(SNR)_i = \frac{\eta P_r}{h v \Delta f}. \tag{7.42b}$$

This is similar to (7.1), except that now P_r is the total input signal power. $(SNR)_i$ is referred to as the input signal-to-noise ratio to the square-law device following the photodetector.

By use of a blocking capacitor, the dc part of the photodetector response r_{dc} can be filtered out. The signal, which then has zero mean, is sent to a full-wave square-law device. If we let $s_a(t) = 2\beta A_1 A_L \cos[(\omega_1 - \omega_L)t + (\phi_1 - \phi_L)]$ and $s_b(t) = 2\beta A_2 A_L \cos[(\omega_2 - \omega_L)t + (\phi_2 - \phi_L)]$, and let $n(t)$ be the noise, then using a generalization of the "direct method" of *Davenport* and *Root* [7.55] for the sum of three signals, we can write the input to the square-law device $x(t)$ as

$$x(t) = s_a(t) + s_b(t) + n(t). \tag{7.43}$$

The output of the square-law device $y(t)$ is then given by

$$\begin{aligned} y(t) &= \alpha x^2(t) \\ &= \alpha[s_a^2(t) + s_b^2(t) + n^2(t) + 2s_a(t)s_b(t) \\ &\quad + 2s_a(t)n(t) + 2s_b(t)n(t)], \end{aligned} \tag{7.44}$$

where α is a scaling constant. For a stationary random process, the expectation value of $y(t)$ is

$$E(y) = \alpha[E(s_a^2) + E(s_b^2) + E(n^2)]$$
$$= \alpha(\sigma_a^2 + \sigma_b^2 + \sigma_n^2) \tag{7.45}$$

for all t, where E denotes the expectation value. In (7.45), we have set $\sigma_a^2 = E(s_a^2)$, $\sigma_b^2 = E(s_b^2)$, and $\sigma_n^2 = E(n^2)$. Furthermore,

$$y^2(t) = \alpha^2[s_a(t) + s_b(t) + n(t)]^4, \tag{7.46}$$

from which the mean-square value of $y(t)$ is evaluated to be

$$E(y^2) = \alpha^2[E(s_a^4) + E(s_b^4) + E(n^4)$$
$$+ 6\sigma_a^2\sigma_b^2 + 6\sigma_a^2\sigma_n^2 + 6\sigma_b^2\sigma_n^2]. \tag{7.47}$$

In obtaining (7.45) and (7.47), we have assumed that $s_a(t)$, $s_b(t)$, and $n(t)$ are all independent of each other, and that $E(s_a) = E(s_b) = E(n) = 0$.

The autocorrelation function of the output of the square-law device is

$$R_y(t_1, t_2) = E(y_1 y_2) = \alpha^2 E[(s_{a1} + s_{b1} + n_1)^2 (s_{a2} + s_{b2} + n_2)^2]. \tag{7.48}$$

For stationary processes, setting $\tau = t_1 - t_2$, we obtain

$$R_y(\tau) = R_{a \times a}(\tau) + R_{b \times b}(\tau) + R_{n \times n}(\tau) + R_{a \times b}(\tau)$$
$$+ R_{a \times n}(\tau) + R_{b \times n}(\tau), \tag{7.49}$$

in which

$$R_{a \times a}(\tau) = \alpha^2 R_{a2}(\tau), \tag{7.50a}$$
$$R_{b \times b}(\tau) = \alpha^2 R_{b2}(\tau), \tag{7.50b}$$
$$R_{n \times n}(\tau) = \alpha^2 R_{n2}(\tau), \tag{7.50c}$$
$$R_{a \times b}(\tau) = 4\alpha^2 R_a(\tau) R_b(\tau) + 2\alpha^2 \sigma_a^2 \sigma_b^2, \tag{7.50d}$$
$$R_{a \times n}(\tau) = 4\alpha^2 R_a(\tau) R_n(\tau) + 2\alpha^2 \sigma_a^2 \sigma_n^2, \tag{7.50e}$$
$$R_{b \times n}(\tau) = 4\alpha^2 R_b(\tau) R_n(\tau) + 2\alpha^2 \sigma_b^2 \sigma_n^2, \tag{7.50f}$$

with $R_{a2}(\tau) = E(s_{a1}^2 s_{a2}^2)$, $R_a(\tau) = E(s_{a1} s_{a2})$, etc.

If we know the exact forms of these correlation functions, we can use the Fourier transform to obtain the power spectral density of the output which will, in turn, enable us to evaluate the final output signal-to-noise power ratio for the three-frequency system.

7.3.2 Application to cw Radar with Sinewave Input Signals

We now assume that the two inputs to the photodetector are pure sinusoidal waves with constant phase over the spatial extent of the photodetector. This would be the case, for example, when the combining beam splitting mirror is optically flat and all broadening effects may be neglected. We let $A_a = 2\beta A_1 A_L$, $A_b = 2\beta A_2 A_L$, $\omega_a = \omega_1 - \omega_L$, $\omega_b = \omega_2 - \omega_L$, $\phi_a = \phi_1 - \phi_L$, and $\phi_b = \phi_2 - \phi_L$. The signal input to the square-law device is then

$$s(t) = s_a(t) + s_b(t)$$
$$= A_a \cos(\omega_a t + \phi_a) + A_b \cos(\omega_b t + \phi_b). \tag{7.51}$$

The amplitudes A_a and A_b are in this case constant, and the phases ϕ_a and ϕ_b are taken to be random variables uniformly distributed over the interval $(0, 2\pi)$ and independent of each other. We easily obtain

$$R_a(\tau) = E(s_{a1} s_{a2}) = A_a^2 E[\cos(\omega_a t_1 + \phi_a) \cos(\omega_a t_2 + \phi_a)]$$
$$= \tfrac{1}{2} A_a^2 \cos \omega_a \tau, \tag{7.52}$$

with $\tau = t_1 - t_2$. Similarly

$$R_b(\tau) = E(s_{b1} s_{b2}) = \tfrac{1}{2} A_b^2 \cos \omega_b \tau. \tag{7.53}$$

The total correlation function of the input signal $R_s(\tau)$ is the sum of the individual correlation functions

$$R_s(\tau) = R_a(\tau) + R_b(\tau) = \tfrac{1}{2} A_a^2 \cos \omega_a \tau + \tfrac{1}{2} A_b^2 \cos \omega_b \tau. \tag{7.54}$$

Taking the Fourier transform, we obtain the power spectral density for the input signal:

$$S_s(f) = \frac{A_a^2}{4} [\delta(f - f_a) + \delta(f + f_a)] + \frac{A_b^2}{4} [\delta(f - f_b) + \delta(f + f_b)], \tag{7.55}$$

where $f_a = \omega_a/2\pi = f_1 - f_L$ and $f_b = \omega_b/2\pi = f_2 - f_L$. The shot noise arising from the strong LO is taken to be white Gaussian over the frequency band $[0, f_n]$. Thus, the noise spectrum is

$$S_n(f) = \begin{cases} N, & \text{for } 0 < |f| < f_n, \\ 0, & \text{elsewhere}. \end{cases} \tag{7.56}$$

The total input power spectral density $S_x(f)$ including the noise is shown in Fig. 7.4. We arbitrarily assume that $f_1 > f_2$ or $f_a > f_b$, and $A_b > A_a$.

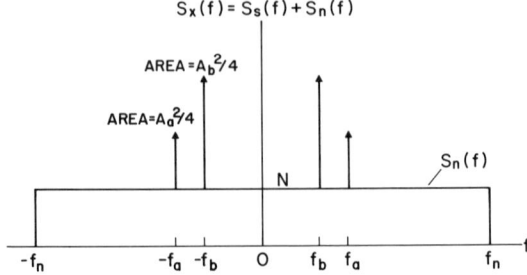

Fig. 7.4. The power spectral density seen at the input to the square-law device for the sine-wave case. If $A_a \neq A_b$, we arbitrarily choose $A_b > A_a$ as shown

From (7.50), we obtain

$$R_{a \times a}(\tau) = \alpha^2 E(s_{a1}^2 s_{a2}^2) = \alpha^2 E[A_a^4 \cos^2(\omega_a t_1 + \phi_a)\cos^2(\omega_a t_2 + \phi_a)]$$

$$= \frac{\alpha^2}{4} A_a^4 + \frac{\alpha^2}{8} A_a^4 \cos 2\omega_a \tau. \tag{7.57a}$$

Similarly

$$R_{b \times b}(\tau) = \frac{\alpha^2}{4} A_b^4 + \frac{\alpha^2}{8} A_b^4 \cos 2\omega_b \tau. \tag{7.57b}$$

Also

$$R_{a \times b}(\tau) = 4\alpha^2 E(s_{a1} s_{a2}) E(s_{b1} s_{b2}) + 2\alpha^2 \sigma_a^2 \sigma_b^2$$

$$= \frac{\alpha^2}{2} A_a^2 A_b^2 \cos(\omega_a - \omega_b)\tau + \frac{\alpha^2}{2} A_a^2 A_b^2 \cos(\omega_a + \omega_b)\tau$$

$$+ \frac{\alpha^2}{2} A_a^2 A_b^2. \tag{7.57c}$$

We note that because of the zero means,

$$\sigma_a^2 = R_a(0) = \frac{A_a^2}{2} \quad \text{and} \quad \sigma_b^2 = R_b(0) = \frac{A_b^2}{2}. \tag{7.58}$$

The total signal-by-signal correlation function $R_{s \times s}(\tau)$ is therefore given by

$$R_{s \times s}(\tau) = R_{a \times a}(\tau) + R_{b \times b}(\tau) + R_{a \times b}(\tau)$$

$$= \frac{\alpha^2}{4}(A_a^2 + A_b^2)^2 + \frac{\alpha^2}{8} A_a^4 \cos 2\omega_a \tau + \frac{\alpha^2}{8} A_b^4 \cos 2\omega_b \tau$$

$$+ \frac{\alpha^2}{2} A_a^2 A_b^2 \cos(\omega_a - \omega_b)\tau + \frac{\alpha^2}{2} A_a^2 A_b^2 \cos(\omega_a + \omega_b)\tau. \tag{7.59}$$

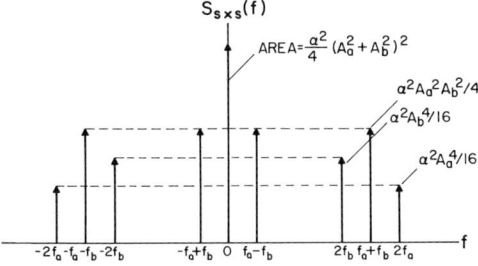

Fig. 7.5. The signal-by-signal power spectral density at the output of the square-law device (sinewave case)

The signal-by-signal part of the power spectral density $S_{s \times s}(f)$ is then, by taking the Fourier transform of $R_{s \times s}(\tau)$,

$$S_{s \times s}(f) = \frac{\alpha^2}{4}(A_a^2 + A_b^2)^2 \delta(f)$$

$$+ \frac{\alpha^2 A_a^4}{16}[\delta(f - 2f_a) + \delta(f + 2f_a)]$$

$$+ \frac{\alpha^2 A_b^4}{16}[\delta(f - 2f_b) + \delta(f + 2f_b)]$$

$$+ \frac{\alpha^2 A_a^2 A_b^2}{4}[\delta(f - f_a + f_b) + \delta(f + f_a - f_b)]$$

$$+ \frac{\alpha^2 A_a^2 A_b^2}{4}[\delta(f - f_a - f_b) + \delta(f + f_a + f_b)]. \quad (7.60)$$

Equation (7.60) is shown in Fig. 7.5.

For the signal-by-noise part, we have from (7.50e) and (7.50f) that

$$R_{s \times n}(\tau) = R_{a \times n}(\tau) + R_{b \times n}(\tau)$$

$$= 4\alpha^2 R_n(\tau)[R_a(\tau) + R_b(\tau)] + 2\alpha^2 \sigma_n^2(\sigma_a^2 + \sigma_b^2)$$

$$= 2\alpha^2 A_a^2 R_n(\tau)\cos\omega_a\tau + 2\alpha^2 A_b^2 R_n(\tau)\cos\omega_b\tau$$

$$+ \alpha^2(A_a^2 + A_b^2)\sigma_n^2. \quad (7.61)$$

The corresponding power spectral density is then

$$S_{s \times n}(f) = \alpha^2 A_a^2[S_n(f - f_a) + S_n(f + f_a)]$$

$$+ \alpha^2 A_b^2[S_n(f - f_b) + S_n(f + f_b)] + \alpha^2(A_a^2 + A_b^2)\sigma_n^2 \delta(f), \quad (7.62)$$

where $S_n(f - f_0)$ indicates that f_0 replaces 0 as the center frequency. This spectral density is plotted in Fig. 7.6.

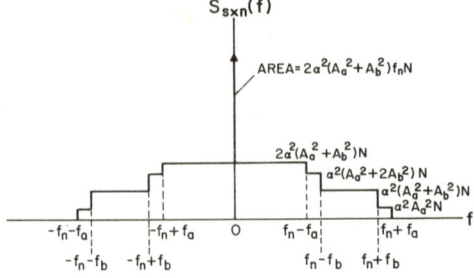

Fig. 7.6. The signal-by-noise power spectral density at the output of the square-law device (sinewave case)

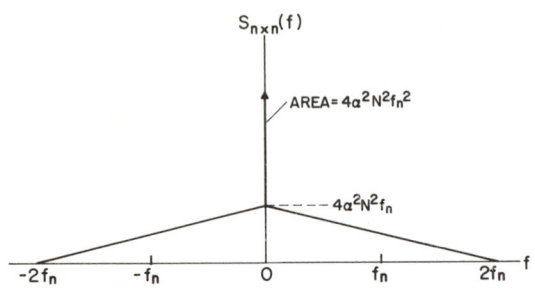

Fig. 7.7. The noise-by-noise power spectral density at the output of the square-law device

Since $R_{n^2}(\tau) = 2R_n^2(\tau) + \sigma_n^4$ for Gaussian noise [7.56], (7.50c) for the noise-by-noise part becomes

$$R_{n \times n}(\tau) = 2\alpha^2 R_n^2(\tau) + \alpha^2 \sigma_n^4, \tag{7.63}$$

whence

$$S_{n \times n}(f) = 2\alpha^2 \int_{-\infty}^{\infty} S_n(f') S_n(f-f') df' + \alpha^2 \sigma_n^4 \delta(f). \tag{7.64}$$

For the input noise spectrum described by (7.56), we have

$$\sigma_n^2 = \int_{-\infty}^{\infty} S_n(f) df = 2f_n N. \tag{7.65}$$

Equation (7.64) can therefore be simplified to

$$S_{n \times n}(f) = 4\alpha^2 f_n^2 N^2 \delta(f) + \begin{cases} 2\alpha^2 N^2 (2f_n - |f|), & \text{for } |f| < 2f_n, \\ 0, & \text{elsewhere}, \end{cases} \tag{7.66}$$

which is shown in Fig. 7.7.

The total output power spectral density $S_y(f)$ is the sum of $S_{s \times s}(f)$, $S_{s \times n}(f)$, and $S_{n \times n}(f)$ and is plotted in Fig. 7.8, where we have assumed that $f_c = f_a - f_b = f_1 - f_2$ lies in the region between the origin and $f_n - f_a$. In fact, f_c could be anywhere over the band f_n. We will consider two extreme cases: (a) $0 < f_c < f_n - f_a$ and (b) $f_n - f_b < f_c < f_n$.

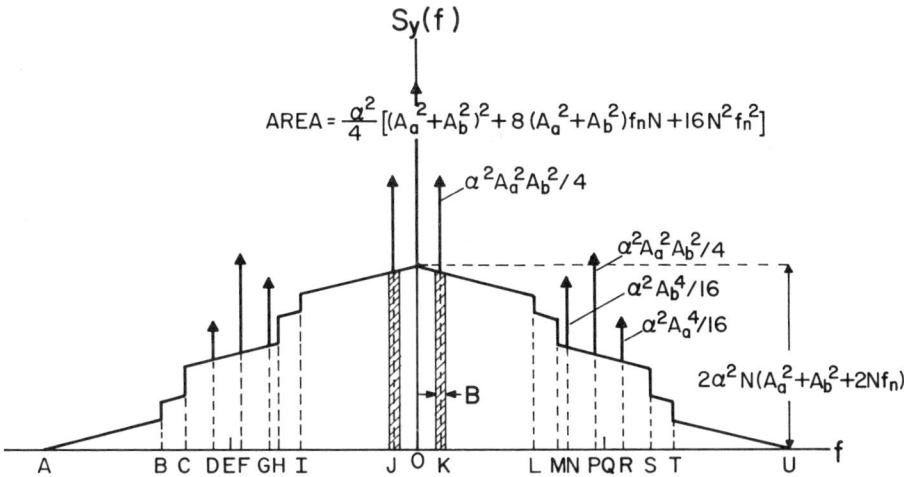

Fig. 7.8. The total power spectral density at the output of the square-law device for the sinewave case (not to scale). Correspondence between letters in figure and abscissa frequencies are: (A) $-2f_n$ (B) $-f_n - f_a$ (C) $-f_n - f_b$ (D) $-2f_a$ (E) $-f_n$ (F) $-f_a - f_b$ (G) $-2f_b$ (H) $-f_n + f_b$ (I) $-f_n + f_a$ (J) $-f_a + f_b$ (K) $f_a - f_b$ (L) $f_n - f_a$ (M) $f_n - f_b$ (N) $2f_b$ (P) $f_a + f_b$ (Q) f_n (R) $2f_a$ (S) $f_n + f_b$ (T) $f_n + f_a$ (U) $2f_n$. The crosshatched area represents the final bandpass filter, of bandwidth B

Since $f_c = f_a - f_b$ is known with great accuracy we can place a bandpass filter, with center frequency f_c, after the square-law device and obtain an output signal at this frequency. We wish to obtain the output signal-to-noise ratio $(SNR)_o$ in terms of the input signal-to-noise ratio $(SNR)_i = (SNR)_{power}$ for the two extreme cases indicated in the previous section.

From (7.60) along with Fig. 7.5 (or Fig. 7.8), we see that the output signal power S_0 at the frequency f_c is

$$S_0 = \frac{\alpha^2 A_a^2 A_b^2}{2}. \tag{7.67}$$

For a bandpass filter with bandwidth B, and for $0 < f_c < f_n - f_a$, the output noise power is the area under the power spectral density curve enclosed by B (see Figs. 7.6–8), which we choose to be rectangular for simplicity. Although strictly speaking, the rectangular function B (as well as Δf) is not realizable, this is not critical since it is the integrated area under the curve which is important rather than the detailed shape. The result is

$$N_0 = 4\alpha^2 NB(A_a^2 + A_b^2) + 4\alpha^2 N^2 B(2f_n - f_a + f_b),$$
$$0 < f_c < f_n - f_a. \tag{7.68}$$

The first term is due to the $s \times n$ interaction, while the second term is due to the $n \times n$ interaction. Equation (7.68) can also be obtained from (7.62–66).

The final signal-to-noise ratio at the output of the bandpass filter for this first case is therefore given by

$$(\text{SNR})_0 = \frac{S_0}{N_0} = \frac{A_a^2 A_b^2}{8NB[(A_a^2 + A_b^2) + N(2f_n - f_a + f_b)]}, \quad 0 \leq f_c < f_n - f_a. \quad (7.69)$$

The input signal power S_i and noise power N_i are, from Fig. 7.4,

$$S_i = R_s(0) = \tfrac{1}{2}(A_a^2 + A_b^2), \quad (7.70a)$$

$$N_i = \sigma_n^2 = 2f_n N. \quad (7.70b)$$

Thus, in terms of the input signal-to-noise ratio $(\text{SNR})_i = S_i/N_i = (A_a^2 + A_b^2)/4f_n N$, (7.69) can be written as

$$(\text{SNR})_0 = \frac{k_P (\text{SNR})_i^2}{\left(1 - \dfrac{f_a - f_b}{2f_n}\right) + 2(\text{SNR})_i}, \quad 0 < f_c < f_n - f_a, \quad (7.71)$$

with

$$k_P = \frac{f_n A_a^2 A_b^2}{B(A_a^2 + A_b^2)^2} = \frac{f_n A_1^2 A_2^2}{B(A_1^2 + A_2^2)^2} = \frac{f_n}{B}\left\{\frac{\xi_P}{(1+\xi_P)^2}\right\}, \quad (7.72)$$

where ξ_P represents the ratio of the signal power levels in the two beams, i.e., $\xi_P = A_2^2/A_1^2$.

The output signal-to-noise ratio is therefore inversely proportional to B, indicating that a small value for the final bandwidth is desired. Actually, B should be chosen much smaller than f_n in order that the above results be exactly correct, although results for an arbitrary value of B can easily be obtained from Fig. 7.8 and its associated equations.

If $f_c = f_a - f_b \to 0$, the minimum value for $(\text{SNR})_0$ is obtained:

$$(\text{SNR})_0^{\min} = \frac{k_P (\text{SNR})_i^2}{1 + 2(\text{SNR})_i}. \quad (7.73)$$

A log plot of (7.73) is shown in Fig. 7.9. Since $(\text{SNR})_0$ will increase as f_c increases, the curve will be shifted up as f_c increases from zero. The degenerate case $f_a \equiv f_b$ should be avoided in practice because of additional noise contributions.

We now consider the second case, where $f_n - f_b < f_c < f_n$. Following the same procedure as above, we obtain

$$N_0 = 2\alpha^2 N B(A_a^2 + A_b^2) + 4\alpha^2 N^2 B(2f_n - f_a + f_b), \quad f_n - f_b < f_c < f_n, \quad (7.74)$$

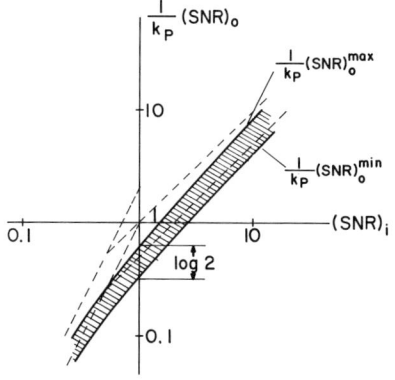

Fig. 7.9. The $(1/k_P)(SNR)_0$ vs. $(SNR)_i$ curve. The actual signal-to-noise ratio depends upon the relative values of $(f_a - f_b)$ and f_n, and lies in the shaded region. This curve also applies to the Gaussian signal case, provided that the quantity k_P is replaced by k_G or k_L, for Gaussian and Lorentzian spectra, respectively (see text)

so that

$$(SNR)_0 = \frac{k_P (SNR)_i^2}{\left(1 - \dfrac{f_a - f_b}{2f_n}\right) + (SNR)_i}, \quad f_n - f_b < f_c < f_n. \tag{7.75}$$

If we choose $f_c = f_a - f_b = f_n$, a maximum value for the $(SNR)_0$ is obtained:

$$(SNR)_0^{max} = \frac{2k_P (SNR)_i^2}{1 + 2(SNR)_i}. \tag{7.76}$$

The log plot of (7.76) is also given in Fig. 7.9; the result is the $(SNR)_0^{min}$ curve shifted vertically upward by $\log 2$. It now becomes clear that, for intermediate cases, i.e., $0 < f_c < f_n$, the $(SNR)_0$ curve will lie in between the curves $(SNR)_0^{min}$ and $(SNR)_0^{max}$. Thus, from a signal-to-noise ratio point of view, it is preferable to maintain the known difference frequency Δv at a maximum value close to f_n. For all cases, decreasing B will always yield improvement.

We can easily show that equal received power in each beam leads to optimum operation. From (7.72), we see that $k_P \propto A_a^2 A_b^2 / (A_a^2 + A_b^2)^2$. Because of the symmetry between A_a and A_b, we maximize k_P by calculating the derivative $(\partial k_P / \partial A_a)|_{A_b = const.} = 0$. This leads to the relation $A_a = A_b$.

From (7.73) and (7.76) the output signal-to-noise ratio $(SNR)_0$ is seen to be bounded as follows:

$$\frac{k_P (SNR)_i^2}{1 + 2(SNR)_i} \leq (SNR)_0 \leq \frac{2k_P (SNR)_i^2}{1 + 2(SNR)_i}, \tag{7.77}$$

again assuming $f_L < f_1', f_2'$ or $f_L > f_1', f_2'$. For LO frequencies between the two signal frequencies, however, the output of the square-law device at $|f_1' - f_2'|$ arises from the sum-frequency rather than the difference-frequency term and therefore falls in a region of lower noise. This, then, is the most desirable configuration from an SNR point of view, and allows us to comfortably realize the upper bound. (We recall, however, that the degenerate case should be avoided.) Nevertheless, we proceed under the more conservative assumption that the LO frequency is either greater than or less than both received signal frequencies. If we examine the specific (and optimum) case of $A_a = A_b$ so that $k_p = f_n/4B$, with $(SNR)_i = \eta P_r/hvf_n$ as given by (7.42b), we obtain

$$\frac{f_n}{4B}\left[\frac{(\eta P_r/hvf_n)^2}{1+2\eta P_r/hvf_n}\right] \leq (SNR)_0 \leq \frac{f_n}{2B}\left[\frac{(\eta P_r/hvf_n)^2}{1+2\eta P_r/hvf_n}\right]. \tag{7.78}$$

This yields an approximate average value given by

$$(SNR)_0 \simeq \frac{f_n}{3B}\left[\frac{(\eta P_r/hvf_n)^2}{1+2\eta P_r/hvf_n}\right]. \tag{7.79}$$

The case where $A_a \neq A_b$ will be considered in Section 7.4.

To obtain the minimum detectable total power $(MDP)_0$ for this three-frequency system, we set $(SNR)_0 = 1$ and solve for P_r. Thus,

$$\eta \frac{(MDP)_0}{hvf_n} = \frac{3B}{f_n} + \left[\frac{3B}{f_n}\left(1+\frac{3B}{f_n}\right)\right]^{1/2}. \tag{7.80a}$$

Since $B \ll f_n$,

$$\eta \frac{(MDP)_0}{hvf_n} \simeq \frac{3B}{f_n} + \left(\frac{3B}{f_n}\right)^{1/2}\left(1+\frac{3B}{2f_n}\right) \simeq \left(\frac{3B}{f_n}\right)^{1/2} \tag{7.80b}$$

and therefore

$$(MDP)_0 \simeq \sqrt{3Bf_n}(hv/\eta) = \sqrt{\frac{3B}{f_n}}P_r(\min), \tag{7.81}$$

with $P_r(\min) = f_n hv/\eta$ as the standard minimum detectable power for the conventional heterodyne system with Doppler uncertainty f_n [7.4, 5]. By choosing $B \ll f_n$, therefore, it is possible to achieve a reduced minimum detectable power

$$(MDP)_0 \ll P_r(\min), \quad B \ll f_n \tag{7.82}$$

using three-frequency nonlinear heterodyne detection. *Equations (7.77–82) represent the key results obtained in this analysis.*

In the limit of large $(SNR)_i$ (strong input signals and/or small Doppler uncertainty), $P_r/f_n \gg hv/\eta$, and (7.78) yields the relationship

$$\frac{\eta P_r}{8hvB} \leq (SNR)_0 \leq \frac{\eta P_r}{4hvB}. \tag{7.83}$$

This can be approximated, then, as

$$(\text{SNR})_0 \simeq \frac{\eta P_r}{6h\nu B} = \frac{f_n}{6B}\left(\frac{\eta P_r}{h\nu f_n}\right)$$

$$= \frac{f_n}{6B}(\text{SNR})_{\text{power}}, \qquad (7.84)$$

where $(\text{SNR})_{\text{power}} = \eta P_r/h\nu f_n$ is the signal-to-noise ratio for the conventional heterodyne system as given in (7.42b). Again, we can provide that

$$(\text{SNR})_0 \gg (\text{SNR})_{\text{power}} \qquad (7.85)$$

by choosing $B \ll f_n$.

7.3.3 Application to cw Radar with Gaussian Input Signals (Gaussian Spectra)

The sinusoidal signal assumption is, in many cases, an idealization which provides simple physical and mathematical insights into a problem. In most real situations, however, the heterodyne signal will have a narrowband character [7.2, 5, 9, 10, 14]. This may be due to the surface roughness of a scattering target in a radar system, or due to the modulation imposed on the carrier in a communications system. For a scatterer returning Gaussian radiation, experiments show that the power spectral density is also frequently in the form of a Gaussian [7.9, 14, 57, 58]. It is the purpose of this section to investigate this case.

We assume that the two signal inputs to the photodetector, $E_1(t)$ and $E_2(t)$, take the form of narrowband Gaussian processes:

$$E_1(t) = A_1(t)\cos(\omega_1 t + \phi_1), \qquad (7.86\text{a})$$

$$E_2(t) = A_2(t)\cos(\omega_2 t + \phi_2), \qquad (7.86\text{b})$$

where, for any given t, the two independent amplitudes $A_1(t)$ and $A_2(t)$ are random variables with Rayleigh distributions, and the two independent phases ϕ_1 and ϕ_2 are uniformly distributed over the interval $(0, 2\pi)$. It should be pointed out that the independent amplitude case considered here is appropriate only for a sufficiently large value of $\Delta\nu$ and for sufficiently large targets. After mixing with a stable LO and filtering out the dc portion, the signal input to the square-law device is easily found to be [7.9, 10, 14]

$$s(t) = 2\beta A_L A_1(t)\cos[(\omega_1 - \omega_L)t + (\phi_1 - \phi_L)]$$
$$+ 2\beta A_L A_2(t)\cos[(\omega_2 - \omega_L)t + (\phi_2 - \phi_L)]. \qquad (7.87)$$

Since β and A_L are constant, the new amplitudes $A_a(t) = 2\beta A_L A_1(t)$ and $A_b(t) = 2\beta A_L A_2(t)$ remain Rayleigh distributed. Similarly, the new phases $\phi_a = \phi_1 - \phi_L$

and $\phi_b = \phi_2 - \phi_L$ can be easily shown to possess uniform distributions over $(0, 2\pi)$. Therefore the narrowband Gaussian nature of the signals is preserved, provided that the envelope variations are slower than the intermediate frequencies $\omega_1 - \omega_L$ and $\omega_2 - \omega_L$ [7.9, 10, 14] and we write

$$s(t) = A_a(t)\cos(\omega_a t + \phi_a) + A_b(t)\cos(\omega_b t + \phi_b). \tag{7.88}$$

The power spectral densities for the narrowband Gaussian inputs are also taken to be Gaussian. Thus

$$S_s(f) = S_a(f) + S_b(f), \tag{7.89a}$$

with

$$S_a(f) = P_a \exp\left[-\frac{(f-f_a)^2}{2\gamma_a^2}\right] + P_a \exp\left[-\frac{(f+f_a)^2}{2\gamma_a^2}\right], \tag{7.89b}$$

$$S_b(f) = P_b \exp\left[-\frac{(f-f_b)^2}{2\gamma_b^2}\right] + P_b \exp\left[-\frac{(f+f_b)^2}{2\gamma_b^2}\right], \tag{7.89c}$$

where P_a and P_b represent the peak values of the Gaussian distributions, and γ_a and γ_b are their standard deviations. The signal powers are then given by

$$\sigma_a^2 = \int_{-\infty}^{\infty} S_a(f)df = 2\sqrt{2\pi}\gamma_a P_a = \frac{\langle A_a^2 \rangle}{2}, \tag{7.90a}$$

$$\sigma_b^2 = \int_{-\infty}^{\infty} S_b(f)df = 2\sqrt{2\pi}\gamma_b P_b = \frac{\langle A_b^2 \rangle}{2}. \tag{7.90b}$$

The noise input once again is assumed to be white Gaussian over the real frequency band $[0, f_n]$, and therefore has the same spectral density as the sinewave input case. The total power spectral density at the input to the square-law device is presented in Fig. 7.10 (compare with Fig. 7.4 for the sinewave case).

Because the signals are stationary Gaussian processes, the signal-by-signal correlation functions at the output of the square-law device are given by [see (7.63)]

$$R_{a \times a}(\tau) = 2\alpha^2 R_a^2(\tau) + \alpha^2 \sigma_a^4, \tag{7.91a}$$

$$R_{b \times b}(\tau) = 2\alpha^2 R_b^2(\tau) + \alpha^2 \sigma_b^4. \tag{7.91b}$$

From (7.64), the Fourier transform of $R_{a \times a}(\tau)$ is

$$S_{a \times a}(f) = 2\alpha^2 \int_{-\infty}^{\infty} S_a(f')S_a(f-f')df' + \alpha^2 \sigma_a^4 \delta(f)$$

$$= 2\sqrt{\pi}\alpha^2 \gamma_a P_a^2 \left[e^{-(f-2f_a)^2/4\gamma_a^2} + e^{-(f-2f_a)^2/4\gamma_a^2} + 2e^{-f^2/4\gamma_a^2}\right]$$

$$+ \alpha^2 \sigma_a^4 \delta(f), \tag{7.92}$$

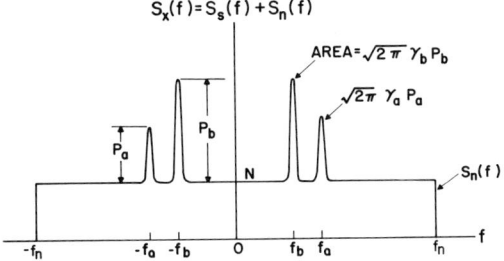

Fig. 7.10. The power spectral density at the input to the square-law device for the Gaussian signal case. We arbitrarily choose $P_b > P_a$

with the identical result for $S_{b \times b}(f)$ (b replaces a). Furthermore, using (7.50d) we obtain

$$S_{a \times b}(f) = 4\sqrt{2\pi}\alpha^2 \frac{(\gamma_a P_a)(\gamma_b P_b)}{\sqrt{\gamma_a^2 + \gamma_b^2}} \left(\exp\left\{ -\frac{[f-(f_a+f_b)]^2}{2(\gamma_a^2+\gamma_b^2)} \right\} \right.$$

$$+ \exp\left\{ -\frac{[f+(f_a+f_b)]^2}{2(\gamma_a^2+\gamma_b^2)} \right\} + \exp\left\{ -\frac{[f-(f_a-f_b)]^2}{2(\gamma_a^2+\gamma_b^2)} \right\}$$

$$\left. + \exp\left\{ -\frac{[f+(f_a-f_b)]^2}{2(\gamma_a^2+\gamma_b^2)} \right\} \right) + 2\alpha^2 \sigma_a^2 \sigma_b^2 \delta(f). \tag{7.93}$$

The total signal-by-signal power spectral density $S_{s \times s}(f)$ is, of course, given by

$$S_{s \times s}(f) = S_{a \times a}(f) + S_{b \times b}(f) + S_{a \times b}(f), \tag{7.94}$$

and is shown in Fig. 7.11 (compare with Fig. 7.5 for the sinewave case).

For the signal-by-noise part, we have

$$R_{s \times n}(\tau) = R_{a \times n}(\tau) + R_{b \times n}(\tau), \tag{7.95}$$

and, from (7.50e), we write

$$R_{a \times n}(\tau) = 4\alpha^2 R_a(\tau) R_n(\tau) + 2\alpha^2 \sigma_a^2 \sigma_n^2, \tag{7.96}$$

which Fourier transforms to

$$S_{a \times n}(f) = 4\alpha^2 \int_{-\infty}^{\infty} S_a(f') S_n(f-f') df' + 2\alpha^2 \sigma_a^2 \sigma_n^2 \delta(f). \tag{7.97}$$

This can be readily evaluated to yield

$$S_{a \times n}(f) = 4\sqrt{2\pi}\alpha^2 N P_a \gamma_a \left[\Phi\left(\frac{f+f_n-f_a}{\gamma_a}\right) - \Phi\left(\frac{f-f_n-f_a}{\gamma_a}\right) \right.$$

$$\left. + \Phi\left(\frac{f+f_n+f_a}{\gamma_a}\right) - \Phi\left(\frac{f-f_n+f_a}{\gamma_a}\right) \right] + 2\alpha^2 \sigma_a^2 \sigma_n^2 \delta(f), \tag{7.98}$$

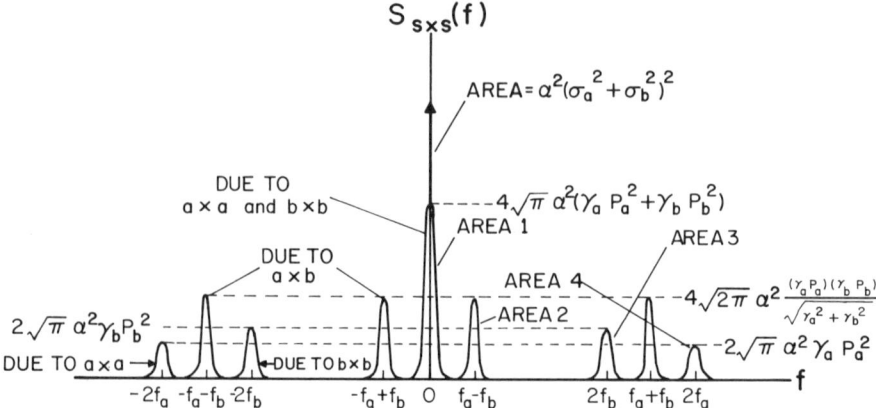

Fig. 7.11. The signal-by-signal power spectral density at the output of the square-law device (Gaussian signal case). Areas under curves: AREA 1 $= 8\pi\alpha^2(\gamma_a^2 P_a^2 + \gamma_b^2 P_b^2)$; AREA 2 $= 8\pi\alpha^2(\gamma_a P_a)(\gamma_b P_b)$; AREA 3 $= 4\pi\alpha^2 \gamma_b^2 P_b^2$; AREA 4 $= 4\pi\alpha^2 \gamma_a^2 P_a^2$

where $\Phi(x)$ is the normal distribution function

$$\Phi(x) = \frac{1}{\sqrt{2\pi}} \int_{-\infty}^{x} e^{-x'^2/2} dx'. \tag{7.99}$$

Similarly, we obtain the identical result for $S_{b \times n}$ (b replaces a). The total signal-by-noise power spectral density $S_{s \times n}(f)$ is just the sum of $S_{a \times n}(f)$ and $S_{b \times n}(f)$, or

$$S_{s \times n}(f) = S_{a \times n}(f) + S_{b \times n}(f). \tag{7.100}$$

A sketch of $S_{s \times n}(f)$ is given in Fig. 7.12. Assuming that the standard deviations γ_a and γ_b are small in comparison with the width of the plateau regions in Fig. 7.6, the plot will be very similar to that of the sinewave case. The only notable difference is the rounding of sharp corners. Small values of γ_a and γ_b also guard against spectrum overlap which would make the solution of the problem more difficult.

The noise-by-noise power spectral density is the same as that for the pure sinewave case (Fig. 7.7). The total output power spectral density $S_y(f)$ is plotted in Fig. 7.13 where we arbitrarily have assumed that $f_a > f_b$ and $f_a - f_b < f_n - f_a < f_n$.

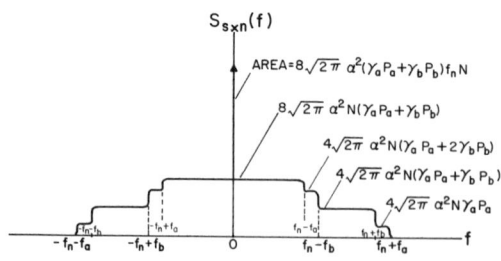

Fig. 7.12. The signal-by-noise power spectral density at the output of the square-law device (Gaussian signal case)

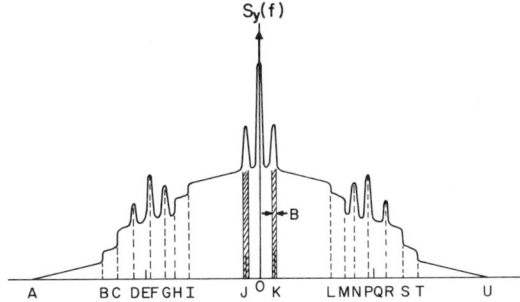

Fig. 7.13. The total power spectral density at the output of the square-law device for the Gaussian signal case. Note the smoothing of all sharp edges. This figure is similar to Fig. 7.6 except that the delta functions are replaced by Gaussians

Here again, we place a bandpass filter of bandwidth B and center frequency $f_c = f_a - f_b$ after the square-law device. Referring to Fig. 7.11 and (7.93), we find

$$S_0 = \frac{8\alpha^2 \sqrt{2\pi}(\gamma_a P_a)(\gamma_b P_b)}{\sqrt{\gamma_a^2 + \gamma_b^2}} \left[\int_{-B/2}^{B/2} e^{-f^2/2(\gamma_a^2 + \gamma_b^2)} df \right]$$

$$= 16\pi\alpha^2 (\gamma_a P_a)(\gamma_b P_b) \left[2\Phi\left(\frac{B}{2\sqrt{\gamma_a^2 + \gamma_b^2}}\right) - 1 \right]. \quad (7.101)$$

The input signal power is, from (7.90), given by

$$S_i = 2\sqrt{2\pi}(\gamma_a P_a + \gamma_b P_b). \quad (7.102)$$

The input noise power is the same as for the sinewave case. Referring to Figs. 7.7 and 7.12, the output noise power can be very well approximated by

$$N_0 = 16\sqrt{2\pi}\alpha^2 NB(\gamma_a P_a + \gamma_b P_b) + 4\alpha^2 N^2 B(2f_n - f_a + f_b),$$
$$0 < f_a - f_b < f_n - f_a, \quad (7.103a)$$

$$N_0 = 8\sqrt{2\pi}\alpha^2 NB(\gamma_a P_a + \gamma_b P_b) + 4\alpha^2 N^2 B(2f_n - f_a + f_b),$$
$$f_n - f_b < f_a - f_b < f_n. \quad (7.103b)$$

The input signal-to-noise ratio $(SNR)_i$ is simply

$$(SNR)_i = \frac{\sqrt{2\pi}}{Nf_n}(\gamma_a P_a + \gamma_b P_b) \quad (7.104)$$

while the output signal-to-noise ratio $(SNR)_0$ can be easily evaluated:

$$(SNR)_0 = \frac{k_G (SNR)_i^2}{\left(1 - \frac{f_a - f_b}{2f_n}\right) + 2(SNR)_i}, \quad 0 < f_a - f_b < f_n - f_a, \quad (7.105a)$$

$$(SNR)_0 = \frac{k_G (SNR)_i^2}{\left(1 - \frac{f_a - f_b}{2f_n}\right) + (SNR)_i}, \quad f_n - f_b < f_a - f_b < f_n. \quad (7.105b)$$

These equations are identical to (7.71) and (7.75) with the exception that the factor k_P [(7.72)] has been replaced by the factor k_G given by

$$k_G = \frac{f_n}{2\sqrt{\gamma_a^2+\gamma_b^2}} \frac{(\gamma_a P_a)(\gamma_b P_b)}{(\gamma_a P_a + \gamma_b P_b)^2} \left\{\frac{1}{u}[2\Phi(u)-1]\right\}$$

$$= \frac{f_n}{2\sqrt{\gamma_1^2+\gamma_2^2}} \frac{(\gamma_1 P_1)(\gamma_2 P_2)}{(\gamma_1 P_1 + \gamma_2 P_2)^2} \left\{\frac{1}{u}[2\Phi(u)-1]\right\}$$

$$= \frac{f_n}{2\sqrt{\gamma_1^2+\gamma_2^2}} \left\{\frac{\xi_G}{(1+\xi_G)^2}\right\} \left\{\frac{1}{u}[2\Phi(u)-1]\right\}. \tag{7.106a}$$

Here

$$u = \frac{B}{2\sqrt{\gamma_a^2+\gamma_b^2}} = \frac{B}{2\sqrt{\gamma_1^2+\gamma_2^2}} \tag{7.106b}$$

and ξ_G, which is the ratio of the beam powers, is

$$\xi_G = \gamma_2 P_2/\gamma_1 P_1. \tag{7.106c}$$

Now, if $f_a - f_b \to 0$ in (7.105a) and choosing $f_a - f_b = f_n$ in (7.105b), we obtain the following bounds for $(SNR)_0$

$$(SNR)_0^{min} = \frac{k_G(SNR)_i^2}{1+2(SNR)_i}, \tag{7.107a}$$

$$(SNR)_0^{max} = \frac{2k_G(SNR)_i^2}{1+2(SNR)_i}, \tag{7.107b}$$

in analogy with (7.73) and (7.76). Again we note that when the LO frequency is between the received signal frequencies, the upper bound can be safely used. The results presented in Fig. 7.9 are therefore also appropriate to this case with the substitution of k_G for k_P.

Since $(SNR)_i$ is the same for the Gaussian case as it is for the pure sinewave case [7.10], the results presented in Section 7.3.2 apply directly with the simple replacement of k_P by k_G; thus

$$\frac{1}{B} \to \frac{1}{2\sqrt{\gamma_a^2+\gamma_b^2}} \left\{\frac{1}{u}[2\Phi(u)-1]\right\} = \frac{2\Phi(u)-1}{B}, \tag{7.108a}$$

with

$$A_a^2 \to 4\sqrt{2\pi}\gamma_a P_a, \tag{7.108b}$$

$$A_b^2 \to 4\sqrt{2\pi}\gamma_b P_b. \tag{7.108c}$$

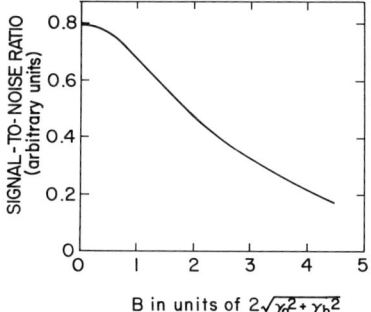

Fig. 7.14. Plot of the function $z(u)$ vs u, which represents the B dependence of the signal-to-noise ratio for Gaussian signals

For equal powers in both beams, therefore, the final signal-to-noise ratio for the Gaussian case is degraded in comparison with the sinewave case by the factor $[2\Phi(u)-1]=[\Phi(u)-\Phi(-u)]<1$. It is clear that the spectral width, as determined by the quantity $\sqrt{\gamma_a^2+\gamma_b^2}$, should be minimized if possible. An expression for the MDP is given in Section 7.4.3.

We now consider the variation of the quantity k_G with the bandwidth B, assuming that the parameters f_n, γ_a, γ_b, P_a, and P_b are fixed. Since $u \propto B$ and $k_G \propto (1/u)[2\Phi(u)-1] \equiv z(u)$, it is sufficient to examine the function $z(u)$ sketched in Fig. 7.14. It is apparent that k_G, and therefore $(SNR)_0$, increases with decreasing B. This evidences the fact that the noise power decreases more rapidly than the signal power as the bandwidth B is narrowed on $f_c = f_a - f_b$. This effect can be observed in Fig. 7.13. For both sinewave and Gaussian signal inputs, therefore, it is desirable to minimize B. The limitation, of course, is provided by the unequal Doppler shifts of the two input signals. For a given bandwidth B, furthermore, it is understood that all of the signal will be detected in the sinewave case, while only a portion of it will be detected in the Gaussian case.

If we let γ_a and $\gamma_b \to 0$ while keeping the power constant (i.e., $\sqrt{2\pi}\gamma_a P_a = A_a^2/4$ and $\sqrt{2\pi}\gamma_b P_b = A_b^2/4$), the Gaussian spectra shrink to delta functions. In this limit, we observe that

$$\lim_{\substack{\gamma_a \to 0 \\ \gamma_b \to 0}} S_0(\text{Gaussian}) = \frac{\alpha^2 A_a^2 A_b^2}{2}[2\Phi(\infty)-1] = \frac{\alpha^2 A_a^2 A_b^2}{2}, \qquad (7.109)$$

which is, as expected, the expression obtained for the output power for the sinewave case [(7.67)].

7.3.4 Application to cw Radar with Gaussian Input Signals (Lorentzian Spectra)

In those cases where the narrowband Gaussian signal inputs possess Lorentzian power spectra rather than Gaussian power spectra, the input power spectral densities are given by

$$S_a(f) = \frac{D_a \Gamma_a}{2[(f-f_a)^2 + \Gamma_a^2]} + \frac{D_a \Gamma_a}{2[(f+f_a)^2 + \Gamma_a^2]}, \quad (7.110\text{a})$$

$$S_b(f) = \frac{D_b \Gamma_b}{2[(f-f_b)^2 + \Gamma_b^2]} + \frac{D_b \Gamma_b}{2[(f+f_b)^2 + \Gamma_b^2]}. \quad (7.110\text{b})$$

Here D_a and D_b are arbitrary constants, while Γ_a and Γ_b are constants reflecting the spectral width. Performing the inverse Fourier transform, the autocorrelation functions are found to be

$$R_a(\tau) = \pi D_a e^{-2\pi \Gamma_a |\tau|} \cos 2\pi f_a \tau, \quad (7.111\text{a})$$

$$R_b(\tau) = \pi D_b e^{-2\pi \Gamma_b |\tau|} \cos 2\pi f_b \tau, \quad (7.111\text{b})$$

whereas the input signal powers are given by

$$\sigma_a^2 = \int_{-\infty}^{\infty} S_a(f) df = \pi D_a = \frac{\langle A_a^2 \rangle}{2}, \quad (7.112\text{a})$$

$$\sigma_b^2 = \int_{-\infty}^{\infty} S_b(f) df = \pi D_b = \frac{\langle A_b^2 \rangle}{2}. \quad (7.112\text{b})$$

With the same noise input as considered in the previous section, and making use of (7.50), (7.91), (7.97), and (7.111), the output power spectral densities are

$$S_{a \times a}(f) = \alpha^2 \pi D_a^2 \left[\frac{2\Gamma_a}{f^2 + 4\Gamma_a^2} + \frac{\Gamma_a}{(f-2f_a)^2 + 4\Gamma_a^2} + \frac{\Gamma_a}{(f+2f_a)^2 + 4\Gamma_a^2} \right]$$
$$+ \alpha^2 \sigma_a^4 \delta(f), \quad (7.113\text{a})$$

$$S_{b \times b}(f) = S_{a \times a}(f)|_{\text{sub. a} \to \text{sub. b}}, \quad (7.113\text{b})$$

$$S_{a \times b}(f) = \alpha^2 \pi D_a D_b \left[\frac{\Gamma_a + \Gamma_b}{(f-f_a-f_b)^2 + (\Gamma_a+\Gamma_b)^2} + \frac{\Gamma_a + \Gamma_b}{(f+f_a+f_b)^2 + (\Gamma_a+\Gamma_b)^2} \right.$$
$$\left. + \frac{\Gamma_a + \Gamma_b}{(f-f_a+f_b)^2 + (\Gamma_a+\Gamma_b)^2} + \frac{\Gamma_a + \Gamma_b}{(f+f_a-f_b)^2 + (\Gamma_a+\Gamma_b)^2} \right]$$
$$+ 2\alpha^2 \sigma_a^2 \sigma_b^2 \delta(f), \quad (7.113\text{c})$$

$$S_{a \times n}(f) = 2\alpha^2 N D_a \tan^{-1} \left\{ \frac{4 f_n \Gamma_a (f^2 + f_a^2 + \Gamma_a^2 - f_n^2)}{[(f-f_a)^2 + \Gamma_a^2 - f_n^2][(f+f_a)^2 + \Gamma_a^2 - f_n^2] - 4 f_n^2 \Gamma_a^2} \right\}$$
$$+ 2\alpha^2 \sigma_a^2 \sigma_n^2 \delta(f), \quad (7.113\text{d})$$

$$S_{b \times n}(f) = S_{a \times n}|_{\text{sub. a} \to \text{sub. b}}, \quad (7.113\text{e})$$

with $0 < \tan^{-1} x < \pi$ (negative angles excluded). For narrow spectral widths (Γ_a, Γ_b small), it is clear that the power spectral densities for the Lorentzian case should look just about the same as those for the Gaussian case. We may simply replace $2\sqrt{2\pi\gamma_a P_a}$ and $2\sqrt{2\pi\gamma_b P_b}$, respectively, by πD_a and πD_b [see (7.90) and (7.112)].

We may calculate the output signal and noise to obtain $(SNR)_0$ as follows. Let

$$S_0 = 2\pi\alpha^2 D_a D_b \int_{-B/2}^{B/2} \frac{\Gamma_a + \Gamma_b}{f^2 + (\Gamma_a + \Gamma_b)^2} df$$

$$= 2\pi\alpha^2 D_a D_b \tan^{-1}\left[\frac{4B(\Gamma_a + \Gamma_b)}{4(\Gamma_a + \Gamma_b)^2 - B^2}\right], \tag{7.114}$$

with $0 < \tan^{-1} x < \pi$. For small B, the output noise power is

$$N_0 = \int_{f_a - f_b - B/2}^{f_a - f_b + B/2} [S_{a \times n}(f) + S_{b \times n}(f) + S_{n \times n}(f)] df$$

$$+ \int_{-f_a + f_b - B/2}^{-f_a + f_b + B/2} [S_{a \times n}(f) + S_{b \times n}(f) + S_{n \times n}(f)] df$$

$$\simeq B[S_{a \times n}(f_a - f_b) + S_{a \times n}(-f_a + f_b) + S_{b \times n}(f_a - f_b) + S_{b \times n}(-f_a + f_b)]$$

$$+ 4\alpha^2 N^2 B(2f_n - f_a + f_b). \tag{7.115}$$

As an approximation, we use the replacements $2\sqrt{2\pi\gamma_a P_a} \to \pi D_a$ and $2\sqrt{2\pi\gamma_b P_b} \to \pi D_b$ in Fig. 7.12 to obtain

$$N_0 = 8\pi\alpha^2 NB(D_a + D_b) + 4\alpha^2 N^2 B(2f_n - f_a + f_b),$$
$$0 < f_a - f_b < f_n - f_a, \tag{7.116a}$$

$$N_0 = 4\pi\alpha^2 NB(D_a + D_b) + 4\alpha^2 N^2 B(2f_n - f_a + f_b),$$
$$f_n - f_b < f_a - f_b < f_n. \tag{7.116b}$$

Using an input signal-to-noise ratio $(SNR)_i$ [see (7.112)] given by

$$(SNR)_i = \frac{\pi(D_a + D_b)}{2Nf_n}, \tag{7.117}$$

and using (7.114), (7.116), and (7.117), we find the output signal-to-noise ratio $(SNR)_0$ to be

$$(SNR)_0 = \frac{k_L (SNR)_i^2}{\left(1 - \frac{f_a - f_b}{2f_n}\right) + 2(SNR)_i}, \quad 0 < f_a - f_b < f_n - f_a, \tag{7.118a}$$

$$(SNR)_0 = \frac{k_L (SNR)_i^2}{\left(1 - \frac{f_a - f_b}{2f_n}\right) + (SNR)_i}, \quad f_n - f_b < f_a - f_b < f_n. \tag{7.118b}$$

This result is therefore the same as that for both the sinewave case and the Gaussian spectrum case [see (7.71), (7.75), (7.105)] except that we now use the factor k_L given by

$$k_L = \frac{f_n}{\pi(\Gamma_a + \Gamma_b)} \frac{D_a D_b}{(D_a + D_b)^2} \frac{\tan^{-1}\left(\frac{4v}{4-v^2}\right)}{v}$$

$$= \frac{f_n}{\pi(\Gamma_1 + \Gamma_2)} \frac{D_1 D_2}{(D_1 + D_2)^2} \frac{\tan^{-1}\left(\frac{4v}{4-v^2}\right)}{v}$$

$$= \frac{f_n}{\pi(\Gamma_1 + \Gamma_2)} \left\{ \frac{\xi_L}{(1+\xi_L)^2} \right\} \frac{\tan^{-1}\left(\frac{4v}{4-v^2}\right)}{v} \quad (7.119a)$$

where

$$v = B/(\Gamma_a + \Gamma_b) = B/(\Gamma_1 + \Gamma_2) \quad (7.119b)$$

and

$$\xi_L = D_2/D_1. \quad (7.119c)$$

Thus the results presented in Fig. 7.9 apply also to this case with k_P replaced by k_L.

For small v (small B), the quantity $4v/(4-v^2) \simeq v$. It is not difficult to show that the behavior of the function $(\tan^{-1} v)/v$ is similar to that of $z(u)$ (see Fig. 7.14). The maximum value occurs at the origin so that in this case too, optimum operation occurs for $B \to 0$.

Furthermore, using the replacements $\pi D_a \to A_a^2/2$ and $\pi D_b \to A_b^2/2$ and letting $\Gamma_a, \Gamma_b \to 0$, (7.119) reduces to

$$k_L = \frac{f_n}{B} \frac{A_a^2 A_b^2}{(A_a^2 + A_b^2)^2} \left\{ \lim_{\Gamma_a + \Gamma_b \to 0} \frac{1}{\pi} \tan^{-1}\left[\frac{4B(\Gamma_a + \Gamma_b)}{4(\Gamma_a + \Gamma_b)^2 - B^2}\right] \right\}$$

$$= \frac{f_n}{B} \frac{A_a^2 A_b^2}{(A_a^2 + A_b^2)^2} = k_P. \quad (7.120)$$

Thus, the present case also reduces to the sinewave case as the spectral width approaches zero with the power fixed. Similarly it is clear from the above that for fixed B the spectral width, as determined by the quantity $(\Gamma_a + \Gamma_b)$, should be minimized if possible.

7.3.5 Application to an Analog Communications System

Use of the three-frequency method for a communications system (in which the transmitter and receiver may be moving relative to each other) is similar to the radar already described, and is indicated in Fig. 7.15. Note, however, that only

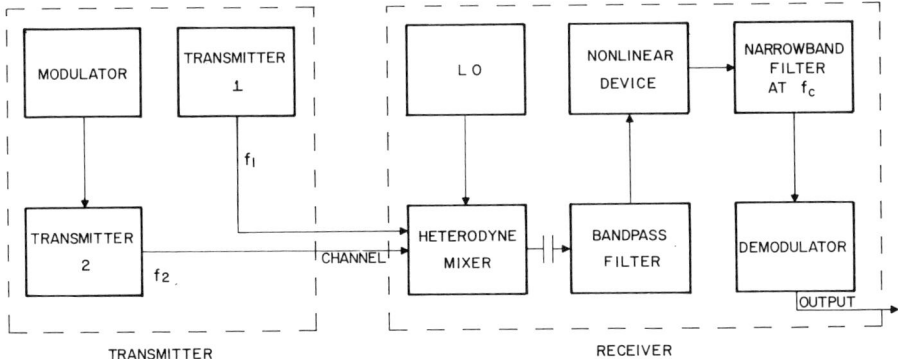

Fig. 7.15. A three-frequency nonlinear heterodyne analog communications system

one of the carrier waves (f_2) is modulated, and that a demodulator is included. By modulating only one of the beams, the s × s component reaching the demodulator results from the convolution of a delta function (at f_1) with the modulated signal (centered at f_2), which is simply the original undistorted spectral information ready for demodulation by a suitable device such as a mixer, an envelope detector, or a discriminator. The maximum rate at which modulation may be decoded (or the information capacity) of the system will, of course, depend upon the time response of the system which is usually governed by the final bandwidth B. We could, in the alternative, construct an analog FM communication system in which a single frequency laser beam is split into two frequencies by a modulator (e.g., an acoustooptic modulator [7.49]). In that case, the frequency difference f_c will carry the information. In the absence of information (i.e., f_c constant), f_c will be spectrally very pure (it should be as narrow as the spectral width of the modulator drive signal). Thus the three-frequency nonlinear system could provide the advantage of lower deviation and thereby provide bandwidth compression.

7.3.6 Operation at Low Frequencies and in Various Configurations

Thus far, we have been especially concerned with absorption detectors operating in the optical and infrared regions of the electromagnetic spectrum ($h\nu \gg kT$, where k is the Boltzmann constant and T is the detector temperature). In this case, the intensity of the incoming wave is obtained from the analytic signal and excludes double- and sum-frequency components [7.10, 12–14]. Nevertheless, (7.77) is a general result for intensity detection which applies also to the microwave and radiowave regions ($h\nu \ll kT$).

For low frequencies, the intensity is related to the square of the electric field, $I \propto E^2$. For a diode mixer which is either operating in the square-law regime or in the linear regime followed by a square-law device, the detector response r for a three-frequency system is then given by the classical expression

$$r = \beta' E_t^2 = \beta'[A_1 \cos(\omega_1 t + \phi_1) + A_2 \cos(\omega_2 t + \phi_2) + A_L \cos(\omega_L t + \phi_L)]^2$$
$$= \beta'\{A_1^2 \cos^2(\omega_1 t + \phi_1) + A_2^2 \cos^2(\omega_2 t + \phi_2) + A_L^2 \cos^2(\omega_L t + \phi_L)$$
$$+ A_1 A_2 \cos[(\omega_1 - \omega_2)t + (\varphi_1 - \varphi_2)] + A_1 A_2 \cos[(\omega_1 + \omega_2)t + (\varphi_1 + \varphi_2)]$$
$$+ A_1 A_L \cos[(\omega_1 - \omega_L)t + (\phi_1 - \phi_L)] + A_1 A_L \cos[(\omega_1 + \omega_L)t + (\phi_1 + \phi_L)]$$
$$+ A_2 A_L \cos[(\omega_2 - \omega_L)t + (\phi_2 - \phi_L)] + A_2 A_L \cos[(\omega_2 + \omega_L)t + (\phi_2 + \phi_L)]\}.$$
(7.121)

Note that $A^2 \cos^2(\omega t + \phi) = (A^2/2)[1 + \cos(2\omega t + 2\phi)]$. Now, since the detector generally does not follow the instantaneous intensity at double- and sum-frequencies ($2\omega_1$, $\omega_1 + \omega_2$, ...), only dc and difference-frequency terms remain. Hence (7.121) will in practice reduce to (7.36). The calculations leading to (7.121) will remain correct, provided of course, that we insert the proper relation for $(SNR)_i$ in the classical low frequency detection regime. Generally, this is obtained by replacing $h\nu$ by kT and η by $1/F_T$, where F_T is the noise figure of the receiver.

Once the target is ascertained to be present, a wide bandpass filter can be gradually narrowed about $2|f_1' - f_L|$ or $2|f_2' - f_L|$ and thereby used to obtain Doppler information. Alternatively one could, of course, switch to a conventional configuration.

It is of interest to examine the operation of the three-frequency nonlinear heterodyne system in a variety of configurations [7.59] different from those assumed earlier. In this section, we consider the behavior of the system under the following conditions: 1) at zero frequency (dc), 2) without a final bandpass filter, 3) with increased Doppler information, 4) as an optimum system with no uncertainty in Doppler shift, and 5) with a νth law nonlinear device other than square-law. We also consider the consequences of four-frequency nonlinear heterodyne detection; this will be examined in greater detail in Section 7.4.

Assuming that $B \rightarrow 0$, a calculation of $(SNR)_0$ for the final filter centered at zero frequency rather than at f_c yields the result given by (7.73) with $k_P = 1$, thus independent of f_n, A_a, A_b, and B. The advantageous factor f_n/B therefore does not appear in the equation. For $B > 0$, the noise increases while the signal does not so that the above result is optimum for zero frequency. It must be kept in mind, however, that this result has been obtained for a system containing a blocking capacitor (see Fig. 7.3).

If we altogether omit the final bandpass filter, a rather complicated expression for $(SNR)_0$ obtains. Assuming that f_a and f_b are much smaller than f_n, and with $A_a = A_b$, the MDP is calculated to be $(MDP)_0 \simeq 2.9 h\nu f_n/\eta$, approximately a factor three worse than the conventional system.

We now consider the case in which the bandpass filter barely encompasses f_a and f_b so that $f_i = f_b$ and $f_n = f_a$, with $f_a > f_b$. Clearly, decreasing this bandwidth should decrease the overall noise and thereby improve system performance, but this requires knowledge of the Doppler shifts involved. A calculation of the average MDP yields the result $(MDP)_0 \simeq 3hvB/\eta$, which is considerably lower than the result given in (7.81), as it should be.

A calculation for the case in which complete Doppler information is available displays clearly the additional noise introduced by the three-frequency system over the conventional system, by virtue of the nonlinear processing. The result for the MDP in this case is $(MDP)_0 \simeq 1.8hvB/\eta$, which is a factor of 1.8 greater than the MDP for the conventional heterodyne receiver with bandwidth B.

General considerations [7.60] show that all vth law detectors behave in an essentially similar manner to the full-wave square-law detector in terms of the ratio of $(SNR)_0$ and $(SNR)_i$. For large values of $(SNR)_i$, therefore, $(SNR)_0$ is expected to be directly proportional to $(SNR)_i$. Thus, if a half-wave linear device were used instead of the full-wave square-law device considered previously, we would expect results similar to those obtained earlier. This provides a wide choice for designing a heterodyne-nonlinear detector combination, perhaps in a single package.

Finally, we consider the consequences of four-frequency mixing. This would have the advantage of making the transmitter and LO identical. Assuming that only one of the sidebands of the LO is strong (A_{L1}), the $(SNR)_0$ is still given by (7.73) and (7.76), but in this case

$$^4k_P = \frac{f_n A_a^2 A_b^2}{B(A_a^2 + A_b^2 + A_c^2)^2} = \frac{f_n A_1^2 A_2^2}{B(A_1^2 + A_2^2 + A_{L2}^2)^2}, \qquad (7.122)$$

with $A_c/A_a = A_{L2}/A_1$. If $A_1 = A_2 = A_{L2}$, we obtain $^4k_P = f_n/9B$ which is to be compared with the value $^3k_P = f_n/4B$ for the three-frequency case. Thus a single unit may be used both as transmitter and local oscillator without a great deal of loss, provided that one of the two sidebands of the LO beam is attenuated down to the level of the received signal. From (7.122), it is seen that the LO optimally consists of a single frequency, however.

The worst case, in which the LO consists of two strong frequencies separated by f_c, gives rise to very large s × n terms arising from the beating between the two LO frequencies. Aside from other terms not contained in (7.122), this case would make the A_c^4 term in the denominator of (7.122) very large, leading to a factor $(P_r/P_L)^2$ in $(SNR)_0$ (P_r and P_L are typical signal and LO powers, respectively). Thus, in other than very large input SNR situations, the transmitter cannot be directly used as an LO without attenuation of one of its sidebands.

7.3.7 Numerical Example: A CO_2 Laser Radar

As an example of three-frequency nonlinear heterodyne detection, we consider a CO_2 laser radar operating at 10.6 µm in the infrared [7.14, 61, 62] (see Fig. 7.3). If we assume that we wish to acquire and track a 1-m-radius satellite with a rotation rate of 1 rpm, the expected bandwidth (resulting from rotation) of the radar return is of order $4R\omega_\perp/\lambda \sim 40\,\text{kHz}$ [7.9]. We therefore choose a difference frequency f_c at a (convenient) value of 1 MHz, which eliminates spectrum overlap. If the satellite has a radial velocity $\sim 10\,\text{km/s}$, the Doppler frequency is $\sim 2\,\text{GHz}$, yielding a value $f_\text{c}' - f_\text{c} = 2v_\parallel f_\text{c}/c \sim 60\,\text{Hz}$. This shift is very small indeed compared with general frequency modulations in an ordinary heterodyne system, justifying the assumption that $f_\text{c}' - f_\text{c} \to 0$. Thus, assuming we have only an upper bound on the satellite velocity, i.e., its velocity may be anywhere in the range 0–10 km/s, we choose $\Delta f = f_\text{n} \sim 2\,\text{GHz}$ and $B \sim 20\,\text{kHz}$. The MDP for this system would, therefore, be $\sim (\sqrt{3} h\nu/\eta)(Bf_\text{n})^{1/2}$ which is equivalent to the MDP obtained from a conventional setup with a bandwidth of approximately 10 MHz. If the Doppler shift is more confined, the MDP is correspondingly reduced. For strong returns, of course, the SNR will show an enhancement commensurate with the bandwidth B. Thus, the advantages of the three-frequency heterodyne system may be secured with such a radar. Similar results would be obtained at other frequencies; in the microwave, for example, f_c may be made as small as tens of Hz. In some cases, it may be possible to reduce clutter by the insertion of an extremely sharp notch filter at exactly f_c.

7.3.8 Application to Binary Communications and Pulsed Radar (Vacuum Channel)

The previous subsections were primarily concerned with the behavior of the three-frequency nonlinear heterodyne system for applications in cw radar and analog communications. As such, a determination of the output signal-to-noise ratio $(\text{SNR})_0$ was adequate to characterize the system. In this subsection, we investigate applications in digital communications and pulsed radar, and therefore examine system performance in terms of the error probability P_e. Evaluation of the probability of error under various conditions requires a decision criterion as well as a knowledge of the signal statistics; we now investigate operation of the three-frequency nonlinear heterodyne scheme in the time domain rather than in the frequency domain.

Because of the added complexity of dealing in the time domain, we limit our investigation to sinewave signals, Gaussian local oscillator (LO) noise, and envelope detection. The configuration of such a receiver is therefore similar to that considered previously, with the addition of an envelope detector (see Figs. 7.3 and 7.15). We therefore examine the case of a particular "square-law envelope detector", consisting of a square-law device, a narrowband filter, and an envelope detector [7.63]. Although envelope detection is generally suboptimum because it is insensitive to phase, it is easy to implement practically and is therefore widely used [7.64].

We begin with an investigation of binary communications and pulsed radar for both nonorthogonal and orthogonal signaling formats in the vacuum channel. In Section 7.3.9, we examine envelope probability distributions and binary signaling for sinewave signals in the lognormal channel (clear air turbulent atmosphere). The advantages of the three-frequency nonlinear heterodyne scheme in the digital communications/pulsed radar configuration are similar to those cited for cw radar/analog communications.

We assume here, as previously, that when a signal is present the fields incident on the mixer are parallel, plane polarized, and spatially first-order coherent over the detector aperture. In general, therefore, the input to the square-law device, as previously [see (7.51)], will be two narrowband signals plus white Gaussian noise with zero mean resulting from the LO, over the band $[0, f_n]$. Thus

$$s(t) = A_a \cos(\omega_a t + \phi_a) + A_b \cos(\omega_b t + \phi_b), \tag{7.123}$$

with A_a, A_b, ϕ_a, and ϕ_b stochastic processes. The amplitudes are assumed to be independent of the phases. We first treat the specific case of sinusoidal signals, i.e., A_a and A_b constant and ϕ_a, ϕ_b independent random variables uniformly distributed over $(0, 2\pi)$.

In the time domain, the white Gaussian noise, which arises from the LO, can be expressed as [7.65]

$$n(t) = \sum_{k=1}^{\infty} u_k \cos \omega_k t + \sum_{k=1}^{\infty} v_k \sin \omega_k t. \tag{7.124}$$

Here, $\omega_k = k\omega_0$ with $\omega_0 = 2\pi/2T$. If the input signal is a pulse, the pulse duration is the time interval $(-T, T)$. The coefficients u_k and v_k may therefore be written as

$$u_k = \frac{1}{T} \int_{-T}^{T} n(t) \cos \omega_k t \, dt, \tag{7.125a}$$

and

$$v_k = \frac{1}{T} \int_{-T}^{T} n(t) \sin \omega_k t \, dt. \tag{7.125b}$$

(In the alternative, a narrowband representation could be used.) Since u_k and v_k are linear transformations of the Gaussian random variable $n(t)$, they are also Gaussian random variables [7.66]; furthermore it can be shown that for T large, all u_k's and v_k's are uncorrelated and independent of one another [7.67]. Since the mean of $n(t)$ is taken to be zero, we find

$$\langle u_k \rangle = \left\langle \frac{1}{T} \int_{-T}^{T} n(t) \cos \omega_k t \, dt \right\rangle = 0 \tag{7.126a}$$

and similarly

$$\langle v_k \rangle = 0, \tag{7.126b}$$

while the variance $\langle u_k^2 \rangle$ is given by

$$\langle u_k^2 \rangle = \frac{1}{T^2} \int_{-T}^{T} \int_{-T}^{T} \langle n(t)n(t') \rangle \cos\omega_k t \cos\omega_k t' dt dt' = \frac{N}{T}. \tag{7.127a}$$

Similarly,

$$\langle v_k^2 \rangle = \frac{N}{T}. \tag{7.127b}$$

In calculating these quantities, we have assumed that the Gaussian noise $n(t)$ is stationary, and that the band $[f_i, f_n]$ is sufficiently large so that the noise can be approximated to be completely white (over an infinite band) leading to an autocorrelation function $R_n(t-t') \simeq N\delta(t-t')$. Here N is the height of the white noise spectrum.

The input $x(t)$ to the square-law device can now be written as

$$\begin{aligned} x(t) &= s(t) + n(t) \\ &= A_a \cos(\omega_a t + \phi_a) + A_b \cos(\omega_b t + \phi_b) \\ &\quad + \sum_k u_k \cos\omega_k t + \sum_k v_k \sin\omega_k t. \end{aligned} \tag{7.128}$$

We note that since ω_0 is small, it is always possible to find integers m and n such that $m\omega_0$ and $n\omega_0$ are very close to ω_a and ω_b, respectively. This implies that T is much larger than $2\pi/\omega_a$ and $2\pi/\omega_b$.

By direct substitution, we find the output of the square-law device $y(t)$ to be

$$\begin{aligned} y(t) &= \alpha x^2(t) \\ &= \alpha(\tfrac{1}{2} \sum_k u_k^2(1 + \cos 2\omega_k t) + \tfrac{1}{2} \sum_k v_k^2(1 - \cos 2\omega_k t) \\ &\quad + \sum_k u_k v_k \sin 2\omega_k t + \sum\sum_{i>j} u_i u_j [\cos(\omega_i - \omega_j)t + \cos(\omega_i + \omega_j)t] \\ &\quad + \sum\sum_{i>j} v_i v_j [\cos(\omega_i - \omega_j)t - \cos(\omega_i + \omega_j)t] \\ &\quad + \sum\sum_{i>j} u_i v_j [\sin(\omega_i + \omega_j)t - \sin(\omega_i - \omega_j)t] \\ &\quad + \sum\sum_{i<j} u_i v_j [\sin(\omega_i + \omega_j)t + \sin(\omega_j - \omega_i)t] \\ &\quad + \tfrac{1}{2} A_a^2 [1 + \cos(2\omega_a t + 2\phi_a)] + \tfrac{1}{2} A_b^2 [1 + \cos(2\omega_b t + 2\phi_b)] \\ &\quad + A_a A_b \{\cos[(\omega_a + \omega_b)t + \phi_a + \phi_b] + \cos[(\omega_a - \omega_b)t + \phi_a - \phi_b]\} \\ &\quad + A_a \sum_k u_k \{\cos[(\omega_k + \omega_a)t + \phi_a] + \cos[(\omega_k - \omega_a)t - \phi_a]\} \\ &\quad + A_b \sum_k u_k \{\cos[(\omega_k + \omega_b)t + \phi_b] + \cos[(\omega_k - \omega_b)t - \phi_b]\} \\ &\quad + A_a \sum_k v_k \{\sin[(\omega_k + \omega_a)t + \phi_a] + \sin[(\omega_k - \omega_a)t - \phi_a]\} \\ &\quad + A_b \sum_k v_k \{\sin[(\omega_k + \omega_b)t + \phi_b] + \sin[(\omega_k - \omega_b)t - \phi_b]\}), \end{aligned} \tag{7.129}$$

where we have used the following symmetrical relations:

$$\tfrac{1}{2}\sum\sum_{i>j} u_i u_j [\cos(\omega_i+\omega_j)t + \cos(\omega_i-\omega_j)t]$$
$$= \tfrac{1}{2}\sum\sum_{i<j} u_i u_j [\cos(\omega_i+\omega_j)t + \cos(\omega_i-\omega_j)t], \qquad (7.130\text{a})$$

and

$$\tfrac{1}{2}\sum\sum_{i>j} v_i v_j [\cos(\omega_i+\omega_j)t + \cos(\omega_i-\omega_j)t]$$
$$= \tfrac{1}{2}\sum\sum_{i<j} v_i v_j [\cos(\omega_i+\omega_j)t + \cos(\omega_i-\omega_j)t]. \qquad (7.130\text{b})$$

Since it is the effective bandwidth rather than the shape of the final narrowband filter which is important, we choose a realizable impulse response for this filter given by

$$h(t) = 2B\cos 2\pi f_c t \qquad 0 < t < \frac{1}{B}. \qquad (7.131)$$

This choice facilitates the computation in the time domain and provides accord with signal-to-noise ratios calculated previously. Assuming B is very small, the time output from the bandpass filter $z(t)$ is given by

$$z(t) = \int_0^{1/B} h(t-t') y(t') dt$$
$$= A\cos(\omega_c t + \phi) + u\cos\omega_c t + v\sin\omega_c t. \qquad (7.132)$$

Here

$$A = \alpha A_a A_b, \qquad (7.133\text{a})$$
$$\phi = \phi_a - \phi_b, \qquad (7.133\text{b})$$

and after a great deal of calculation, u and v turn out to be the sum of an infinite number of random variables, and therefore Gaussian. The means and variances of u and v are found to be [7.59]

$$\langle u \rangle = \langle v \rangle = 0 \qquad (7.134)$$

and

$$\langle u^2 \rangle = \langle v^2 \rangle \simeq 4\alpha^2 \frac{N}{T} [2f_n N + (\langle A_a^2 \rangle + \langle A_b^2 \rangle)], \qquad (7.135)$$

assuming $f_0, f_c \ll f_n$. It is also found that

$$\langle uv \rangle = \langle u \rangle \langle v \rangle = 0, \qquad (7.136)$$

indicating that u and v are uncorrelated and independent processes. Equations (7.134), (7.135), and (7.136) indicate that the last two terms in (7.132), $u\cos\omega_c t + v\sin\omega_c t$, constitute a narrowband Gaussian random process with zero mean and center frequency ω_c. In fact, (7.135) represents the output noise power N_0.

We can corroborate this rather broad result [7.59] for a specific case by generalizing the results obtained by *Kac* and *Siegert* [7.68] and *Emerson* [7.69], who have treated a related problem. We assume the output of the heterodyne mixer to consist of two sinewave signals plus uncorrelated (white) Gaussian noise. The system, in this case, consists of a realizable IF Gaussian bandpass filter, with arbitrary width Δf and a center frequency around f_a or f_b (which is large in comparison with f_c), the usual square-law device, and a realizable final narrowband filter with bandwidth B. Under the restrictions $f_c \ll f_a, f_b$ and $B \ll f_c \ll \Delta f$, it may be shown that the output of the final narrowband filter will be a sinewave signal plus a Gaussian random process. For noise alone, the output will simply be Gaussian. Thus the envelope distribution for noise will be Rayleigh, while that for signal-plus-noise will be Rician. This is we might add, the same result obtained for conventional two-frequency heterodyne detection, although the means and variances will not have the same relationship in that case.

For $f_c \ll f_n$, as prescribed previously, it is not difficult to verify that the above description in the time domain is in accord with the frequency-domain results presented previously. Since the relationship between the pulse width T and the minimum bandwidth of the final filter is governed by the Fourier transform property $TB \sim 1$ [7.70], (7.135) for the noise power in this regime may be written as

$$\langle u^2 \rangle = \langle v^2 \rangle = N_0 \simeq 4\alpha^2 NB[2f_n N + (\langle A_a^2 \rangle + \langle A_b^2 \rangle)]. \tag{7.137}$$

Using (7.133a), we therefore obtain for the output signal-to-noise ratio

$$(\text{SNR})_0 = \frac{S_0}{N_0} = \frac{\langle A^2 \rangle}{2\langle u^2 \rangle} = \frac{\langle A_a^2 A_b^2 \rangle}{8NB[(\langle A_a^2 \rangle + \langle A_b^2 \rangle)2f_n N]}. \tag{7.138}$$

Using an input signal-to-noise ratio given by

$$(\text{SNR})_i = \frac{\langle A_a^2 \rangle + \langle A_b^2 \rangle}{4f_n N}, \tag{7.139}$$

we finally obtain

$$(\text{SNR})_0 = \frac{k_Q(\text{SNR})_i^2}{1 + 2(\text{SNR})_i}, \tag{7.140}$$

with

$$k_Q = \frac{f_n \langle A_a^2 A_b^2 \rangle}{B(\langle A_a^2 \rangle + \langle A_b^2 \rangle)^2} = \frac{f_n \langle A_1^2 A_2^2 \rangle}{B(\langle A_1^2 \rangle + \langle A_2^2 \rangle)^2}. \tag{7.141}$$

These expressions are valid in the regime $f_c \ll f_n$, and are analogous to (7.72) and (7.73) of Section 7.3.2. Our treatment is therefore consistent with that presented previously.

According to (7.132) and the discussion following, in the presence of signal plus noise the output of the narrowband final filter $z(t)$, after being passed through the envelope detector, is given by the Rician distribution [7.71]

$$f_1(r) = \frac{r}{\sigma^2} I_0\left(\frac{Ar}{\sigma^2}\right) \exp\left(-\frac{r^2 + A^2}{2\sigma^2}\right). \tag{7.142}$$

Here, r represents the envelope of $z(t)$, $\sigma^2 = \langle u^2 \rangle = 4\alpha^2 NB[2f_n N + (\langle A_a^2 \rangle + \langle A_b^2 \rangle)]$, and $I_0(x)$ is the modified Bessel function of the first kind and zero order, also expressible as

$$I_0(x) = \frac{1}{2\pi} \int_0^{2\pi} \exp(x \cos\theta) d\theta. \tag{7.143a}$$

We may use the asymptotic expansion for $x \ll 1$ [7.72],

$$I_0(x) = 1 + \frac{x^2}{4} + \ldots \simeq e^{x^2/4}, \tag{7.143b}$$

while for $x \gg 1$,

$$I_0(x) \simeq \frac{e^x}{\sqrt{2\pi x}}. \tag{7.143c}$$

In the presence of noise alone, i.e., for $A_a = A_b = 0$, the probability density function for the envelope $f_0(r)$ is the Rayleigh distribution

$$f_0(r) = \frac{r}{\sigma_0^2} \exp\left(-\frac{r^2}{2\sigma_0^2}\right). \tag{7.144}$$

Here σ_0^2 is the noise power in the absence of signal, i.e.,

$$\sigma_0^2 = \langle u^2 \rangle|_{A_a = A_b = 0} = 8\alpha^2 B f_n N^2. \tag{7.145}$$

We note that in our nonlinear problem $\sigma^2 \neq \sigma_0^2$ because of the presence of $s \times n$ terms in σ^2. In the usual linear systems problem, these terms do not appear, and $\sigma^2 = \sigma_0^2$.

Given the probability distributions for the output signals, we can proceed to investigate binary communications and pulsed radar systems performance upon choosing a decision rule. In the following, we consider both orthogonal and nonorthogonal formats for digital signaling.

Nonorthogonal Signaling Formats

We first consider pulse-code modulation where it is the intensity which is modulated. This simple nonorthogonal scheme is frequently referred to as PCM/IM [7.73]. The signal is considered to be present when a 1 is transmitted, and absent when a 0 is transmitted. To evaluate system performance, we choose the likelihood-ratio criterion [7.72, 73]. If Q represents the *a priori* probability that a 1 is transmitted, the signal is judged to be present if

$$Qf_1(r) \geqq (1-Q)f_0(r). \tag{7.146}$$

For simplicity, we assume throughout that the different types of errors are equally costly. Since the signals are pulse coded, the value of r chosen is the average value over the pulse width. The decision threshold r_D is the value of r for which the equality in (7.146) holds. Using (7.142), (7.143a), and (7.144) for sinewave signals and Gaussian noise, r_D is therefore the solution to the transcendental equation

$$\frac{1}{2\pi} \int_0^{2\pi} \exp\left(\frac{Ar}{\sigma^2}\cos\theta\right) d\theta$$
$$= \left(\frac{1-Q}{Q}\right) \frac{\sigma^2}{\sigma_0^2} e^{A^2/2\sigma^2} \exp\left(-\frac{\sigma^2-\sigma_0^2}{2\sigma_0^2\sigma^2}r^2\right). \tag{7.147}$$

Using (7.137) and (7.141), it is clear that

$$\sigma^2 = \sigma_0^2(1+2\sqrt{\xi_0/k_Q}), \tag{7.148}$$

where

$$\xi_0 \equiv \frac{\langle A^2 \rangle}{2\sigma_0^2}. \tag{7.149}$$

For sinewave inputs, A, A_a, and A_b are constant and the quantity k_Q is identical with the quantity k_P introduced earlier [see (7.72), Sec. 7.3.2]. Defining $r_0 \equiv r/\sigma_0$, (7.147) can be rewritten as

$$\frac{1}{2\pi} \int_0^{2\pi} \exp\left(\frac{\sqrt{2\xi_0} r_0 \cos\theta}{1+2\sqrt{\xi_0/k_P}}\right) d\theta$$
$$= \left(\frac{1-Q}{Q}\right)\left(1+2\sqrt{\frac{\xi_0}{k_P}}\right) \exp\left(\frac{\xi_0}{1+2\sqrt{\xi_0/k_P}}\right)$$
$$\cdot \exp\left(-\frac{r_0^2 \sqrt{\xi_0/k_P}}{1+2\sqrt{\xi_0/k_P}}\right). \tag{7.150}$$

Therefore, with k_P and Q fixed, the solution to (7.150) for r_0, which we call $\hat{r}_0$, is a function only of ξ_0. If we further define

$$\xi' \equiv (\text{SNR})_0 = \frac{\langle A^2 \rangle}{2\sigma^2}, \tag{7.151}$$

then the quantity $\xi' = (\sigma_0^2/\sigma^2)\xi_0 = \xi_0/(1 + 2\sqrt{\xi_0/k_P})$ is also a function only of ξ_0. Thus, $\hat{r}_0$ is a function only of ξ'. The decision threshold $r_D \equiv \sigma_0 \hat{r}_0$ is therefore a function of both ξ' and σ_0.

The probability of a decoding error P_e is given by

$$P_e = Q \int_0^{r_D} f_1(r) dr + (1-Q) \int_{r_D}^{\infty} f_0(r) dr, \tag{7.152}$$

which in the present case, may be written as

$$P_e = Q \int_0^{r_D} \frac{r}{\sigma^2} I_0\left(\frac{Ar}{\sigma^2}\right) e^{-\frac{r^2 + A^2}{2\sigma^2}} dr$$

$$+ (1-Q) \int_{r_D}^{\infty} \frac{r}{\sigma_0^2} e^{-\frac{r^2}{2\sigma_0^2}} dr. \tag{7.153}$$

Replacing r/σ by r', we can rewrite the first integral I_1 in (7.152) as follows

$$I_1 = \int_0^{\hat{r}_0 \sigma_0/\sigma} r' I_0(\sqrt{2\xi'} r') e^{-\left(\frac{r'^2}{2} + \xi'\right)} dr'. \tag{7.154}$$

Since $\hat{r}_0 \sigma_0/\sigma = \hat{r}_0/(1 + 2\sqrt{\xi_0/k_P})^{1/2}$ is a function only of ξ_0 which, in turn, is a function only of the output signal-to-noise ratio ξ', this integral is a function only of ξ'. The second integral in (7.153) can be easily evaluated as follows:

$$\int_{r_D}^{\infty} \frac{r}{\sigma_0^2} e^{-r^2/2\sigma_0^2} dr = \int_{r_D/\sigma_0}^{\infty} x e^{-x^2/2} dx = -e^{-x^2/2}\Big|_{r_D/\sigma_0}^{\infty}$$

$$= e^{-r_D^2/2\sigma_0^2} = e^{-\hat{r}_0^2/2}, \tag{7.155}$$

which is also a function only of ξ'.

Therefore, with fixed $k_P (\propto f_n/B)$ and fixed Q, the probability of error P_e is a function only of the output signal-to-noise ratio $\xi' \equiv (\text{SNR})_0$. By use of (7.141), in turn, P_e can be written in terms of $(\text{SNR})_i$. Computer results for the probability of error are presented in Fig. 7.16 [(7.143) has been used for the computer calculation], in which P_e is plotted against $(\text{SNR})_i$ for several values of f_n/B, with the usual choice $Q = 0.5$ and $A_a = A_b$. The solid curves represent this PCM/IM scheme. For fixed f_n, the advantage of using small B is obvious.

Also shown in Fig. 7.16 is the P_e versus SNR curve for the conventional two-frequency heterodyne system in which no square-law device is used and f_n must be narrowed to Δf to provide a detectable signal. The output for this case is again a sinewave signal plus a narrowband (Δf) Gaussian noise [7.73]. Thus the

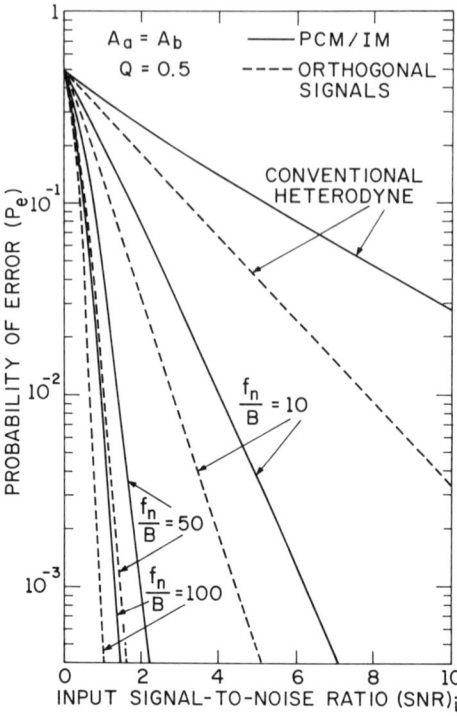

Fig. 7.16a. Probability of error vs (SNR)$_i$ for the three-frequency binary communication system in the vacuum channel. The input signals are assumed to be sinusoidal while the noise is Gaussian. The result for the conventional heterodyne system is shown for comparison (log vs linear plot)

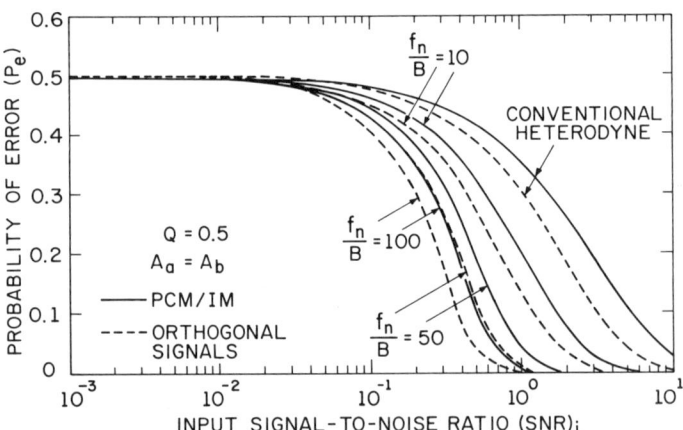

Fig. 7.16b. Same curves as Fig. 7.16a on a linear vs log plot

computation is the same as for the three-frequency heterodyne case with $\sigma^2 = \sigma_0^2$ and $\xi' = \xi_0 = (SNR)_i$. The probability of error at a given signal-to-noise ratio for the ordinary heterodyne system is seen to be higher than for the three-frequency system. This results from the exclusion of noise demanded by the final bandpass filter where $f_n/B > 1$, thus providing higher (SNR)$_o$ and lower P_e for the three-frequency system. Since the Rician and Rayleigh distributions have been calculated only for $B \ll f_c$ (hence $B \ll f_n$) and for white noise, the optimum three-frequency case considered previously is not shown in Fig. 7.16.

Pulsed Radar Application

The three-frequency nonlinear heterodyne system can also be used for pulsed radar applications. The configuration is similar to that considered previously. Pulses are sent to the target and the maximum-likelihood test is used to determine whether the target is or is not present (reflected or scattered signal deemed present or absent). For a detailed treatment of conventional range-gated pulsed radar applications, the reader is referred to the book by *Davenport* and *Root* [7.74].

Orthogonal Signaling Formats

We consider a number of orthogonal signaling formats—we begin with frequency shift keying (FSK) which is also referred to as PCM/FM. In such a scheme, the frequency of one of the transmitted beams is fixed at the value f_1, while the frequency of the other is caused to shift between two values, f_2 and f_2' (not to be confused with the Doppler shifted f_2' considered earlier). When a 1(0) is transmitted, the second carrier will be at frequency $f_2(f_2')$. The difference frequency will therefore shift between $f_c = f_1 - f_2$ and $f_c' = f_1 - f_2'$ (assuming $f_1 > f_2, f_2'$). The frequencies $|f_1 - f_L|$, $|f_2 - f_L|$, and $|f_2' - f_2|$ will all lie within the band f_n. A block diagram for such a system is shown in Fig. 7.17. Two narrow bandpass final filters with center frequencies at f_c and f_c' (not to be confused with the Doppler shifted f_c' considered earlier) are used. Following each bandpass filter is an envelope detector. If a 1(0) is transmitted, the signal will ideally pass through the top (bottom) narrow bandpass filter along with the noise; only noise will be present at the other filter.

For such an orthogonal format, the optimum single detector receiver chooses the largest signal as the correct one. Let the outputs of the first and second envelope detectors be represented by r_1 and r_2, respectively, while the probability density functions for r_1 and r_2 are $h_1(r_1)$ and $h_2(r_2)$, respectively. If we assume that a 1 is transmitted, we have

$$h_1(r_1) = f_1(r_1), \tag{7.156a}$$

$$h_2(r_2) = f_0(r_2), \tag{7.156b}$$

where $f_1(\cdot)$ and $f_0(\cdot)$ are given by (7.142) and (7.144), respectively. Using the decision rule of the largest, error occurs during times when $r_2 > r_1$. The error probability P_{e_1} is, therefore,

$$P_{e_1} = \int_0^\infty dr_1 [f_1(r_1) \int_{r_1}^\infty dr_2 f_0(r_2)]$$

$$= \frac{\sigma_0^2}{\sigma_0^2 + \sigma^2} e^{-A^2/2(\sigma_0^2 + \sigma^2)}. \tag{7.157}$$

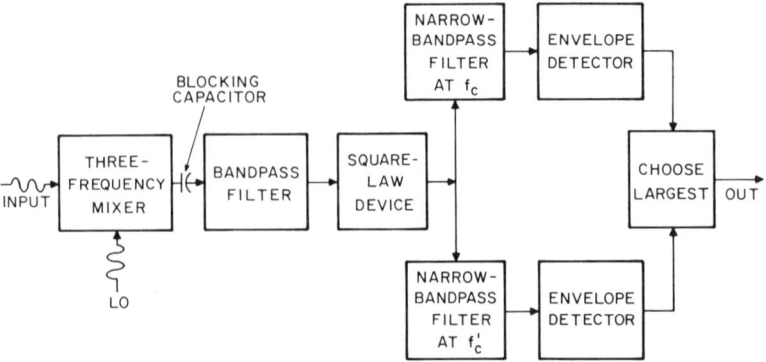

Fig. 7.17. Block diagram for the PCM/FM three-frequency nonlinear heterodyne receiver

This can be readily shown to be a function only of ξ'. In exactly the same manner, the error probability P_{e0}, when 0 is transmitted, is given by the same expression; thus $P_{e0} = P_{e_1}$. The overall probability of error P_e is therefore given by

$$P_e = QP_{e_1} + (1-Q)P_{e0} = P_{e_1}, \qquad (7.158)$$

which is presented in Fig. 7.16 in dashed form with the same parameters as for the PCM/IM case. The conventional heterodyne case is also shown [7.73]. The improvement obtained by using the orthogonal PCM/FM signaling format is seen to be substantial.

Another binary orthogonal pulse-code modulation scheme is polarization modulation (PCM/PL). Thus the bit 1(0) is represented by right (left) circular or vertical (horizontal) linear polarization. At the transmitter, a polarization modulator converts the laser beam into one of two polarization states. At the receiver (see Fig. 7.18), the circularly polarized beam may be passed through an optical filter and then be converted to horizontal or vertical linear polarization by a quarter-wave plate. The linear polarization components are spatially separated (e.g., by a Wollaston prism) so that the vertically polarized component will strike the upper photodetector and the horizontally polarized component will strike the lower photodetector. With 100 % modulation, when the bit 1 is transmitted, only vertical polarization will appear at the receiver and the radiation will ideally strike only the upper detector. When a 0 is transmitted, only horizontal polarization will appear and a signal will ideally strike only the lower detector. The "choice of largest" decision rule is used for decoding. It is not difficult to see that the results for P_e in this case are identical to those for the PCM/FM system. Depolarization effects of the atmosphere, which are not generally large, will result in a decrease of $(SNR)_i$ and thus $(SNR)_0$ [7.75–78].

The final orthogonal format which we consider is binary pulse-position modulation (PPM/IM). In this scheme, each bit period is divided into two equal subintervals. If a 1(0) is transmitted, the pulse is caused to occur in the first (second) subinterval. A block diagram for one implementation of such a system

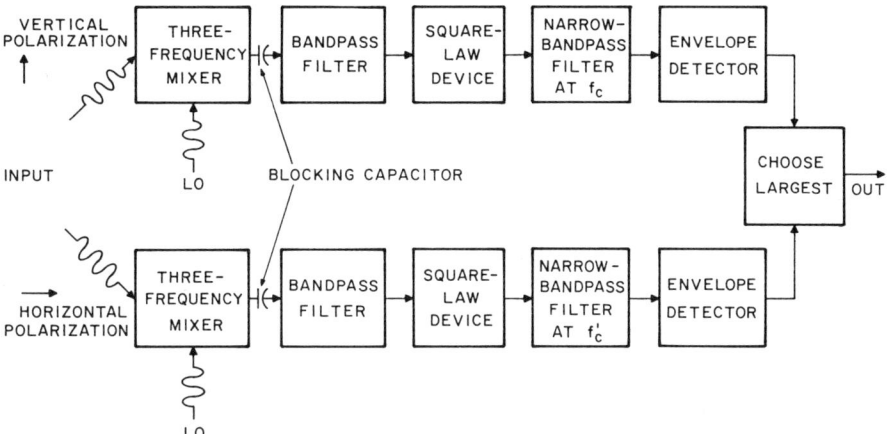

Fig. 7.18. Block diagram for the PCM/PL three-frequency nonlinear heterodyne receiver

is presented in Fig. 7.19. The upper (lower) gate is open for every initial (final) subinterval, and closed for every final (initial) subinterval. A time delay equal to the subinterval length is provided for the signal in the upper gate so that the outputs for both intervals can be compared at the same time. The rule of largest decision is used for decoding. The results for the probability of error are again the same as those for the PCM/FM system.

The input signals for the PCM/FM, PCM/PL, and PPM/IM systems possess the orthogonality property

$$\int_{-T}^{T} S_1(t)S_0(t)dt = 0, \tag{7.159}$$

where $S_1(t)$ is the signal waveform representing a 1 state, and $S_0(t)$ is the signal waveform representing a 0 state. Depending on specific definitions, such orthogonal modulation schemes are generally superior to nonorthogonal

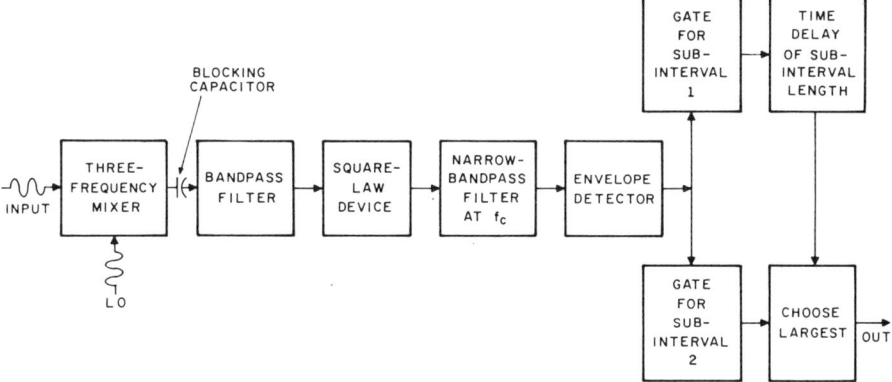

Fig. 7.19. Block diagram for the PPM/IM three-frequency nonlinear heterodyne receiver

schemes in terms of error probability performance [7.76–78] and have the further advantage of requiring no more than a simple comparison for optimum reception. The M-ary signaling case is a straightforward generalization of the binary case [7.79].

7.3.9 Application to Binary Communications and Pulsed Radar (Lognormal Atmospheric Channel)

Whereas the previous section (7.3.8) was concerned with the calculation of system performance for the vacuum channel, we now turn to the error probabilities for three-frequency nonlinear heterodyne detection for the atmospheric channel. The behavior of the clear-air turbulent atmosphere as a lognormal channel for optical radiation has been well documented both theoretically and experimentally [7.76–78, 80–82]. We therefore choose the amplitudes A_1 and A_2 to be lognormally distributed, and the phases ϕ_1 and ϕ_2 to be uniformly distributed over $(0, 2\pi)$. Since $A_a \propto A_1$ and $A_b \propto A_2$, while $\phi_a = \phi_1 - \phi_L$ and $\phi_b = \phi_2 - \phi_L$, we can write

$$A_a = u_a B_a, \qquad (7.160a)$$

$$A_b = u_b B_b, \qquad (7.160b)$$

where B_a and B_b are constants, and u_a and u_b have the same lognormal distribution

$$P_N(u_i) = \frac{1}{\sigma_\chi u \sqrt{2\pi}} \exp\left[-\frac{1}{2\sigma_\chi^2}(\ln u_i - m)^2\right], \quad i = a, b. \qquad (7.161)$$

Here σ_χ is the logarithmic-amplitude standard deviation which is related to the logarithmic-irradiance standard deviation σ by the formula $4\sigma_\chi^2 = \sigma^2$ [7.82]. Assuming energy is conserved and that there is no scattering of radiation out of the beam, we choose

$$\langle u_i^2 \rangle = 1 \qquad (7.162)$$

which is equivalent to setting $m = -\sigma_\chi^2$.

Using (7.133a) the output amplitude A is given by

$$A = \alpha A_a A_b = \alpha B_a B_b u_a u_b. \qquad (7.163)$$

If u_a and u_b are independent, we obtain

$$\langle A^2 \rangle = \alpha^2 B_a^2 B_b^2 \langle u_a^2 \rangle \langle u_b^2 \rangle = \alpha^2 B_a^2 B_b^2, \qquad (7.164a)$$

or

$$\alpha B_a B_b = \sqrt{\langle A^2 \rangle} = \sqrt{\alpha^2 \langle A_a^2 A_b^2 \rangle}. \tag{7.164b}$$

Furthermore,

$$\ln A = \ln u_a + \ln u_b + \ln \alpha B_a B_b. \tag{7.165}$$

Since the quantities $y_a \equiv \ln u_a$ and $y_b \equiv \ln u_b$ will both be normally distributed as

$$f_N(y_i) = \frac{1}{\sqrt{2\pi}\sigma_\chi} \exp\left[-\frac{1}{2\sigma_\chi^2}(y_i + \sigma_\chi^2)^2\right], \tag{7.166}$$

if u_a and u_b are independent, the variable $y_L = \ln A$ will have the normal distribution

$$f_L(y_L) = \frac{1}{2\sigma_\chi \sqrt{\pi}} \exp\left[-\frac{1}{4\sigma_\chi^2}(y_L + 2\sigma_\chi^2 - \ln \alpha B_a B_b)^2\right], \tag{7.167}$$

from which we obtain the probability density for A

$$f_A(A) = \frac{1}{2\sigma_\chi \sqrt{\pi} A} \exp\left[-\frac{1}{4\sigma_\chi^2}\left(\ln \frac{A}{\sqrt{\langle A^2 \rangle}} + 2\sigma_\chi^2\right)^2\right], \tag{7.168}$$

u_a, u_b independent, where we have made use of (7.164b).

We also consider the situation $u_a = u_b = u$, which would arise if both incoming signals were sufficiently close in frequency and space such that they suffered precisely the same fluctuations at each instant of time [7.83]. This case is more likely to occur in a practical situation than the independent case. For dependent fluctuations, then,

$$A = \alpha A_a A_b = \alpha B_a B_b u^2, \tag{7.169}$$

whence

$$\langle A \rangle = \alpha \langle A_a A_b \rangle = \alpha B_a B_b, \tag{7.170}$$

and

$$\ln A = \ln \langle A \rangle + 2 \ln u. \tag{7.171}$$

Since $\ln u$ has the normal distribution $f_L(u)$ as given by (7.166), we find that the variable $y_L = \ln A$ has the normal probability density function

$$f_N(y_L) = \frac{1}{2\sigma_\chi \sqrt{2\pi}} \exp\left[-\frac{1}{8\sigma_\chi^2}(y_L + 2\sigma_\chi^2 - \ln \langle A \rangle)^2\right]. \tag{7.172}$$

By variable transformation, we obtain the probability density function for A as

$$f_A(A) = \frac{1}{2\sigma_\chi \sqrt{2\pi} A} \exp\left[-\frac{1}{8\sigma_\chi^2}\left(\ln \frac{A}{\langle A \rangle} + 2\sigma_\chi^2\right)^2\right], \quad u_a = u_b. \tag{7.173}$$

This equation appears similar to (7.168); we note that $\sqrt{\langle A^2 \rangle}$ is replaced by $\langle A \rangle$ and the effective variance has been doubled. This results in a flattening and broadening of the probability density for the case of identical disturbance to both beams, $u_a = u_b$.

For atmospheric fluctuations which vary slowly in comparison with the pulse time T (this is the usual case, see [7.76–78, 81–83]) the three-frequency system envelope output will be Rician during each time interval. The over-all envelope distribution in the presence of the atmosphere $f_{1A}(r)$ will therefore be a Rician smeared over all possible values of A,

$$f_{1A}(r) = \int_0^\infty f_1(r|A) f_A(A) dA, \tag{7.174}$$

where $f_1(r|A)$ is given by (7.142). In the absence of signal, the envelope probability density remains as it was before [see (7.144)] since the noise alone arises from the local oscillator which is unaffected by atmospheric fluctuations. Thus,

$$f_{0A}(r) = f_0(r). \tag{7.175}$$

Under the assumptions leading to (7.174), and considering the various modulation schemes discussed previously, the probability of error in the presence of the lognormal turbulent atmosphere is given by

$$P_e(\text{turbulent}) = \int_0^\infty P_e(\text{quiescent}) f_A(A) dA. \tag{7.176}$$

This quantity was calculated using the Columbia University IBM-OS 360 computer, and the results are presented in Figs. 7.20–23. In Figs. 7.20 and 7.21, the quantities A_a and A_b were assumed to be independent with the same signal power $\langle A_a^2 \rangle = \langle A_b^2 \rangle$. The error probability curves displayed in these figures correspond to two values of the log-amplitude variance, $\sigma_\chi^2 = 0.25$ and $\sigma_\chi^2 = 0.57$. These correspond approximately to $\sigma = 1$ and $\sigma = 1.5$ (saturation value) [7.81, 82]. Other parameters are identical to those for the quiescent atmosphere as shown in Fig. 7.16. Figures 7.22 and 7.23 are analogous to Figs. 7.20 and 7.21, with the exception of the fact that $A_a = A_b$. For all cases, the results for conventional heterodyne operation are also shown in Figs. 7.20 and 7.21. For $\sigma_\chi \to 0$, the results properly reduce to the quiescent atmosphere data presented in Fig. 7.16. Computer results also indicate that the probability of error curves depend only on the signal-to-noise ratio and not on the absolute noise level in the presence of the lognormal channel, as well as in its absence.

From the graphical data presented in Figs. 7.16, 7.20–7.23, it is clear that orthogonal signaling formats yield better performance than nonorthogonal PCM/IM (this is also the case for direct detection [7.76–78]). Error probabilities

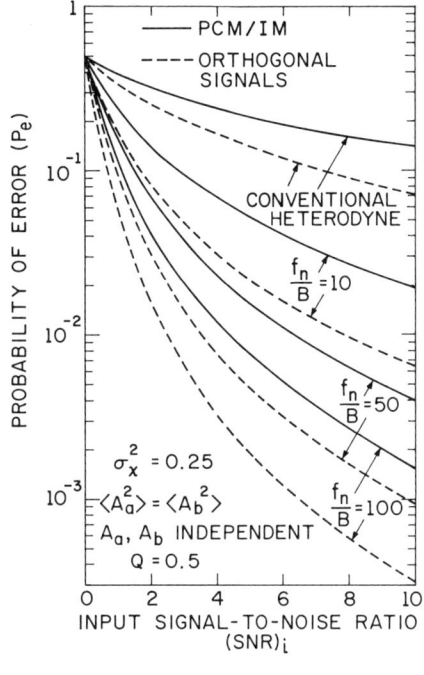

Fig. 7.20. Probability of error vs $(SNR)_i$ for the three-frequency binary communication system with atmospheric turbulence at the level $\sigma_\chi^2 = 0.25$. The input amplitudes A_a and A_b are assumed to be independent, and the noise is Gaussian. The result for the conventional heterodyne system is shown for comparison

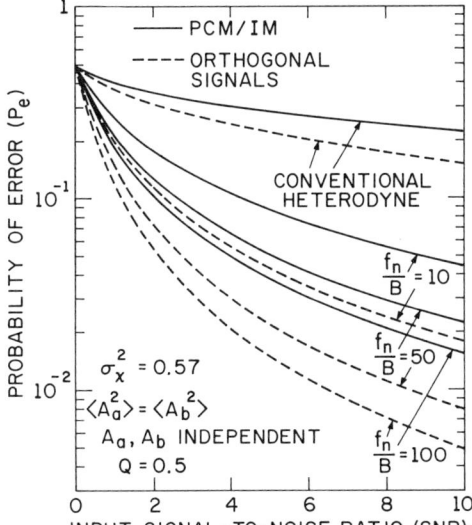

Fig. 7.21. Probability of error vs $(SNR)_i$ for the three-frequency binary communication system with atmospheric turbulence at the level $\sigma_\chi^2 = 0.57$. The input amplitudes A_a and A_b are assumed to be independent, and the noise is Gaussian. The result for the conventional heterodyne system is shown for comparison

are seen to increase with increasing atmospheric turbulence levels. Independent fluctuations in the two signal beams serve as a kind of diversity and thereby improve receiver performance. In all cases, furthermore, it is evident that three-frequency nonlinear heterodyne detection can provide improved performance over conventional heterodyne detection, particularly as the ratio f_n/B increases.

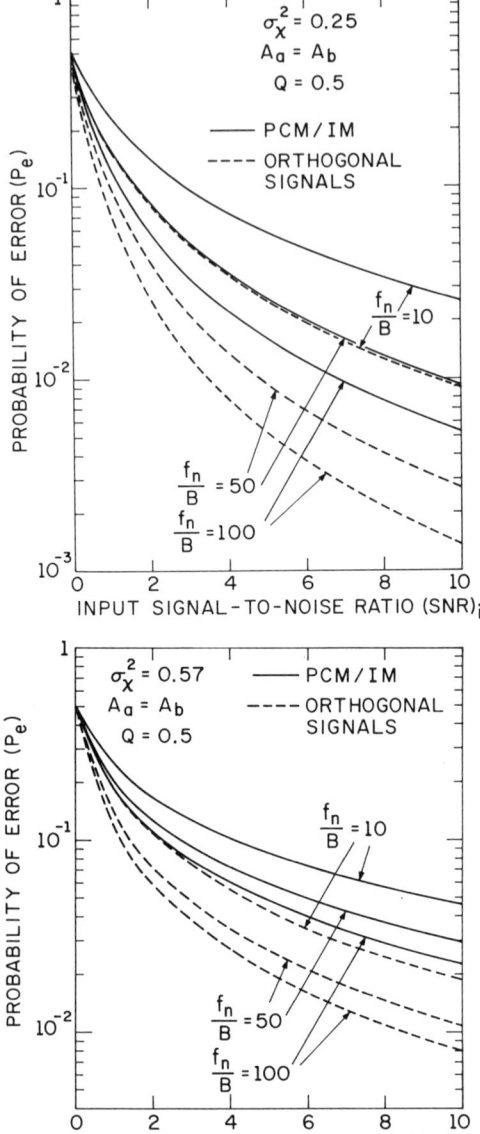

Fig. 7.22. Probability of error vs $(SNR)_i$ for the three-frequency binary communication system with atmospheric turbulence at the level $\sigma_\chi^2 = 0.25$. We assume $A_a = A_b$

Fig. 7.23. Probability of error vs $(SNR)_i$ for the three-frequency binary communication system with atmospheric turbulence at the level $\sigma_\chi^2 = 0.57$. We assume $A_a = A_b$

Finally, receiver performance for the cases of phase detection with a maximum-likelihood criterion [7.84] and phase-shift keying (PSK) have also been obtained [7.59]. While PSK is definitely superior to phase detection, neither scheme provides very satisfactory error probabilities.

The next section (7.3.10) presents an over-all discussion of the usefulness of the three-frequency nonlinear heterodyne technique for radar and communications applications.

7.3.10 Discussion

The three-frequency nonlinear heterodyne detection scheme has certain advantages over the conventional single-photon two-frequency configuration for both analog and digital systems. One advantage is the possibility of increasing the detection sensitivity, and minimizing the probability of error, particularly when little Doppler information is available. It provides an output signal at a well-known difference frequency regardless of the Doppler shift of the transmitted signals. Frequency scanning of the LO or receiver may therefore be eliminated. It allows a target to be continuously observed with Doppler shifts of greater magnitude and range than previously possible. The system is also angle independent in the sense that the Doppler shift is proportional to the radial velocity and therefore is generally a function of angle. A wide bandpass filter following the square-law device can be gradually narrowed about $2|f_1' - f_L|$ or $2|f_2' - f_L|$ in order to obtain Doppler information. If Doppler information is increased, we find that the receiver performance can be improved, in accordance with our expectations.

Since the use of a two-frequency transmitter can be considered as a special case of a modulated single-frequency beam, the system can be thought of as a heterodyne version of signal extraction at a predetermined modulation frequency. Thus, the technique is similar to conventional heterodyne radiometry, but *carefully takes into consideration the effects of Doppler shift and signal statistics*. Since Doppler shift is generally not an important parameter in the usual heterodyne radiometry detection scheme, the final filter bandwidth (associated with the integration time) can almost always be made arbitrarily small; furthermore, it is often possible to maintain a fixed phase relationship between the reference and detected signals, so that an additional factor of 2 (arising from coherent detection) becomes available. These specific benefits are not available for three-frequency nonlinear heterodyne detection.

Under the usual conditions of Doppler uncertainty, optimum operation occurs with the known difference frequency f_c at a maximum value close to f_n, or with the LO frequency between the received signal frequencies, and requires that the radiation power be equally divided between the two received beams. Processing of the dc output from the square-law device was not found to be useful when a blocking capacitor is included in the system. Four-frequency mixing was found to provide acceptable performance only when one of the LO frequencies is substantially attenuated.

Signals of three varieties were considered: a) sinewave input signals, b) Gaussian input signals with Gaussian power spectra, and c) Gaussian input signals with Lorentzian power spectra. Taking the pure sinewave case as a standard, the output signal-to-noise ratio for Gaussian signals is degraded by the factor $[2\Phi(u) - 1] < 1$ (for Gaussian spectra) or by the factor $\{(1/\pi)\tan^{-1}[4v/(4-v^2)]\} < 1$ (for Lorentzian spectra), where u and v are quantities proportional to the bandwidth B of the final narrowband filter. In all cases, decreasing B serves to increase the signal-to-noise ratio and decrease the minimum detectable power. In the Gaussian cases, it was found desirable to keep the width of the spectra as small as possible in order to maximize the signal-to-noise ratio.

The digital results, in particular, may be easily extended in a number of directions. Stochastic signals, rather than sinewave signals, could be treated in the binary communication problem. An extensive treatment of M-ary communications is possible, as is the generalization from a single detector to an array of detectors [7.76–78]. Consideration could be given to the optimum matched filter detector rather than the envelope detector discussed earlier. While the present treatment consists of a per-symbol analysis, prediction could be used to estimate the atmospheric turbulence level over a time period from a particular symbol, for example. In short, the usual variations possible with the conventional heterodyne system may be extended and/or modified for application to the three-frequency nonlinear heterodyne technique.

The principle appears to be applicable in all regions of the electromagnetic spectrum where conventional heterodyne detection is useful. In the next section (7.4), we consider two versions of the system useful for the detection of remote species.

7.4 Multifrequency Single-Photon Selective Heterodyne Radiometry for Detection of Remote Species

The radiation from known remote species, such as extraterrestrial molecules and smokestack effluents, is generally shifted as well as broadened in frequency when detected at a receiving station. Shifts in the center frequency can be attributed to a number of effects, including Doppler shift arising from the mass motion of a group of molecules and red shift arising from emission in the presence of a strong gravitational field. The magnitude of the Doppler shift is proportional to velocity and can be quite large, leading to an uncertainty in the appropriate frequency at which to search for a weak signal. This problem is magnified at high frequencies since Doppler shift is also proportional to frequency.

In this section, we consider using a passive version of the three-frequency single-photon heterodyne technique for partially eliminating the effects of Doppler shift in detecting remotely radiating objects. It is useful where a pair (or pairs) of emission lines exists with a definite and well-known frequency separation, such as those produced by two transitions of a given molecular species or by a given transition of two isotopes of that species. If the two radiated frequencies are close to each other, they are Doppler shifted by essentially the same amount (as with the active system) so that the effects of Doppler shift can be made to nearly cancel in the difference frequency. By employing two signal frequencies instead of one, an effective modulation of the source is achieved so that the bandwidth of the receiver can be narrowed about the difference frequency, in a manner similar to that accomplished by using a radiometer. But

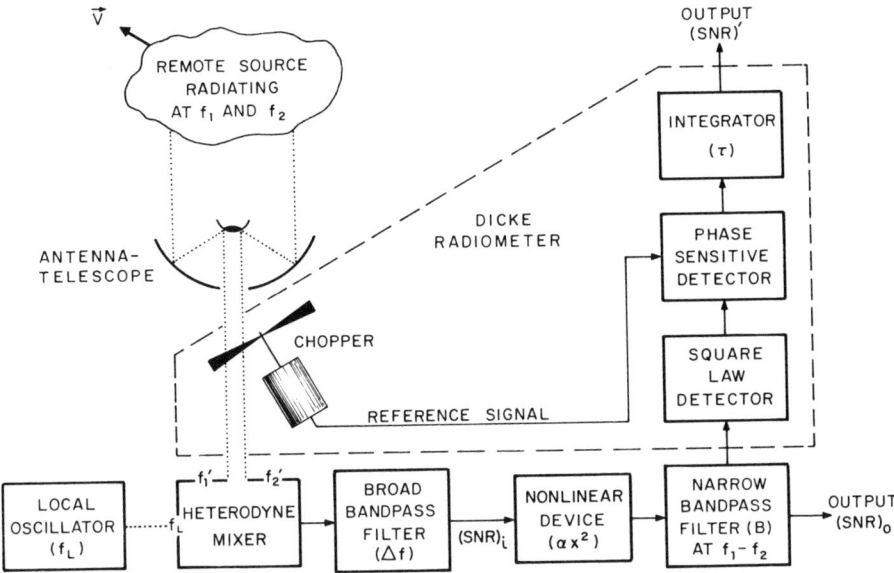

Fig. 7.24. Block diagram for the Doppler-independent three-frequency selective heterodyne radiometer. Dotted lines represent radiation signals, solid lines with arrows represent electrical signals, and dashed lines enclose a Dicke radiometer which can be added to the system if required. For clarity, amplification stages are omitted

whereas the modulation occurs at the detector in the classical radiometer (which is therefore nonspecific), the modulation frequency in the system described here is directly related to the remote species being detected. The system can, furthermore, be coupled with a classical Dicke radiometer [7.85–89] to provide improved performance where warranted. The technique will be most useful in the infrared and optical where the Doppler shifts are large; conventional heterodyne radiometry and spectroscopy have recently begun to find use at these frequencies [7.88–91]. A variation of the system, useful when the Doppler shift is known, is described at the very end of the discussion section (Sec. 7.4.5).

7.4.1 Configuration for Two Received Frequencies

The simplest example of the three-frequency system, useful in the acquisition and tracking of radar and communications signals, has been discussed in Section 7.3. In Fig. 7.24, we show a block diagram for the selective heterodyne radiometry version. The remotely radiating source emits two waves at frequencies f_1 and f_2 whose rest difference frequency $f_c \equiv |f_1 - f_2|$ is known to high accuracy. The waves experience a Doppler shift arising from the mass motion (Doppler feature) of the source. (They are also broadened due to the constituent particle velocity distribution.) Thus, a wave whose center frequency is f is detected at the receiving station with a frequency f'. Assuming that the velocity of the cloud is much smaller than the speed of light c, the nonrelativistic Doppler formula provides

$$f' = f(1 \pm v_{\parallel}/c), \tag{7.177}$$

where $v_{\parallel}$ is the radial component of the over-all velocity vector v. This expression differs from that given in (7.33) by a factor of 2 in the second term; the radiation in the passive system makes only a one-way trip. The frequency difference between the two received waves f'_c is therefore given by

$$f'_c \equiv |f'_1 - f'_2| = f_c \pm (v_{\parallel}/c)f_c \simeq f_c \;; \tag{7.178}$$

thus, the radiated and received difference frequencies are independent of Doppler shift to good approximation when $v_{\parallel}/c \ll 1$.

The two radiation fields f'_1 and f'_2 are again mixed in a heterodyne detector with a strong, coherent, and polarized LO signal (at frequency f_L) yielding two electrical beat signals at $|f'_1 - f_L|$ and $|f'_2 - f_L|$, along with a dc component which is blocked. The third signal at $|f'_1 - f'_2|$, arising without benefit of the LO, is weak and may be neglected. The ac output of the heterodyne mixer is then broadband coupled, through a filter of bandwidth Δf, to a nonlinear device. The value chosen for Δf should be as small as possible in order to maximize the signal-to-noise ratio (SNR), but must encompass the (somewhat unknown) difference frequencies generated in the mixer. The nonlinear device, which also has a response over Δf, then generates a component at the frequency

$$f'_c = |f'_1 - f'_2|. \tag{7.179}$$

Since the output of the nonlinear device is essentially independent of the Doppler shift as well as the LO frequencies, variations in these quantities have little effect on the system output. In many instances, therefore, the necessity for a stable and tunable LO may be eliminated. Again, the reception in this system is angle independent.

The narrowband filter centered at $f_c \simeq f'_c$, and of bandwidth B, placed after the nonlinear device, achieves the low noise bandwidth. Thus, amplifiers and other detection apparatus process electrical signals at (usually) moderate frequencies, which provides ease of matching as well as good receiver noise figure. This, in turn, decreases the LO power necessary for optimum coherent detection. Only the heterodyne mixer and the nonlinear device need have high-frequency response in many instances. For clarity, amplifiers have been eliminated from the block diagram. If warranted, the output of the narrow bandpass filter may be fed into a standard Dicke radiometer (dashed box in Fig. 7.24) consisting of a (third) detector, a phase-sensitive (synchronous) detector, and an integrator with time constant τ. (Although we specify that this detector is square-law in Fig. 7.24, its characteristic is not critical and, in fact, a linear detector will often provide the cleanest signal.) The modulation may be obtained from a chopper as indicated. This technique can sometimes provide improvement in the SNR and has been coupled with a conventional infrared heterodyne radiometer in a number of instances [7.88–91]. It may also be advantageous to use a balanced mixer in this configuration [7.89].

The SNR at the output (o) of the three-frequency nonlinear heterodyne system $(SNR)_o$ is given by [see (7.76) and (7.107b)]

$$(SNR)_o = 2k'(SNR)_i^2/[1 + 2(SNR)_i], \qquad (7.180)$$

assuming the use of an LO which produces no excess noise (and lies between the two signal frequencies), a bandpass filter close to low-pass $(\Delta f \to f_n)$, and a square-law nonlinear device for which it is easy to carry out the calculation. Here $(SNR)_i$ represents the SNR at the input (i) to the square-law device (see Fig. 7.24) which will generally be $\ll 1$. The factor k' appearing in (7.180) is discussed in the next section.

7.4.2 n Received Frequencies and the Factor k'

For the system involving two signal frequencies, the quantity k' has been previously shown to depend on the magnitude and on the statistical and spectral nature of the received radiation, as well as on the widths of the two bandpass filters. Inasmuch as the radiation from remote molecular species may contain multiple frequencies, we consider operation of the system in the more general case when $n(\geq 2)$ lines with equal frequency spacing are passed through the broad bandpass filter *and detected*. Again, a number of cases are of interest: sinusoidal signals (P), independent Gaussian signals with Gaussian spectra (G), and independent Gaussian signals with Lorentzian spectra (L). The Gaussian signal case is considered in detail since radiation from astronomical sources is generally Gaussian [7.92]. The effect of multiple lines is included in the parameter k' by generalizing the previously obtained expressions for k [see (7.72), (7.106), and (7.119)]. For the cases considered above, this quantity can be written as

$$k'_P \simeq \frac{f_n}{B} \left\{ \frac{\sum_{j=1}^{n-1} A_j^2 A_{j+1}^2}{(\sum_{j=1}^{n} A_j^2)^2} \right\}, \qquad (7.181a)$$

$$k'_G \simeq \frac{f_n}{\sqrt{8\gamma}} \left\{ \frac{\sum_{j=1}^{n-1} (\gamma P_j)(\gamma P_{j+1})}{(\sum_{j=1}^{n} \gamma P_j)^2} \right\} \left[\frac{2\Phi(B/\sqrt{8\gamma}) - 1}{(B/\sqrt{8\gamma})} \right], \qquad (7.181b)$$

and

$$k'_L \simeq \frac{f_n}{2\Gamma} \left\{ \frac{\sum_{j=1}^{n-1} D_j D_{j+1}}{(\sum_{j=1}^{n} D_j)^2} \right\}$$
$$\times \left[\frac{\pi^{-1} \tan^{-1}\{(2B/\Gamma)[4 - (B^2/4\Gamma^2)]^{-1}\}}{B/2\Gamma} \right]. \qquad (7.181c)$$

Here, A_j represents the amplitude of the jth line in the sinusoidal case, P_j represents the peak value of the Gaussian spectral distribution and γ is its standard deviation, whereas D_j and Γ represent the height and width of the Lorentzian spectrum, respectively. The quantity Φ is the error function. It has been assumed for simplicity that all spectral widths are identical, i.e., $\gamma_j = \gamma_{j+1} = \gamma$ and $\Gamma_j = \Gamma_{j+1} = \Gamma$; similar but more complex expressions are obtained when this is not the case.

Inasmuch as the quantities in large square brackets in (7.181b) and (7.181c) above are of order unity for $B \lesssim \gamma(\Gamma)$, it is the *larger* of B and $\gamma(\Gamma)$ which limits k' and therefore the SNR in the Gaussian signal case. In particular, for the Gaussian spectrum case with $B = \sqrt{8}\gamma$, $[2\Phi(1) - 1] = 0.68$ whereas for the Lorentzian spectrum case with $B = 4(\sqrt{2} - 1)\Gamma \simeq 1.66\Gamma$, $\pi^{-1} \tan^{-1} 1 = 1/4$. Thus the SNRs for the Gaussian and Lorentzian cases are reduced below that for the sinewave case (delta-function spectrum), for the same bandwidth B. This is understood to arise from the fact that some signal is being excluded in the Gaussian and Lorentzian cases in comparison with the delta-function case, but the noise is approximately the same. For fixed $\gamma(\Gamma)$, the best SNR for the Gaussian and Lorentzian cases is obtained as $B \to 0$, since the noise decreases faster than the signal, as B decreases, in the approximation $v_{\parallel} \to 0$. Of course, B cannot be decreased below the Doppler shift of the frequency difference $|f_c' - f_c| = (v_{\parallel}/c)f_c$, which is unknown but can generally be estimated. For $B \gg \gamma(\Gamma)$, essentially all of the signal is included, and the results reduce to those obtained in the sinewave case. If possible, therefore, lines should be chosen for which the (Doppler) width and the Doppler shift are minimized, i.e., the lines should be narrow and closely spaced in frequency.

In the case where all such lines are of equal spacing, power, and width ($A_j = A_{j+1}$; $P_j = P_{j+1}$, $\gamma_j = \gamma_{j+1} = \gamma$; $D_j = D_{j+1}$, $\Gamma_j = \Gamma_{j+1} = \Gamma$), the braces in (7.181) can be replaced by

$$\{\cdot\}_{P,G,L} \to (n-1)/n^2, \quad n = 2, 3, 4, \ldots. \tag{7.182}$$

For fixed input radiation power, the best operation is clearly achieved for $n = 2$ (so that $\{\cdot\}_{P,G,L} = 1/4$), since additional lines increase the (signal-by-noise contribution to the) total noise more than they do the signal. When increased radiation power becomes available by virtue of the additional lines, however (e.g., the detection of more than one Doppler feature), $n > 2$ can be advantageous.

We also consider the case in which n equal-power, equal-width lines are allowed through the broad bandpass filter, these not being equally spaced, however, so that only one pair of lines contributes to the output signal. In this case, the braces in (7.181) must be replaced by

$$\{\cdot\}_{P,G,L} \to 1/n^2, \quad n = 2, 3, 4, \ldots. \tag{7.183}$$

Performance in this case is degraded for $n > 2$ since the additional lines contribute only to the noise.

Finally, we consider the case in which only two lines ($n=2$) of arbitrary width are received and detected. Recalling that ξ represents the ratio of received power in these two lines, i.e., $\xi_P = A_2^2/A_1^2$, $\xi_G = \gamma_2 P_2/\gamma_1 P_1$, and $\xi_L = D_2/D_1$, the expressions in braces in (7.181) become

$$\{\cdot\}_{P,G,L} = \xi(1+\xi)^{-2}, \tag{7.184}$$

which is again equal to 1/4 for equal-power received signals ($\xi = 1$).

7.4.3 SNR and MDP for Two Gaussian Signals

The expression for the SNR at the output of the three-frequency system $(SNR)_0$ for two Gaussian signals with Gaussian spectra (standard deviations γ_1 and γ_2) is obtained by using (7.180), (7.181b), (7.184), and (7.106). To good approximation, assuming $(SNR)_i \ll 1$, this is given by

$$(SNR)_0 \simeq \frac{f_n}{(\gamma_1^2 + \gamma_2^2)^{1/2}} \left\{ \frac{\xi_G}{(1+\xi_G)^2} \right\}$$

$$\cdot \left[\frac{2\Phi\{B/[2(\gamma_1^2+\gamma_2^2)^{1/2}]\} - 1}{B/[2(\gamma_1^2+\gamma_2^2)^{1/2}]} \right] (SNR)_i^2. \tag{7.185}$$

For quantum-noise limited detectors such as photoemitters and reverse-biased photodiodes operating in the infrared and optical [7.4–7, 10, 14, 15], assuming that the incident radiation and the coherent LO are polarized in the same plane, the input SNR to the nonlinear device is [see (7.1) and (7.42b)]

$$(SNR)_i = \eta P_r/h\nu\Delta f, \quad h\nu \gg kT. \tag{7.186}$$

Here $\eta = \eta_1$ is the detector quantum efficiency. P_r is the total received signal radiation power, and kT is the thermal excitation energy (k is Boltzmann's constant and T is the detector temperature). For photovoltaic and photoconductive detectors, the input SNR is generally one-half that given in (7.186) [7.5, 14].

Heterodyne detectors in the microwave and millimeter regions ($h\nu \ll kT$) include square-law mixers such as the crystal diode detector [7.93], the InSb photoconductive detector [7.94–96], the Golay cell [7.95], the pyroelectric detector [7.95], the metal-oxide-metal diode, and the bolometer [7.87]. The latter three types of detectors have also been used successfully in the middle infrared (at 10.6 μm) [7.97–100]. For this type of detector Johnson noise generally predominates, and the input SNR is given by [7.100]

$$(SNR)_i = P_r/kT_{eff}\Delta f. \tag{7.187}$$

For simplicity, we have lumped a number of detector parameters and operating conditions into the receiver effective temperature T_{eff}. Of particular interest in the mm and far-infrared regions are the low-noise fast Schottky-barrier diodes recently used in a number of experiments for astronomical observations [7.96, 101].

Inserting (7.186) or (7.187) into (7.185), and letting $(SNR)_0 = 1$, we obtain a minimum detectable total power (MDP) at the output of the three-frequency system given by

$$(MDP)_0 \simeq \frac{h\nu}{\eta} \left\{ \frac{1+\xi_G}{\sqrt{\xi_G}} \right\}$$

$$\cdot \left[\frac{B/[2(\gamma_1^2+\gamma_2^2)^{1/2}]}{2\Phi\{B/[2(\gamma_1^2+\gamma_2^2)^{1/2}]\}-1} \right]^{1/2} f_n^{1/2}(\gamma_1^2+\gamma_2^2)^{1/4} \quad (7.188a)$$

for quantum-noise limited detection, and

$$(MDP)_0 \simeq kT_{eff} \left\{ \frac{1+\xi_G}{\sqrt{\xi_G}} \right\}$$

$$\cdot \left[\frac{B/[2(\gamma_1^2+\gamma_2^2)^{1/2}]}{2\Phi\{B/[2(\gamma_1^2+\gamma_2^2)^{1/2}]\}-1} \right]^{1/2} f_n^{1/2}(\gamma_1^2+\gamma_2^2)^{1/4} \quad (7.188b)$$

for Johnson-noise limited detection.

The quantities in braces and in square brackets in (7.188) are both typically of order unity. Since $f_n^{1/2}(\gamma_1^2+\gamma_2^2)^{1/4} \sim (v_\parallel^{max}/c)^{1/2}(f\gamma)^{1/2}$ for $\gamma > B$ while it is $\sim (v_\parallel^{max}/c)[f(f_1-f_2)]^{1/2}$ for $\gamma < B$, where $v_\parallel^{max}$ is the maximum expected radial velocity, the system provides increasing advantage at higher radiation frequencies f (since the effective bandwidth $\sim f^{1/2}$) for fixed γ and $(f_1 - f_2)$. Small linewidths and close spacing of the lines are also important. For certain choices of parameters, which are determined by the species which it is desired to detect, the SNR at the output of the three-frequency selective system will provide a sufficient confidence level for detection. For situations in which this is not the case, further improvement in the SNR could be obtained by using a multichannel receiver and/or a classical radiometer, as mentioned previously.

7.4.4 Numerical Example: Astronomical Radiation from CN

As an example of the use of the system in the mm region, we calculate the MDP for astronomical radiation arising from the following $N=1\rightarrow 0$, $J=3/2\rightarrow 1/2$ hyperfine transitions of the CN radical: $F=5/2\rightarrow 3/2$ ($f_1 = 113\,490.9 \pm 0.2$ MHz) and $F=3/2\rightarrow 1/2$ ($f_2 = 113\,488.1 \pm 0.3$ MHz) [7.102]. Recent radiometric observations of this radiation made use of the simple and sharply defined velocity structure of the Orion-A molecular cloud; a measurement of the $N=1\rightarrow 0$ line of $^{13}C^{16}O$ provided the Doppler effect correction due to the cloud's motion. Using Doppler-independent heterodyne radiometry, on the other hand, requires only a *bound* on the velocity range. A radial velocity within the (substantial) range -200 km/s $\leq v_{\|} \leq 200$ km/s, for example, yields a Doppler shift uncertainty of $2|v_{\|}|f/c \simeq 151.2$ MHz. In this case, the detected frequencies would be bounded by $113\,415.3$ MHz $\leq f_1' \leq 113\,566.5$ MHz and $113\,412.5$ MHz $\leq f_2' \leq 113\,563.7$ MHz. Choosing f_L somewhere between the rest frequencies, e.g., at $113\,490.0$ MHz, we obtain $|f_1' - f_L| \leq 76.5$ MHz and $|f_2' - f_L| \leq 77.5$ MHz, inducing us to choose $\Delta f = f_n = 78$ MHz. Depending on the actual velocity of the cloud, this might allow the beat signal of the LO with other hyperfine lines to be passed to the nonlinear device, which will not impair operation if these other lines are relatively weak. The narrowband filter is centered at the rest difference frequency $f_c = |f_1 - f_2| = 2.8$ MHz, with a minimum width $B = 2|v_{\|}|f_c/c \simeq 1.87$ kHz. Since $B \ll 2(\gamma_1^2 + \gamma_2^2)^{1/2}$, the MDP is essentially determined by f_n and γ (we choose $\gamma = 1.5$ MHz since $\gamma_1 \simeq \gamma_2 \simeq 1.5$ MHz). Inasmuch as $\xi_G \equiv \gamma_2 P_2 / \gamma_1 P_1 \simeq 1/2$ for these lines [7.102], the MDP given in (7.188b) becomes MDP $\simeq kT_{\text{eff}}\{2.12\}[1.11](8.80 \times 10^3)(1.45 \times 10^3) \simeq kT_{\text{eff}}\delta F$, with $\delta F \simeq 30$ MHz representing the effective bandwidth for the calculation. Using the conventional system with this uncertainty in Doppler shift, and assuming that a one-channel receiver is used, the MDP would be $kT_{\text{eff}}\Delta f$ with $\Delta f \simeq 78$ MHz, indicating that improvement is possible with the proposed system.

For situations in which $B > \gamma$, a multichannel receiver using a bank of narrow bandwidth filters could be used in place of the narrow bandpass filter (B), compressing the number of channels below that required in the conventional system. In the infrared and optical, an unknown Doppler shift provides a greater range of uncertainty in the received frequencies than at longer wavelengths; this system should therefore be useful in detecting atomic and molecular radiation at these higher frequencies, particularly in those wavelength regions where atmospheric windows exist. For example, strong CN optical transitions from interstellar sources were first observed in 1940 [7.103, 104]; one could attempt to definitively detect the presence of the CN $R(2)$ line, which would provide an improved estimate for the cosmic blackbody radiation at 1.32 mm [7.104]. Particular attention might also be given to possible infrared emission from CO, which exists in relatively high densities and with a very broad range of velocities in interstellar regions, as determined by its mm-wave emission [7.105, 106]. Clearly, the same considerations apply to the detection of maser radiation from astronomical sources [7.103, 107–109], and to the detection of remote pollutants [7.110, 111].

7.4.5 Discussion

We have described a selective heterodyne radiometer potentially useful in the detection of remote species such as pollutants and interstellar molecules. The system operates on the basis of the difference frequency between two radiated lines which, for closely spaced lines, is relatively insensitive to Doppler shift. This allows for the sensitive detection of known species moving at unknown velocities. The two frequencies may be obtained from individual transitions or from two isotopes of the same species. The system introduces little loss over the conventional heterodyne radiometer and has a number of specific advantages. In particular, it requires knowledge only of rest difference frequencies and not of line rest frequencies which are sometimes difficult to determine [7.112], and it requires neither a stabilized nor a tunable LO. Clearly it requires little knowledge of the source velocity and consequently is generally unsuitable for spectroscopy. Changes in the source velocity or direction do not alter system detectability appreciably. This is particularly important in the infrared and optical where Doppler shifts are generally large.

The SNR and MDP at the output of the system have been obtained for a number of cases of interest including sinusoidal signals and Gaussian signals with both Gaussian and Lorentzian spectra. A configuration involving multiple ($n \geq 2$) signal frequencies has also been considered. Other desirable operating conditions are as follows: 1) The LO frequency should be chosen to be nearly between the signal frequencies, 2) Lines with minimum broadening (low γ) and minimum frequency separation (low B) are most desirable, 3) $\Delta f(f_n)$ should be minimized by bounding the expected Doppler shift as closely as possible, and 4) The strongest pair of lines consistent with the above conditions should be chosen.

The detection of CN radiation provided an example of the use of the technique in the mm region: an indication of possible uses at higher frequencies was provided. For the submillimeter region, it may be possible to use a combination Schottky barrier diode/harmonic mixer which would provide an output at low frequencies as long as the high-frequency beat signals are generated and mixed within the detector. LO harmonics are also readily generated in these devices [7.101] so that harmonic-mixing selective heterodyne radiometry could be performed [7.113]. Josephson junctions, which can sometimes be made to produce their own LO power [7.96], and metal-oxide-metal diodes could also be used. An IMPATT solid state oscillator could conveniently be used as an LO in these regions since frequency stabilization, which is difficult to achieve in these devices [7.96], is not required. At higher frequencies, some fixed-line lasers could possibly be used since the LO frequency need not be tunable.

Disadvantages of the system include the lack of Doppler information, the difficulty of observing absorption lines and continuum radiation, and the added complexity. Use of a calibration load is also more complicated than in the conventional case. Finally, there will be an uncertainty that the detected difference-frequency signal can be properly identified, in analogy with the identification problem for the Doppler shifted signal in the conventional configuration. Thus, the system should be used for the application in which it is most effective: the search for a known emitting weak remote species with an unknown Doppler feature.

Finally, we draw attention to a variation of this scheme, called heterodyne correlation radiometry [7.114] that should be useful for the sensitive detection of radiating species whose Doppler shift is known, but whose presence we wish to affirm. Such radiation (which may be actively induced) can arise, for example, from remote molecular emitters, impurities and pollutants, trace minerals, chemical agents, or a general multiline source. A radiating sample of the species to be detected is physically made a part of the laboratory receiver, and serves as a kind of frequency-domain template with which the remote radiation is correlated, after heterodyne detection. This system is expected to be especially useful for the detection of sources whose radiated energy is distributed over a large number of lines, with frequencies that are not necessarily known. Neither a stable nor a tunable local oscillator is required. The minimum detectable power is expressible in a form similar to that for conventional heterodyning (for both quantum-noise-limited and Johnson-noise-limited detectors). The notable distinction is that the performance of the proposed system improves with increasing number of remotely radiating signal lines and increasing locally produced radiation power. Performance degradation due to undesired impurity radiation is not a problem in general.

Acknowledgement. I am grateful to the John Simon Guggenheim Memorial Foundation and to the National Science Foundation for generous financial support. I am also indebted to Rainfield Y. Yen for permitting me to present many of the calculations (in Sec. 7.3) that he carried out as a part of his Ph.D. thesis, and which we later published jointly.

References

7.1   A. T. Forrester, R. A. Gudmundsen, P. O. Johnson: Phys. Rev. **99**, 1691 (1955)
7.2   A. Javan, E. A. Ballik, W. L. Bond: J. Opt. Soc. Am. **52**, 96 (1962)
7.3   B. J. McMurtry, A. E. Siegman: Appl. Opt. **1**, 51 (1962)
     A. E. Siegman, S. E. Harris, B. J. McMurtry: Optical heterodyning and optical demodulation at microwave frequencies. In: *Optical Masers*, ed. by J. Fox (Wiley-Interscience, New York 1963) pp. 511–527
7.4   M. C. Teich, R. J. Keyes, R. H. Kingston: Appl. Phys. Lett. **9**, 357 (1966)
7.5   M. C. Teich: Proc. IEEE **56**, 37 (1968) [Reprinted in *Infrared Detectors*, ed. by R. D. Hudson, Jr., J. W. Hudson (Dowden, Hutchinson and Ross, Stroudsburg 1975)]
7.6   B. M. Oliver: Proc. IRE **49**, 1960 (1961)
     H. A. Haus, C. H. Townes, B. M. Oliver: Proc. IRE **50**, 1544 (1962)
7.7   A. E. Siegman: Proc. IEEE **54**, 1350 (1966)
7.8   W. S. Read, D. L. Fried: Proc. IEEE **51**, 1787 (1963)
7.8a M. M. Abbas, M. J. Mumma, T. Kostiuk, D. Buhl: Appl. Opt. **15**, 427 (1976)
7.8b E. Jakeman, C. J. Oliver, E. R. Pike: Advances in Phys. **24**, 349 (1975)
7.9   M. C. Teich: Proc. IEEE **57**, 786 (1969)
7.10 M. C. Teich, R. Y. Yen: J. Appl. Phys. **43**, 2480 (1972)
7.11 U. M. Titulaer, R. J. Glauber: Phys. Rev. **140**, B676 (1965); Phys. Rev. **145**, 1041 (1966)
7.12 M. C. Teich: Appl. Phys. Lett. **14**, 201 (1969)
7.13 M. C. Teich: Quantum theory of heterodyne detection. In: *Proc. Third Photoconductivity Conf.*, ed. by E. M. Pell (Pergamon, New York 1971) pp. 1–5
7.14 M. C. Teich: Coherent detection in the infrared. In: *Semiconductors and Semimetals*, ed. by R. K. Willardson and A. C. Beer (Academic, New York 1970) **5**, *Infrared Detectors*, Chap. 9, pp. 361–407
7.15 L. Mandel, E. Wolf: J. Opt. Soc. Am. **65**, 413 (1975)
7.15a G. Lachs: Phys. Rev. **138**, B 1012 (1965)
7.15b J. Peřina: Phys. Lett. **24A**, 333 (1967)

7.15c J. Peřina, R. Horak: J. Phys. A **2**, 702 (1969)
7.15d E. Jakeman, E. R. Pike: J. Phys. A **2**, 115 (1969)
7.15e J. Peřina, V. Peřinová, L. Mišta: Opt. Acta **19**, 579 (1972)
7.15f M. C. Teich, W. J. McGill: Phys. Rev. Lett. **36**, 754 (1976)
7.16   M. C. Teich: IEEE J. Quant. Electron. **QE-11**, 595 (1975)
7.17   A. Einstein: Ann. Physik **17**, 132 (1905) [translation: Am. J. Physics **33**, 367 (1965)]
7.18   L. Mandel: Proc. Phys. Soc. (London) **74**, 233 (1959); see also
       E. M. Purcell: Nature **178**, 1449 (1956)
       L. Mandel: Proc. Phys. Soc. (London) **72**, 1037 (1958)
7.19   R. J. Glauber: Phys. Rev. **130**, 2529 (1963); Phys. Rev. **131**, 2766 (1963)
7.20   P. L. Kelley, W. H. Kleiner: Phys. Rev. **136**, A 316 (1964)
7.21   M. C. Teich, G. J. Wolga: J. Opt. Soc. Am. **57**, 542 (1967)
7.22   M. C. Teich, J. M. Schroeer, G. J. Wolga: Phys. Rev. Lett. **13**, 611 (1964)
7.23   H. Sonnenberg, H. Heffner, W. Spicer: Appl. Phys. Lett. **5**, 95 (1964)
7.24   M. C. Teich: Two quantum photoemission and dc photomixing in sodium, Ph.D. thesis (Cornell University 1966) unpublished [also Report no. 453, Materials Science Center, Cornell University, Ithaca, New York, February 1966]
7.25   M. C. Teich, G. J. Wolga: Phys. Rev. **171**, 809 (1968)
7.26   F. Shiga, S. Imamura: Phys. Lett. **25A**, 706 (1967)
7.27   E. M. Logothetis, P. L. Hartman: Phys. Rev. Lett. **18**, 581 (1967)
7.28   E. M. Logothetis, P. L. Hartman: Phys. Rev. **187**, 460 (1969)
7.29   R. E. B. Makinson, M. J. Buckingham: Proc. Phys. Soc. A (London) **64**, 135 (1951)
7.30   R. L. Smith: Phys. Rev. **128**, 2225 (1962)
7.31   H. C. Bowers: Theoretical and experimental considerations of the double-quantum photoelectric effect, M.S. thesis (Cornell University 1964) unpublished
7.32   I. Adawi: Phys. Rev. **134**, A 788 (1964)
7.33   P. Bloch: J. Appl. Phys. **35**, 2052 (1964)
7.34   M. C. Teich, G. J. Wolga: Phys. Rev. Lett. **16**, 625 (1966)
7.35   P. Lambropoulos, C. Kikuchi, R. K. Osborn: Phys. Rev. **144**, 1081 (1966)
7.36   B. R. Mollow: Phys. Rev. **175**, 1555 (1968)
7.37   G. S. Agarwal: Phys. Rev. A **1**, 1445 (1970)
7.38   M. C. Teich, R. L. Abrams, W. B. Gandrud: Opt. Communic. **2**, 206 (1970)
7.39   P. Diament, M. C. Teich: J. Opt. Soc. Am. **59**, 661 (1969)
7.40   C. Freed, H. A. Haus: Phys. Rev. **141**, 287 (1966)
       G. Lachs: J. Appl. Phys. **39**, 4193 (1968)
7.41   M. C. Teich, P. Diament: J. Appl. Phys. **40**, 625 (1969)
7.42   P. P. Barashev: Zh. Eksper. i Teor. Fiz. (USSR) **59**, 1318 (1970) [translation: Soviet Phys. JETP **32**, 720 (1971)]
7.43   P. P. Barashev: Phys. Stat. Sol. (a) **9**, 9 (Part I) and 387 (Part II) (1972)
7.44   L. Mandel: J. Opt. Soc. Am. **57**, 613 (1967)
       L. Mandel, E. Wolf: Rev. Mod. Phys. **37**, 231 (1965)
7.44a M. J. Beran, J. DeVelis, G. Parrent: Phys. Rev. **154**, 1224 (1967)
7.45   M. C. Teich, D. A. Berkley, G. J. Wolga: Rev. Sci. Instr. **36**, 973 (1965)
7.46   R. A. Fox, R. M. Kogan, E. J. Robinson: Phys. Rev. Lett. **26**, 1416 (1971)
       S. Klarsfeld, A. Maquet: Phys. Rev. Lett. **29**, 79 (1972)
7.47   S. J. Ippolito, S. Rosenberg, M. C. Teich: Rev. Sci. Instr. **41**, 331 (1970)
7.48   M. C. Teich: Appl. Phys. Lett. **15**, 420 (1969); U.S. Patent Number 3, 875, 399
7.49   R. L. Abrams, R. C. White, Jr.: IEEE J. Quantum Electron. **QE-8**, 13 (1972)
7.50   M. C. Teich, R. Y. Yen: Appl. Opt. **14**, 666 (1975)
7.51   M. C. Teich, R. Y. Yen: Appl. Opt. **14**, 680 (1975)
7.52   M. C. Teich: Rev. Sci. Instr. **46**, 1313 (1975)
7.53   W. B. Davenport, Jr., W. L. Root: *An Introduction to the Theory of Random Signals and Noise* (McGraw-Hill, New York 1958) p. 112

7.54 D.J.Angelakos, T.E.Everhart: *Microwave Communications* (McGraw-Hill, New York 1968) p. 204
7.55 Ref. [7.53], pp. 257–259
7.56 Ref. [7.53], p. 255
7.57 M.I.Skolnik: *Introduction to Radar Systems* (McGraw-Hill, New York 1958) p. 185
7.58 W.E.Murray, Jr.: Coherent laser radar Doppler signatures. In: *Optics Research* (MIT Lincoln Lab.1969) No. 1, p. 6
7.59 R.Y.Yen: Optical communications: An investigation of several techniques, Ph.D. thesis (Columbia University 1972) unpublished
7.60 Ref. [7.53], p. 308
7.61 H.A.Bostick: IEEE. J. Quantum Electron. **QE-3**, 232 (1967)
7.62 H.A.Bostik, L.J.Sullivan: Laser radar and tracking. In: *Optics Research* (MIT Lincoln Lab. 1969) No. 1, p. 21
7.63 Ref. [7.53], p. 193
7.64 Ref. [7.53], p. 356
7.65 R.G.Gallager: *Information Theory and Reliable Communication* (Wiley, New York 1968) p. 365
7.66 R.J.Schwarz, B.Friedland: *Linear Systems* (McGraw-Hill, New York 1965) p. 299
7.67 Ref. [7.53], pp. 93–101
7.68 M.Kac, A.J.F.Siegert: J. Appl. Phys. **18**, 383 (1947)
7.69 R.C.Emerson: J. Appl. Phys. **24**, 1168 (1953)
7.70 V.Voorhis: *Microwave Receivers* (McGraw-Hill, New York 1948) p. 157
7.71 S.O.Rice: Mathematical Analysis of Random Noise. In: *Selected Papers on Noise and Stochastic Processes*, ed. by N.Wax (Dover, New York 1954) pp. 133–294
7.72 S.Stein, J.J.Jones: *Modern Communication Principles* (McGraw-Hill, New York 1967) pp. 138–139
7.73 W.K.Pratt: *Laser Communication Systems* (Wiley, New York 1969) pp. 224–229
7.74 Ref. [7.53], pp. 352–355
7.75 D.L.Fried, G.E.Mevers: J. Opt. Soc. Am. **55**, 740–741 (1965)
7.76 M.C.Teich, S.Rosenberg: Appl. Opt. **12**, 2616 (1973)
7.77 S.Rosenberg, M.C.Teich: Appl. Opt. **12**, 2625 (1973)
7.78 S.Rosenberg, M.C.Teich: IEEE Trans. Inform. Theory **IT-19**, 807 (1973)
7.79 Ref. [7.72], pp. 286–309
7.80 V.I.Tatarski: *Wave Propogation in a Turbulent Medium* (McGraw-Hill, New York 1961)
7.81 P.Diament, M.C.Teich: J. Opt. Soc. Am. **60**, 1489 (1970)
J.Peřina, V.Peřinová, M.C.Teich, P.Diament: Phys. Rev. **A7**, 1732 (1973)
7.82 R.S.Lawrence, J.W.Strohbehn: Proc. IEEE **58**, 1523 (1970)
7.83 D.L.Fried: Appl. Opt. **10**, 721 (1971)
7.84 Ref. [7.53], p. 167
7.85 R.H.Dicke: Rev. Sci. Instr. **17**, 268 (1946)
7.86 J.D.Kraus: *Radio Astronomy* (McGraw-Hill, New York 1966)
7.87 T.G.Phillips, K.B.Jefferts: IEEE Trans. Microwave Theory Tech. **MTT-22**, 1290 (1974)
7.88 J.Gay, A.Journet, B.Christophe, M.Robert: Appl. Phys. Lett. **22**, 448 (1973)
7.89 H.van de Stadt: Astron. Astrophys. **36**, 341 (1974)
7.90 T.de Graauw, H.van de Stadt: Nature (Phys. Sci.) **246**, 73 (1973)
7.91 D.W.Peterson, M.A.Johnson, A.L.Betz: Nature (Phys. Sci.) **250**, 128 (1974)
7.92 N.J.Evans II, R.E.Hills, O.E.H.Rybeck, E.Kollberg: Phys. Rev. **A6**, 1643 (1972)
7.93 C.H.Townes, A.L.Schawlow: *Microwave Spectroscopy* (McGraw-Hill, New York 1955)
7.94 E.H.Putley: Proc. IEEE **54**, 1096 (1966)
7.95 H.A.Gebbie, N.W.B.Stone, E.H.Putley, N.Shaw: Nature (Phys. Sci.) **214**, 165 (1967)
7.96 A.A.Penzias, C.A.Burrus: Ann. Rev. Astronomy Astrophys. **11**, 51 (1973)
7.97 R.L.Abrams, A.M.Glass: Appl. Phys. Lett. **15**, 251 (1969)
7.98 E.Leiba: Compt. Rend. (Paris) **268**, B 31 (1969)
7.99 R.L.Abrams, W.B.Gandrud: Appl. Phys. Lett. **17**, 150 (1970)
7.100 B.Contreras, O.L.Gaddy: Appl. Phys. Lett. **18**, 277 (1971)

7.101 H.R.Fetterman, B.J.Clifton, P.E.Tannenwald, C.D.Parker, H.Penfield: IEEE Trans. Microwave Theory Tech. **MTT-22**, 1013 (1974)
K.Mizuno, R.Kuwahara, S.Ono: Appl. Phys. Lett. **26**, 605 (1975)
7.102 A.A.Penzias, R.W.Wilson, K.B.Jefferts: Phys. Rev. Lett. **32**, 701 (1974)
7.103 D.M.Rank, C.H.Townes, W.J.Welch: Science **174**, 1083 (1971)
7.104 P.Thaddeus: Ann. Rev. Astronomy Astrophys. **10**, 305 (1972)
7.105 R.W.Wilson, K.B.Jefferts, A.A.Penzias: Astrophys. J. Lett. **161**, L43 (1970)
7.106 P.M.Solomon: Phys. Today **26**, #3, 32 (1973)
7.107 M.M.Litvak: Ann. Rev. Astronomy Astrophys. **12**, 97 (1974)
7.108 L.E.Snyder, D.Buhl: Astrophys. J. Lett. **189**, L31 (1974)
7.109 L.E.Snyder: IEEE Trans. Microwave Theory Tech. **MTT-22**, 1299 (1974)
7.110 E.D.Hinkley, P.L.Kelley: Science **171**, 635 (1971)
7.111 R.Menzies: Appl. Phys. Lett. **22**, 592 (1973)
7.112 B.Zuckerman, P.Palmer: Ann. Rev. Astronomy Astrophys. **12**, 279 (1974)
7.113 P.F.Goldsmith, R.L.Plambeck, R.Y.Chiao: IEEE Trans. Microwave Theory Tech. **MTT-22**, 1115 (1974)
7.114 M.C.Teich: Proc. Society Photo-Optical Instrumentation Engineers **82**, 132 (1976)

8. Recent Advances in Optical and Infrared Detector Technology

R. J. Keyes

Since the publication of the first edition of this book, the progress of a vigorous optical and infrared detector technology has led to advancements in the photon detection process. These advancements are more in the vein of perfecting or putting into practice existing concepts than the introduction of a fundamentally new detection phenomenon. Nevertheless these advances are destined to have a significant impact on the utility of optical and infrared detectors over a wide spectrum of applications, including television, communications, satellite surveillance, night vision, smart military ordinance, and household devices to name a few.

The material presented in this chapter was largely the result of communications with the contributors to this volume, digestion of information from current scientific periodicals and conversations with George Foyt of Lincoln Laboratory. The progress in the perfection of optical and infrared detectors over the past few years has been voluminous and all facets cannot be justly presented in a single chapter. This chapter presents only a few of the advances which the author feels are indicative of the general trend of progress; the many references are designed to provide the reader with a source of more detailed information and a broader view of evolving detection technology.

In order to be consistent with the earlier chapters of this volume the detector advances are segregated into general detector types (thermal, photoconductor and photovoltaic, photoemissive, charge transfer devices, and heterodyne). The final section of this chapter presents future detector and related technology advances (from the author's point of view) that would greatly enhance our ability to use optical and infrared detectors to solve basic problems of our society.

8.1 Thermal Detectors

Although no new types of detector have come to prominence since the original publication, interest in thermal detectors has grown steadily, leading to both improvements in performance and reliability and fuller engineering development. This progress has resulted from pressures from three different types of requirement. First, for higher performance detectors for infrared astronomy, both ground and satellite based. Secondly for high performance uncooled detectors for use in radiometers in both earth satellites and interplanetary probes. The third is for cheap but reliable detectors to meet commercial requirements for burglar alarms and industrial instrumentation. This third requirement has led to large scale manufacture of certain detectors leading to a more widespread exploitation of thermal (8–13 µm) radiation than has been possible hitherto. From this activity the realization is emerging that the uncooled thermal detectors are not just cheap and inferior alternatives to the cooled photon detectors but that their particular combination of properties fits them for tasks which other detectors cannot fulfill. This process is continuing so that we can expect the rapid development of thermal detectors to continue for some time.

8.1.1 The Thermopile

The main emphasis of recent work has been wider exploitation of the evaporated film type. Further details of their use in deep space radiometers have been published [8.1] and in the growing commercial market for cheap 10 µm detectors the thermopile is competing strongly with the pyroelectric detector for the "cheap but reliable" market [8.2, 2a]. Ground based laboratory applications include electrically calibrated radiometers using evaporated chromium–nickel [8.3] or copper constantan wire [8.3a] thermopiles, a total energy sensor used to monitor radiation from Tokamaks and other nuclear fusion experiments [8.4] and an alignment device for use on a CO_2 laser fusion experiment [8.5]. The performance of evaporated film thermopiles has been reviewed by *Korn* and *Shtrikman* [8.6]. *Vigdorovich* et al. [8.6a] have described the use of composite films.

8.1.2 The Bolometer

Attempts are still being made to produce improved bolometers operating at or near room temperature. These include vanadium oxide metal-semiconductor transition devices [8.7], bismuth-lead layers [8.8], metallic nickel [8.8a], and aluminium [8.8b], silicon carbide [8.8c], and doped barium titanate ceramic [8.8d] elements. New designs of radiometers and power meters using thermistor bolometers have been described [8.9–11]. Microelectronic techniques have been used to produce fast but sensitive (NEP $\sim 10^{-9}$ WHz$^{-1/2}$ at 25 MHz and 100 µm wavelengths) uncooled for improved bolometers [8.11a].

Considerable effort has gone into the improvement of cryogenic bolometers required for infrared and sub-mm astronomy. The emphasis has been on the optimization of various types of composite structure [8.12–14]. In these devices a radiation receiver is coupled thermally to the bolometric element which acts as a sensitive thermometer. Both semiconductors [8.13–16] and superconductors [8.12, 17] are being used for the bolometric elements. Whilst the design of efficient absorbers is understood in principle, the problem of coupling the absorber to the sensor is more difficult, especially when device operation below 1 K is desired. The coupling of the sensor to its heat sink presents a similar problem. Proposed solutions to these problems include the use of modern fabrication techniques such

Table 8.1. Composite cryogenic bolometers

Absorber	Sensor	Operating temperature [K]	Electrical NEP [WHz$^{-1/2}$]	Measuring frequency [Hz]	Response time [ms]	Ref.
Doped Si 1 mm D	Ge	1.8	1.3×10^{-14}			[8.15]
Cu foil with ferrite black 2 mm D	Ge	1.7	2×10^{-14}	80		[8.19]
Bi film on sapphire 4 × 4 mm²	Al	1.5	1.7×10^{-15}	52	83	[8.12]
	Ti	0.4	2×10^{-16} (est.)			
Bi film on sapphire 4 × 4 mm²	Ge	1.2	3×10^{-15}	17	25	[8.16]
	Ge	0.35	6×10^{-16}	17	6	
Ion implanted si 5 × 5 mm²	Epitaxial Si on sapphire	0.1	$\sim 10^{-16}$ (est.)			[8.14]
Unsupported Al film (gas kinetics detector)	Al	1.5	4×10^{-17}		10^{-4}	[8.18]

as epitaxial growth of Si on sapphire for the basic element, the introduction of impurities by ion implantation to form the absorbing layers and photolithographic methods to define the structure [8.14]. The results of recent work are summarized in Table 8.1. Most authors discuss in detail the electrical noise equivalent power. To obtain the true radiation NEP (as presented in Fig. 3.8) a realistic correction factor must be introduced. Consideration of the problem of introducing the radiation into the detector indicates that this factor is not likely to be better than 50% [8.13]. Even then, the value for the NEP that several authors predict with the most advanced designs is approaching 10^{-16} WHz$^{-1/2}$. To achieve this, research on low noise cryogenic amplifiers will be required. Even this may not be the limit, for *Chin* [8.18] has obtained an NEP $\sim 4 \times 10^{-17}$ WHz$^{-1/2}$ with a response time of $\sim 10^{-7}$/s using a freely suspended superconducting Al film cooled by He gas. *Chin*'s device was not intended for radiation detection but for studying gas kinetics.

Examples of the use of cryogenic bolometers in ground [8.20], balloon [8.21], aircraft [8.19], and satellite [8.22] based instruments are given in the references quoted. Other papers on cryogenic bolometers include [8.23–25a].

8.1.3 The Golay Cell and Related Detectors

To some extent in modern spectrometers the Golay cell has been replaced by pyroelectric detectors [8.26] but it is still widely used especially for sub-mm applications [8.27] where the design of pyroelectric detectors has not been fully optimized for use at the longer wavelengths. The condenser microphonic type of detector is still widely used in gas analyzers. Its operation has been discussed recently by *Sachdev* et al. [8.28].

8.1.4 Pyroelectric Detector

The exploitation of the pyroelectric detector has grown rapidly. Considerable effort has been put into the development of both simple detectors capable of mass production and of sophisticated designs capable of a performance approaching the fundamental limit. Whilst the performance of the best pyroelectric detectors now available may not be superior to that of the best TGS detectors previously reported, they are made from refractory oxide material such as $LiTaO_3$ so that they represent a considerable advance both in ease of manufacture and reliability in use. This advance in technique has led to renewed interest in pyroelectric detector arrays. When combined with CCDs or other types of multiplex read-out, these arrays offer attractive performance where medium size arrays (less than 1000 elements) are required and ultimately should provide large two-dimensional arrays (10^5–10^6 elements) which will have a superior performance to the pyroelectric vidicon. Since the pyroelectric vidicon itself (see Sect. 3.7) is already proving useful for a wide range of thermal imaging applications, this development will increase significantly the range of applications for which high grade thermal imaging becomes economically viable.

Recent reviews of pyroelectric detectors have been published by *Liu* and *Long* [8.29], *Doyle* [8.30], *Marshall* [8.31], *Kremenchugsky* and *Samoilov* [8.31a]. *Lang* [8.32] has continued to publish his very useful literature guides to pyroelectricity which review all aspects of research on pyroelectricity as well as that on detectors and other applications. Papers relating more generally to the design of pyroelectric devices have been published by *Stokowski* [8.33], *Newnham* et al. [8.34], *Zook* and *Liu* [8.35], *Putley* [8.36], *Byatt* [8.36a], *Elfimov* et al. [8.36b], *Hamid* [8.36c], *Katsube* et al. [8.36d], *Simhony* et al. [8.36e], *Zajosz* [8.36f], and *Tidjani* et al. [8.36g].

Of recent papers on pyroelectric detectors, some of the most interesting are those by *Stokowski* and his colleagues [8.37] who have applied modern fabrication techniques such as ion beam machining to the manufacture of very sensitive $LiTaO_3$ detectors. This work received an award in the Industrial Research Development 1978 competition for the 100 best new products [8.38]. New techniques for fabricating high performance detectors, especially for use in the sub-mm range have also been proposed by *Hadni* and his colleagues [8.39].

They have suggested the possible advantage of cryogenic operation of pyroelectric devices. At first sight this would seem a retrograde step, but for certain sub-mm applications where the only alternative detectors operate at liquid helium temperatures there could be advantages in this suggestion by *Hadni*. One difficulty with sub-mm cryogenic detectors, which applies particularly to semiconductors (both the bolometers and the extrinsic photon detectors) and also to some extent to the superconductors, is the difficulty of obtaining material of the correct composition and of fabricating a delicate structure. Because pyroelectrics utilize intrinsic material properties which are not very structure sensitive this problem becomes much less severe, even for cryogenic operation. There are sub-mm applications which could best be met by use of a detector array. This technology should be much easier to develop based on pyroelectrics than it would on any of the other available types of detector. *Bordoni* et al. [8.40] have shown that the NEP of a PZT detector improves by a factor greater than 250 on cooling to 1.18 K. With the known ease of manufacture of PZT based pyroelectric detectors, this could lead to useful devices for use in sub-mm plasma diagnostics or laser studies.

TGS is still used for the highest performance pyroelectric detectors, although it may soon be largely replaced by $LiTaO_3$. The recent successful operation of radiometers in the Pioneer Venus probe [8.41, 41a] and of met. satellites such as TIROS N [8.42] has demonstrated the capabilities of TGS devices.

The pyroelectric detectors now being manufactured in quantity use either $LiTaO_3$ or a ceramic based on PZT modified to enhance its pyroelectric properties [8.43]. Strontium Barium Niobate (SBN) [8.44] and polyvinylidene fluoride films [8.36a, d] are also used. The use of polymer films could expand as the understanding of these films increases.

Some recent papers describing applications of pyroelectric detectors include *Elbers* et al. [8.45], *Button* and *Wolfe* [8.46], *McColl* [8.47], *Chaloner* et al. [8.48], *Derrie* et al. [8.48a], and *Hartung* and *Jurgeit* [8.48b]. One area of growing interest where pyroelectric devices could be very suitable is the development of a radiometer in the 26–35 μm band for clear air turbulence detection to reduce the hazard to high flying aircraft [8.49]. The successes achieved by pyroelectric detectors both in high technology areas and as simple detectors of thermal radiation should assure them a continuing place amongst infrared devices.

8.1.5 Other Types of Thermal Detectors

Whilst not a detector in the conventional sense, the spectrophone should be mentioned. In this device acoustic pressure waves are generated when modulated radiation is absorbed by molecules (usually gas but also condensed matter) in an absorption cell. The pressure waves are detected by a sensitive microphone placed in the cell. This device stems from the same source as the Golay and Luft cells (see Sect. 3.4). It has attracted a lot of interest recently because being a selective device, signals are only produced by material of interest and therefore it has a very high sensitivity for detection of traces of pollutants of specific molecules in complex media. *Busse* and *Bullemer* [8.50] compared its performance against that of a more conventional detector for analytical spectroscopy, whilst *Colles* et al. [8.51] gave a general review. Work is continuing on Nernst effect detectors [8.50a–c].

8.1.6 The Use of Thermal Detectors in Infrared Imaging Systems

Of the thermal imaging systems discussed in the first edition, only the pyroelectric vidicon has received continuing development. This has reached the point where pyroelectric cameras are becoming fairly widely available with a performance comparable to that of a single element mechanically scanned cooled photon detector system but at a fraction of the cost. Several recent papers have described applications of the pyroelectric camera including thermonuclear fusion experiments [8.52, 53], infrared spectroscopy [8.54, 55], industrial instrumentation and performance checking in surgical operations [8.56, 57, 57a] and high speed infrared photography [8.57b]. Other applications for thermal imaging systems have been described for which the pyroelectric camera would seem the most suitable equipment [8.58]. The current performance of the pyroelectric vidicon has been described in several recent review articles [8.59–63].

Research to obtain further improvements in the performance of pyroelectric imaging systems is proceeding in several directions. Some further improvement in the spatial resolution obtainable with the vidicon will be obtainable with the reticulated targets which are now being fabricated [8.64]. Improper ferroelectrics offer improved materials [8.64a]. However, another approach is to develop smaller and more rugged tubes with lower power consumption aimed at providing cheaper and more convenient tubes for those applications for which the present performance is adequate [8.65, 66].

An alternative to the pyroelectric vidicon is a solid-state device in which an array of pyroelectric detectors is accessed by a CCD or other multiplexer. Because the basic sensitivity of a single element pyroelectric detector is almost two orders higher than that of a resolution element in a pyroelectric vidicon limited by electron beam and amplifier noise, this approach has much more long term potential than the pyroelectric vidicon. However, up till now it has been less attractive than the vidicon because the performance of the multiplexing electronics had not reached a suitable level. Recent progress in electronics has now reached the point where it is very relevant to re-examine this approach [8.31, 37, 67–69]. The ultimate performance is such that an uncooled unscanned all-solid-state imager could compare favourably (for land based applications at any rate) with most advanced cryogenic imaging systems in performance but with a significant reduction in initial cost and logistic support requirements and a significant increase in reliability.

8.2 Photovoltaic, Photoconductive, and Avalanche Diode Detectors

Recently, increasing technology development has been devoted to photovoltaic $Pb_{1-x}Sn_xSe$ and $Pb_{x-1}Sn_xS$ alloys while interest in $Hg_{x-1}Sn_xTe$ has been decreasing. Interest in doped Si as infrared detector material remains high. In this update, we shall briefly review the most significant recent results for these materials. No important additions have been made recently to the theory given in Chapter 4.

8.2.1 Intrinsic Photovoltaic Detectors

It now seems certain that the strong Auger recombination predicted theoretically by *Emtage* in 1976 (see Ref. [4.39] of Chapt. 4) is in fact present in the $Pb_{1-x}Sn_xTe$ alloys, although the evidence is still inferential. This apparent strong Auger recombination, and resulting short carrier lifetime ($\tau \sim 10^{-8}$ s), has been a factor in the recent de-emphasis of $Pb_{1-x}Sn_xTe$ detector development; with strong Auger recombination the D_λ^* potential of photovoltaic $Pb_{1-x}Sn_xTe$ detectors would be more limited than that predicted in Chapter 4, where the possibility of only radiative recombination is assumed. Emtage's theory is such that Auger recombination should be weaker in $Pb_{1-x}Sn_xSe$ and $Pb_{1-x}Sn_xS$ alloys, because their surfaces of constant energy at the valence and conduction band edges are less anisotropic than in $Pb_{1-x}Sn_xTe$; thus these other two IV–VI alloy systems could be preferable to $Pb_{1-x}Sn_xTe$ for high-performance detectors [8.70].

Peterson and *Casselman* [8.71] have been extending *Petersen*'s earlier calculations of Auger recombination in n-type $Hg_{1-x}Cd_xTe$ to p-type material. The purpose is to determine how important Auger recombination in fact is in p-type $Hg_{1-x}Cd_xTe$. Their calculations assume Auger transitions involving the light-hole valence band as well as the heavy-hole valence band. The result will help determine the maximum D_λ^* really possible in $Hg_{1-x}Cd_xTe$ photovoltaic detectors.

Development of $Hg_{1-x}Cd_xTe$ photovoltaic detector technology continued. See, for example, the publications of *Fiorito* and co-workers [8.72a] and of *Becla* and *Pawlikowski* [8.72b], and the review by *Dornhaus* and *Nimtz* [8.73].

8.2.2 Intrinsic Photoconductive Detectors

In recent work *Borrello* and co-workers [8.74] have demonstrated experimentally the dependence of detector material parameters on background radiation described in Appendix E (p. 142); their results are for $Hg_{0.8}Cd_{0.2}Te$ photoconductive detectors sensitive in the 8–14 μm wavelength range. *Kinch* and co-workers [8.75] have demonstrated $Hg_{0.8}Cd_{0.2}Te$ photoconductive detectors of high performance, approaching the limit imposed by Auger recombination.

Kinch and co-workers in another paper [8.76] have proposed and demonstrated a geometric method of enhancing the D_λ^* of $Hg_{1-x}Cd_xTe$. It involves locating the ohmic electrical contacts farther than a carrier diffusion length away from the edges of the active, illuminated detector area; this design prevents rapid loss of photoexcited carriers by recombination at the contacts. These "ohmic" contacts correspond to the $\Delta n = \Delta p = 0$ boundary condition defined on p. 110. This can be a useful approach, but it has a larger volume of detector material which contributes more noise than signal, so that the ultimate D_λ^* is degraded by a geometrical factor [see Ref. 8.77, Eq. (15)].

8.2.3 Extrinsic Photoconductive Detectors

Specially doped Si is the major material in this category. Some data have now been published in the open literature on doped Si as an extrinsic photoconductive infrared detector material. Properties of dopants, not previously well known, have been measured and seem to provide additional possibilities for doped Si as an effective detector material.

Sclar [8.77] has reported an extensive study of extrinsic photoconductive Si detectors for the 3–5 µm and 8–14 µm infrared wavelength ranges. His study included indium, sulfur, and thallium as dopants for 3–5 µm detection and aluminum, gallium, bismuth, and magnesium as dopants for 8–14 µm.

Scott and *Schmit* [8.78] have reported a careful study of the infrared properties of thallium-doped Si, doped as highly as 5×10^{16} atoms/cm^3; thallium acts as an acceptor. They measured an infrared ionization energy of 0.246 eV, corresponding to a long-wavelength cutoff $\lambda_{co} = 5.0$ µm, well suited for 3–5 µm detection. They also estimated a peak optical cross section of 2.6×10^{-17} cm^2 for thallium in Si. Later work by *Brotherton* and *Gill* [8.79] verified the 0.24 eV ionization energy of thallium by means of thermal emission rate measurements.

Vydyanath and co-workers [8.80] have reported the development of selenium-doped Si for 3–5 µm detection; selenium acts as a donor. The ionization energy of selenium was found from photoconductivity measurements to be 0.3 eV, corresponding to $\lambda_{co} = 4.1$ µm. They found the maximum solubility of selenium in Si to be slightly under 10^{17} atoms/cm^3.

8.2.4 Avalanche Photodiodes (APD)

Low-loss, wide-bandwidth optical fiber communication systems have been the main driving force behind the recent progress in the development of high speed, high gain heterostructure avalanche diodes for the spectral region of 1.0 to 1.6 µm where various fiber optic materials exhibit low transmission losses. In addition to fiber optic communications, these heterostructure APDs may add a new dimension to the capabilities of active laser imaging and rangefinding in the 1.5 µm atmospheric window, passive night vision in the region of peak sky-glow emission (~ 1.6 µm), and to laser satellite communication systems. The beauty of APDs formed by layered structures of binary, ternary, and or quarternary compounds is that their peak sensitivity can be positioned at any desired wavelength between 0.4 µm to 1.8 µm by the proper selection of group II–V alloys and their stoichiometric composition.

Hurwitz and *Hsieh* [8.81] have reported avalanche gains for 1.2 µm radiation in excess of 12 with 150 ps rise times and 45% quantum efficiencies in epitaxial layers of GaInAsP on InP substrates. The APD sensitive area is an etched mesa of 150 µm in diameter. The authors feel that this alloy combination can provide avalanche sensitivity with good lattice matching over the spectral range of 0.9 to 1.6 µm by varying the stoichiometric composition of the GaInAsP alloy.

More recently *Katsuhiko Nishida* et al. [8.82] have reported avalanche gains as high as 3000 at 1.25 μm wavelength with dark currents as low as 1 μA/cm^2 at 1/2 the breakdown voltage for a 50 μm diameter InGaAsP/InP device. The high gain and low dark current is attributed to narrow spatial separation of the avalanche $p-n$ junction (formed in the InP window) from the InGaAsP light absorbing regions.

A recent review article by *Law* et al. [8.83] provides some insight into the design criteria of group III–V alloy heterostructure avalanche photodiodes in terms of their speed of response, noise mechanisms and gain. A useful direct comparison of GaAlSb, GaAlAsSb, and InGaAsP APDs is given in terms of their basic operational parameters.

8.3 Photoemissive Detectors

Recent advances in photoemissive detection fall into six general areas. First are general improvements in "classical" emitters, both in processing [8.84] and in improved compositions, as seen in manufacturers new sales literature.

The second area of improvement is with wide bandgap NEA emitters, including GaP and GaAsP for the visible and GaAs for the near IR [8.85, 86]. Third are emitters constructed in the transmissive mode to allow eventual imaging; one published example is [8.87]. Fourth are evolutionary improvements in cathodes optimized for 1.06 μm detection [8.88]. Fifth, and guiding all of the above NEA advances, are greatly improved structural models and theories for the (CsO) covered surface [8.89–95]. (on related oxygen sticking and sputter yields). The sixth area is one of revolutionary advances in field-assisted 1–1.65 μm emitters [8.96–101].

With the exception of the field-assisted devices, quantitative performance improvements are significant but rather moderate. Performance in the "visible" includes greater than 50% quantum efficiency at 0.53 μm and similarly at the bandgap of GaAs, 14% QE at 0.85 μm for transmission mode GaAs, and 9% QE at 1.06 μm for reflection mode InGaAsP.

The very significant advances in modeling the surface include much deeper understanding of the "reconstructed" semiconductor/vacuum surface (the lattice is not simply terminated, but the final layers adjust to form new surface structures without intrinsic surface states or pinning of the Fermi level in pure, unstrained crystals), and understanding of the effect of O and Cs (or other metals) on this surface. The oxygen or perhaps induced strain is found to cause the pinning and the surface states (and similarly for the metal if oxygen is absent); finally, NEA photoemission has been related to specific (CsO) suboxides. Other related work includes efforts with sputtering, oxygen sticking probabilities, and an enormous volume of literature on Schottky barriers (the latter not included in the updated references).

The field-assisted cathodes were being designed at the time of the first edition, and [8.96, 97] contain early reports of success and specific proposals for constructing this photoemissive device. In the most widely reported implementation of this long-suggested device, two layers are added to the "normal" NEA IR photoemitter as it is described in the body of the text. For example, for IR emission from InGaAsP, a layer of wide bandgap InP (known to have a useful high-mass conduction band above the conduction band minimum) is grown on the InGaAsP photoconductor; this surface is then coated with a 100–200 Å layer of Ag to form a Schottky barrier, and this final surface is coated with (CsO) until NEA InP(/Ag) is obtained. In operation, the Schottky is reverse baised to produce a depletion region throughout the InP. Photoemission for the 1.0 to 1.65 μm photon wavelengths requires that electrons transverse the InP while being injected into the higher lying conduction band, from which emission through the Ag/(CsO) must occur as in an an InP NEA emitter. At biases of 3–5 V and with cooling to 125 K, quantum efficiencies in the order of 2×10^{-3} at wavelengths out to 1.65 μm have been reported. Efficiency suffers with operation at room temperature, and dark current rises rapidly for biases above 3–5 V (while emission saturates in this same voltage range). Although a first order design-optimization model for this device has been published, the detailed operation of the surface is not fully understood.

8.4 Charge Transfer Imaging Devices

Since 1970 extensive research and development of the charge coupling concept in a metal-oxide semiconductor (MOS) has revolutionized the data processing field in general and has been widely used for multiplexing and amplifying signals from optical and infrared detector arrays. Because silicon was the basic semiconductor in the charge coupled device (CCD), it was natural that the first imaging CCDs were sensitive in the spectral region of 0.4 μm to 1.0 μm. In the area of photon detection and imaging the use of the charge coupled concept and the related charge injection device (CID) may be the key to significant improvements in the utility and performance of large scale detector arrays. In addition to substantial increase in the ability to detect small signals this concept also promises savings in volume, and weight, ruggedness, and power description; the latter is of particular importance when operation at cryogenic temperatures or at remote sites is required.

8.4.1 Near Infrared and Visible CCD Imagers

Landauer et al. [8.102] have demonstrated an 800 × 800 element silicon buried channel imaging CCD with an rms readout noise of 15 electrons per pixel which was designed for space telescope imaging in the visible and near infrared regions at

170 K. Because of the small pixel size (15 μm × 15 μm) and the buried-channel structure, saturation occurs at approximately 7×10^4 electrons per pixel. In addition to the large number of elements the array structure is fabricated such that noise-free summation of the charge from two or more pixels (prior to the output amplifier) can be carried out. This type of pixel summation should enhance its capability to perform low contrast, diffuse imaging. The device has line and point imperfection of less than 1% of the pixels [8.103].

Dyck [8.104] reported that a 1024 × 64 element silicon imager, which can be used both in both the storage and time-delay-integration (TDI) modes, has shown high saturation capability (10^6 electrons per pixel) with a total surface uniformity of 1–3% and a 50% quantum efficiency when operated in the storage mode. The rms noise was less than 200 electrons per pixel but the actual value was not reported. When operated in the TDI mode as a line scan imager the sensitivity was said to increase by a factor of 64 over a simple line scan image sensor of equal size elements (20 μm × 20 μm).

Recent measurements on a monolithic platinum-silicon Schottky barrier IR-CCD by *Kosonocky* et al. [8.105] has demonstrated the unique uniformity of these devices for a near infrared imager (1.2 to 4 μm). When used as a 250 element linear line sensor, uniformities of 0.55% were measured and when employed as a 25 × 50 element two-dimensional array the uniformity of response was approximately 2%; limited only by the ability to control the geometric area of individual sensor elements. In the latter case the active detection area (46 μm × 46 μm) covered only 16% of the real estate which resulted in poor photon collection efficiency. This, coupled to the fact that the quantum efficiency of the active region is only 1.5%, leads to the unfavorable condition of converting only 1 photon out of ~ 2000 into measurable signal electrons. In spite of the unfavorable exchange the area array demonstrated an effective image NET of 1 °C and an NEP of 3.36×10^{-11} w with a uniformity of 2% when operated at 80 K. The authors feel, that with new designs and processing improvements, higher density Schottky IR-CCDs will be able to discriminate temperature differences above ambient of 0.1 to 0.2 °C for direct readout devices.

8.4.2 Thermal Scene CCD Imagers

Series/Parallel Scan with time delay and integration remains the principal approach to advanced thermal imaging systems. However, for applications where only a small number of resolution elements are needed, two-dimensional staring detector arrays with CCD or CID readout are being considered [8.106]. This does away with the scanner and a focal optics used with conventional systems. However, to compensate for nonuniformities, both dc offset and gain correction must be made on a pixel by pixel basis. Detector responsivity and readout nonlinearities will increase the number of computations needed for sufficient correction and only experience with the stability of different types of arrays will determine how often the correction algorithms must be calibrated [8.107, 108].

The use of chopping is cumbersome, however it can help provide real time calibration.

Photoconductive extrinsic silicon can be used for staring arrays. Because its photoconductive gain is less than unity it can be used to avoid multiplexor saturation during the long integration time. PV hybrids can also be used but they operate with a detector gain of unity. Since direct transfer of photogenerated charge into a CCD well for a TV field time (1/60 s) with unity gain and anything except a spike filter in front of the array will result in CCD well saturation, various combinations of gain reduction and fast framing, or accumulating subframes into a storage register must be used to avoid saturation. This, of course, is a serious problem in the 8–12 μm region where net gain reductions of several hundreds are required.

Remarkable progress has been made in the development of IR-CCDs for the 3–5 μm region in HgCdTe, and GaInSb [8.109–112]. With sufficient fat zero exceptionally high transfer efficiency (0.998 at 1 MHz) has been demonstrated in HgCdTe. Low surface state density has also been achieved. The IR background will of course automatically provide fat zero except at the very beginning of a TCI register. For some applications charge handling capability will still limit the length of the TDI register. Since contrast in the IR scene increases at shorter wavelengths and higher bandgap material has superior charge handling capability, this favors imagers designed to work out to just 4.2 μm with as long a TDI register as permitted. So far CCDs in HgCdTe have only been fabricated on n-type material and no p–n-junction technology compatible with the CCD processing has been found. Thus sophistications such as anti-blooming drains cannot yet be incorporated.

Direct Injection Hybrid and Extrinsic Silicon

For scanning systems in the 3–5 μm region emphasis in PV hybrid technology has been on increasing f^* by lowering detector capacitance [8.113]. Lateral collection is used to reduce the junction area while maintaining the photoactive detector size. There is also interest in 8–12 μm systems using HgCdTe photodiodes. The PV approach does not apply large bias voltages to the IR material and gain reduction circuits can be incorporated before the input to the CCD TDI register. Because extrinsic silicon photoconductive gain is less than unity, multiplexor CCD saturation with 8–12 μm operation can be avoided, however lower focal plane operating temperatures are required.

Infrared CID

Two-dimensional IR CID arrays for the 3–5 μm region have been fabricated in InSb and HgCdTe [8.114–116]. In some cases (in InSb) large bias charges were needed to obtain acceptable transfer efficiency [8.117]. This reduces responsivity since at the time of readout (injection) the signal charge lies under both the row and column gates and only the charge under the column gate contributes to the voltage at the preamplifier. Lag has also been observed [8.117]. In HgCdTe the long lifetimes in n-type material will limit readout speed due to free carrier recapture.

One of the principal remaining questions is whether any kind of MIS focal plane array CCD or CID can be made to work in the 8–12 μm region. The principal problem is the extra dark current caused by interband tunneling during those periods when a large electric field is applied to the narrow bandgap semiconductor (usually when the well is empty). The problem gets progressively worse as the bandgap is reduced [8.118, 119]. To reduce the electric field at the surface low semiconductor doping is desired.

For the 8–12 μm region the CID used in series/parallel scan mode theoretically has an advantage in that the detector integration time is less than a dwell time. The integrated background charge does not have to be as large as in the CCD approach where integration occurs during the whole TDI process. The well capacity can be less which means that lower voltages can be applied to the IR material. In addition, since the background noise is higher in the 8–12 μm region, the preamplifier used with the CID array will not have to be driven as hard to achieve the noise performance required for BLIP operation.

8.5 Heterodyne Detectors

Over the past few years significant progress has been made in the fabrication of large electrical bandwidth detector arrays approaching theoretical sensitivity (hvB) for use in infrared optical radars and coherent radiometry. *Spears* and *Hoyt* [8.120] have fabricated a 12-element HgCdTe array designed for the rapid target acquisition of a CO_2 laser radar. All elements had a heterodyne sensitivity of better than 7×10^{-20} W/Hz at 760 MHz and 1.3×10^{-19} watts/Hz at 1.5 GHz when operated at 77 K without AR coating [8.121]. Blackbody heterodyne radiometry [8.122] was used to determine the heterodyne sensitivity of each of the 550 μm diameter elements. Improvement in the heterodyne sensitivity of these arrays by a factor of ~ 1.4 was achieved with a 1.1 μm thick ZnS antireflection coating. *Shanley* and *Perry* [8.123] have observed 6.2×10^{-20} watt/Hz sensitivity at a frequency of 1.75 GHz for a 170 μm diameter $n^+ - n^- - p^+$ HgCdTe photodiode operated a liquid nitrogen. The surface was not antireflective coated.

From a sensitivity point of view heterodyne detection is approaching the theoretical hvB limit over the visible and infrared spectral regions. The electrical bandwidth over which sensitivity approaching hvB can be realized is presently dictated by the capacitance of the sensing element, the parasitic resistance of the device, and the availability of local oscillator power. Since the capacitance of heterodyne photodiodes is proportional to the sensing area, higher bandwidths can be realized by replacing a single photodiode with an array of smaller units and summing their output signals.

8.6 Desirable Optical and Infrared Detector Technology Advances

Desirable detector technology advance is in the eye of the beholder and what follows is the editor's point of view. Tendered by the recent adverse impact of energy shortages on world economics and quality of life, coupled to the seemingly inexhaustible source of solar energy, it seems that the realization of inexpensive, large area, reliable, photovoltaic diodes would have to loom as the most desirable advance in detector technology. The key word in these adjectives is "inexpensive", because it will almost certainly force a violently different approach to the manner and the materials used in the photodiode fabrication process. In order to achieve manufacturing costs of several dollars per square foot will demand that the photodiode substrate be made of sheets of metals, glass, or plastic. The basic photosensitive material (silicon, GaAs, or other) will have to be evaporated or sputtered onto the substrate in a fashion similar to current techniques for metalizing rolls of plastic film. Methods of making $p-n$ junctions and their appropriate leads will almost surely involve a similar process of evaporating the proper doping metal followed by high intensity, short time, large area heating pulses (probably laser) in order to diffuse the dopant into the substrate to form $p-n$ junction islands without resorting to costly vacuum oven procedures. Those accustomed to the "clean room" environment of semiconductor fabrication facilities may rightly shudder at this mass production approach to solar photovoltaic arrays, but without the perfection of such techniques the cost of producing meaningful amounts of *electric power* from the sun (for industrial, commercial, and residential power) will be less cost effective than competing methods. An encouraging note is that if the large area solar photovoltaic arrays can be made inexpensively, they need not have conversion efficiencies greater than about 5% in order to be cost effective. When the heat generated by solar cell *inefficiency* is coupled to a system that uses this energy for building heating or air conditioning the conversion efficiency requirement is further reduced.

The widespread application of 10.6 μm heterodyne detection for communications, radar and infrared active imaging is presently restricted by the need to cool the detectors below temperatures obtainable from thermal electric coolers (~ 180 K). Presently the main reason for cooling 10.6 μm heterodyne detectors is to reduce the thermal generation–recombination rate in the photoconductor or the diffusion current in diodes to a level that can be overridden by available laser local oscillator power. In general the use of photoconductors (with band gaps tailored for 10.6 μm radiation) seems to be a realistic approach to operation at 180 K or above[1]. Since the thermal generation–recombination rate is of the order of $10^{16}/\tau$ hole–electron pair per cm^3 at 180 K it follows that a 1 μm thick, 10^{-3} cm^2 area photoconductor with a carrier lifetime (τ) of 10^{-8} s would require ~ 1 milliwatt of LO power to approach $h\nu B$ detector sensitivity at bandwidths of

---

[1] Conversations with D. L. Spears at Lincoln Laboratory indicate that he is currently pursuing this general approach.

100 MHz. In this example the temperature rise due to thermal power flow into the heat sink would be insignificant.

A third very desirable development would embrace solid state imaging devices of high quantum efficiency, high gain, and stability for night vision application in the "sky glow" region of this spectrum at 1.6 μm. The perfection of such devices in this region is significant because scene illumination due to natural sources of radiation is generally greater than 5×10^{-8} [8.124] watts/cm^2, scene contrast is high, and atmospheric transmission through haze and fog is considerably better than in the visible region of the spectrum.

References

8.1   S.C.Chase: SPIE **95**, 30–37 (1976)
      S.C.Chase, J.L.Engel, H.W.Eyerly, Hugh H.Kieffer, F.D.Palluconi: Appl. Opt. **17**, 1243–1251 (1978)
8.2   W.T.Baker: Opt. Spectra **11** (March 1977)
8.2a L.R.Wollman: Electro-Opt. Syst. Des. **11**, 37–44 (Sept. 1979)
8.3   L.P.Boivin, T.C.Smith: Appl. Opt. **17**, 3067–3075 (1978)
8.3a A.Ono: Jpn. J. Appl. Phys. **18**, 697–698, 1995–2002 (1979)
8.4   UKAEA Culham: Unpublished report
8.5   M.D.Bausman, I.Liberman, A.T.Swann: J. Opt. Soc. Am. **68**, 1441–1442 (1978)
8.6   U.Korn, S.Shtrikman: Elect. Electron. Engrs. Israel 10th Convention Tel-Aviv, 10–13 Oct. 1977, pp. 110–117
8.6a V.N.Vigdorovich, G.A.Ukhlinov, N.I.Chibotaru: Instrum. Exp. Tech. (USA) **21**, 521–523 (1978)
8.7   R.S.Scott, G.E.Fredericks: Infrared Phys. **16**, 619–626 (1976)
8.8   Yu.Z.Levin, V.A.Maslov, V.A.Danilov, N.A.Suslova: Sov. J. Opt. Technol. **44**, 759–760 (1977)
8.8a V.A.Beskin, L.S.Kremenchugskii, A.F.Mal'nev, S.K.Sklyarenko, V.P.Timonin, G.I.Shuster: Meas. Tech. (USA) **21**, 228–230 (1978)
8.8b J.H.Degnan: Rev. Sci. Instrum. **50**, 1223–1226 (1979)
8.8c K.Wasa, T.Tohda, Y.Kasahara, S.Hayakawa: Rev. Sci. Instrum. **50**, 1084–1088 (1979)
8.8d B.Morten, M.Prudenziati, A.Taroni: Fisica Technologia (Italy) **1**, 75–95 (1978)
8.9   S.Wieder, E.Jaoudi: Am. J. Phys. **46**, 935–937 (1978)
8.10 N.E.Ivanov, I.A.Ivanova, T.A.Prilezhaeva: Soc. J. Opt. Technol. **44**, 504–505 (1977)
8.11 J.C.Johnson, G.A.Massey: Appl. Opt. **17**, 2268–2269 (1978)
8.11a T.L.Hwang, S.E.Schwarz, D.B.Rutledge: Appl. Phys. Lett. **34**, 773–776 (1979)
8.12 J.Clarke, G.I.Hoffer, P.L.Richard, N.H.Yeh: *Low Temperature Physics* LT14, ed. by M.Krusius, M.Vorio (Elsevier, Amsterdam 1975) pp. 226–229; J. Appl. Phys. **48**, 4865–4879 (1977)
      J.Clarke, P.L.Richards, N.H.Yeh: Appl. Phys. Lett. **30**, 664–666 (1977)
8.13 S.A.El-Atawy, P.A.R.Ade: Infrared Phys. **18**, 683–690 (1978)
8.14 G.Chanin, J.P.Torre, L.Peccoud: Infrared Phys. **18**, 657–662 (1978)
8.15 F.J.Low, R.F.Kurtz, W.M.Poteet, T.Nishimura: Astrophys. J. **214**, L115–L118 (1977)
8.16 N.S.Nishioka, P.L.Richards, D.P.Woody: Appl. Opt. **17**, 1562–1567 (1978)
8.17 G.Gallinaro, G.Roba, R.Tatarek: J. Phys. E **11**, 628–630 (1978)
8.18 M.A.Chin: J. Appl. Phys. **48**, 2723–2728 (1977)
8.19 J.P.Baluteau, M.Anderegg, A.F.M.Moorwood, N.Coron, J.E.Beckman, E.Bussoletti, H.H.Hippelein: Appl. Opt. **16**, 1834–1840 (1977)
8.20 P.L.Richards: Infrared Phys. **17**, 241–244 (1977)

8.21 D.J.W.Kendall, T.A.Clark: Appl. Opt. **18**, 346–353 (1979)
8.22 J.C.Mather: Proc. SPIE **105**, 44–50 (1977)
8.23 T.A.Enukova, V.P.Korotkov, G.N.Mikheeva, N.A.Pankratov, G.G.Pitkevich, Yu.I.Polushkin: Sov. J. Opt. Technol. **44**, 476–477 (1977)
8.24 G.A.Zaitsev, I.A.Khrebtov: J. Appl. Spectrosc. **28**, 125–130 (1978)
8.25 P.N.Nikiforov, N.A.Pankratov: Sov. J. Opt. Technol. **45**, 210–212 (1978)
8.25a N.A.Pankratov, Yu.V.Kulikov, N.V.Shchetinina: Sov. J. Opt. Technol. **45**, 435–437 (1978)
8.26 S.T.Dunn: Appl. Opt. **17**, 1367–1373 (1978)
8.27 J.W.Fleming, K.Hursey: J. Phys. E **12**, 91–92 (1979)
8.28 R.N.Sachdev, J.P.Gupta, K.G.Vohra: Appl. Opt. **17**, 3472–3476 (1978)
8.29 S.T.Liu: Proc. IEEE **66**, 14–26 (1978)
8.30 W.M.Doyle: Electro-Opt. Syst. Des. **11**, 12–16 (Nov. 1978)
8.31 D.E.Marshall: SPIE **132**, 110–117 (1978)
8.31a L.S.Kremenchugsky, V.B.Samoilov: Ukr. Fiz. Zh. (USSR) **24**, 274–287 (1979)
8.32 S.B.Lang: Ferroelectric **17**, 553–573 (1978); **19**, 26–60 (1978)
8.33 S.E.Stokowski: Appl. Opt. **15**, 1767–1774 (1976)
8.34 R.E.Newnham, D.P.Skinner, L.E.Cross: Mater. Res. Bull. **13**, 525–536 (1978)
8.35 J.D.Zook, S.T.Liu: J. Appl. Phys. **49**, 4604–4606 (1978)
8.36 E.H.Putley: Infrared Phys. **18**, 373–374 (1978); **20** (1980) in press
8.36a D.W.G.Byatt: Electron. Industry, pp. 21–27 (Sept. 1979)
8.36b O.V.Elfimov, L.S.Kremenchugskii, S.K.Sklyarenko: Sov. J. Opt. Technol. **45**, 663–664 (1978); Prib. Tekh. Eksp. **3**, 211–213 (May-June 1978)
8.36c S.A.Hamid: Phys. Status Solidi (a) **53**, K75–77 (1979)
8.36d T.Katsube, Y.Nakagawa, K.Ohkubo, M.Hara, F.Ohtani: Sci. Eng. Rpt. Saitama Univ., Ser. C (Japan) **12**, 1–6 (1978)
8.36e M.Simhony, M.Bass: Appl. Phys. Lett. **34**, 426–427 (1979)
 M.Simhony, M.Bass, E.W. van Stryland, E.M.Tenescu, B.Levy: IEEE J. QE-**15**, 206–208 (1979)
8.36f H.J.Zajosz: Thin Solid Films **62**, 229–236 (1979)
8.36g M.El-Hadi Tidjani, P.Belland, D.Veron: Infrared Phys. **19**, 677–681 (1979)
8.37 N.E.Byer, S.E.Stokowski: Martin Marietta Laboratories Tech. Rpt. 76–30 (Unclassified) Presented at IRIS Detector Speciality Meeting, March 1976, San Diego
 S.E.Stokowski: Martin Marietta Laboratories Tech. Rpt. 76–31 (Unclassified) Presented at IRIS Detector Specialty Meeting, March 1976, San Diego
8.38 Industrial Research/Development **16**, 18 (October 1978)
8.39 A.Hadni, R.Thomas, J.Mangin, M.Bagard: Infrared Phys. **18**, 663–668 (1978)
 X.Gerbaux, J.M.Waldschmidt, A.Hadni: Appl. Opt. **17**, 1616–1620 (1978)
8.40 F.Bordoni, P.Carelli, I.Modena, G.L.Romani: Infrared Phys. **19**, 653–657 (1979)
8.41 Electro-Opt. Syst. Des. **11**, 13 (February 1979)
8.41a F.W.Taylor, F.E.Vescelus, J.R.Locke, G.T.Foster, F.B.Forney, R.Beer, J.T.Houghton, J.Delderfield, J.T.Schofield: Appl. Opt. **18**, 3893–3900 (1979)
8.42 C.Carter: Caswell News, April (1979)
8.43 L.Williams: Electron **20**, 15–19 (Sept. 1978)
8.44 V.V.Voronov, N.V.Karlov, G.P.Kuz'min, Yu.S.Kuz'minov, B.A.Kuritsyn, S.M.Nikiforov, V.V.Osiko, A.M.Prokhorov: Sov. J. Quantum Electron. **7**, 1062–1086 (1977)
8.45 D.C.Elbers, W.H.Thomason, J.D.Macomber: Appl. Opt. **17**, 308–310 (1978)
8.46 K.J.Button, S.M.Wolfe: SPIE **105**, 72–79 (1977)
8.47 M.McColl: SPIE **105**, 24–34 (1977)
8.48 C.P.Chaloner, J.R.Drummond, J.T.Houghton, R.F.Jarnot, H.K.Roscoe: Proc. R. Soc. London A**364**, 145–159 (1978)
8.48a J.Debrie, A.de Martino, M.Lequime, R.Frey, F.Pradère: Rev. Sci. Instrum. **50**, 330–332 (1979)
8.48b C.Hartung, R.Jurgeit: Sov. J. Quantum Electron. **8**, 1035–1037 (1978)
8.49 P.M.Kuhn, I.G.Nolt, L.P.Stearns, J.V.Radostitz: Opt. Lett. **3**, 130–132 (1978)
8.50 G.Busse, B.Bullemer: Infrared Phys. **18**, 255–256 (1978)
8.50a G.P.Chuiko, N.M.Chuiko: Inorg. Mater. **15**, 23–26 (1979)

8.50b   A.B.Katrich, V.M.Kuz'michev: Sov. J. Quantum Electron. **8**, 1102–1105 (1979)
8.50c   J.E.Müller: Nachrichtentech. Electron. **28**, 143–144 (1978)
8.51   M.J.Colles, N.R.Geddes, E.Mendizadeh: Contemp. Phys. **20**, 11–36 (1979)
8.52   D.R.Kohler, P.B.Weiss: Topical Meeting on Inertial Confinement Fusion, San Diego, CA USA, 7–9 Feb. 1978, pp. TUC 13/1–4 (Opt. Soc. Am., Washington, D.C. 1977)
        R.F.Benjamin, P.D.Goldstone, J.P.Carpenter: Appl. Opt. **17**, 3809–3811 (1978)
        D.R.Kohler, P.B.Weiss: J. Opt. Soc. Am. **67**, 1445 (1977)
8.53   Electro-Opt. Syst. Des. **10**, 35–39 (Jan. 1978)
8.54   Y.Talmi: Appl. Opt. **17**, 2489–2501 (1978)
8.55   W.M.Wreathall: SPIE **110**, 63–69 (1977)
8.56   R.Watton, D.Burgess, B.Harper: J. Appl. Sci. Eng. A**2**, 47–63 (1977)
8.57   J.W.E.Brydon, A.K.Lamie, D.J.Wheatley: J. Med. Eng. Technol. **3**, 77–80 (1979)
8.57a   G.F.Vermeij: J. Med. Eng. Technol. **3**, 5–11 (1979)
8.57b   A.J.Alcock, H.A.Baldis, P.B.Corkum, J.C.Samson, W.J.Sarjeant: SPIE **97**, 264–268 (1976)
        A.J.Alcock, P.B.Corkum: Canadian J. Phys. **57**, 1280–1290 (1979)
8.58   R.Tice, J.Euskirchen: Opt. Spectra **12**, 32–36 (Sept. 1978)
        C.W.Brice III: Industrial & Commercial Power Systems Tech. Conf. 6–8 June 1978, Cincinnati (IEEE NY USA 1978) pp. 118–120
        K.E.G.Pitt: Microelectron. J. **9**, 19–26 (1979)
        R.W.Burton: Proc. 8th European Microwave Conf. Paris, France 4–8 Sept. 1978, pp. 179–182
8.59   B.Singer: *Adv. in Image Pickup and Display*, Vol. 3 (Academic Press, New York 1977) pp. 1–82
8.60   A.G.Shepherd: Electronics **50**, 99–105 (Nov. 1977)
        Electronique & Appl. Ind. No. 250, 1 April, 1978, pp. 51–55
8.61   L.E.Garn: IEEE Trans. ED-**24**, 1221–1228 (1977)
8.62   R.S.Levitt: Electro-Opt. Syst. Des. **9**, 22–30 (Oct. 1977)
8.63   P.C.H.Dickerson: J. R. Electr. Mech. Eng. (GB), No. 28, 46–50 (April 1978)
8.64   R.Watton: Infrared Phys. **18**, 73–87 (1978)
        R.Watton, D.E.Burgess: Infrared Phys. **19**, 683–688 (1979)
8.64a   A.Shaulov, M.I.Bell, W.A.Smith: J. Appl. Phys. **50**, 4913–4919 (1979)
8.65   J.E.Jacobs, S.A.Remily: Infrared Phys. **19**, 1–7 (1979)
8.66   Electronic Warfare Defense Electron. **10**, 46 (April 1978)
8.67   S.Iwasa, J.Gelpey, J.Marciniec, D.Marshall, W.White, D.Lamb, S.Liu, D.Paffel: IEEE Intern. Electron Devices Meeting, Washington, D.C., Dec 4–6, 1978, pp. 522–525
8.68   C.B.Roundy: Appl. Opt. **18**, 943–945 (1979)
8.69   C.B.Roundy: Infrared Phys. **19**, 507–522 (1979)
8.70   H.Preier: Infrared Phys. **18**, 43 (1978)
8.71   P.E.Petersen, T.N.Casselman: To be published
8.72   G.Fiorito, G.Gasparrini, F.Svelto: Infrared Phys. **17**, 25 (1977); **18**, 59 (1978)
        P.Becla, J.M.Pawlikowski: Infrared Phys. **16**, 457 (1976)
8.73   R.Dornhaus, G.Nimtz: Springer Tracts in Modern Physics, Vol. 78 (Springer, Berlin, Heidelberg, New York 1976) pp. 1–119
8.74   S.Borrello, M.Kinch, D.LaMont: Infrared Phys. **17**, 121 (1977)
8.75   M.A.Kinch, S.R.Borrello, A.Simmons: Infrared Phys. **17**, 127 (1977)
8.76   M.A.Kinch, S.R.Borrello, B.H.Breazeale, A.Simmons: Infrared Phys. **17**, 137 (1977)
8.77   N.Sclar: Infrared Phys. **16**, 435 (1976)
8.78   W.Scott, J.L.Schmit: Appl. Phys. Lett. **33**, 294 (1978)
8.79   S.D.Brotherton, A.Gill: Appl. Phys. Lett. **33**, 953 (1978)
8.80   H.R.Vydyanath, W.J.Helm, J.S.Lorenzo, S.T.Hoelke: Infrared Phys. **19**, 93 (1979)
8.81   C.E.Hurwitz, J.J.Hsieh: Appl. Phys. Lett. **22**, 487 (1978)
8.82   Katsuhiko Nishida et al.: Appl. Phys. Lett. **33**, 251 (1979)
8.83   H.D.Law, K.Nakano, L.R.Tomasetta: IEEE J. QE-15, No. 7, July 1979
8.84   G.Ghosh, B.P.Varma: J. Appl. Phys. **49**, 4549 (1978)
8.85   J.S.Escher, G.A.Antypas: Appl. Phys. Lett. **30**, 314 (1977)
8.86   G.H.Olsen, D.J.Szostak, T.J.Zamekowski, M.Ettenberg: J. Appl. Phys. **48**, 1007 (1977)
8.87   D.G.Fisher, G.H.Olsen: J. Appl. Phys. **50**, 2930 (1979)

8.88   J.S. Escher, G. A. Antypas, J. Edgecumbe: Appl. Phys. Lett. **29**, 153 (1976)
8.89   W. E. Spicer: Appl. Phys. **12**, 115 (1977)
8.90   W. E. Spicer, I. Lindau, P. E. Gregory, C. M. Garner, P. Pianetta, P. W. Chye: J. Vac. Sci. Tech. **13**, 780 (1976)
8.91   W. E. Spicer, P. Pianetta, I. Lindau, P. W. Chye: J. Vac. Sci. Tech. **14**, 885 (1977)
8.92   W. E. Spicer, I. Lindau, C. Y. Su, P. W. Chye, P. Pianetta: Appl. Phys. Lett. **33**, 934 (1978)
8.93   I. Lindau, P. W. Chye, C. M. Garner, P. Pianetta, C. Y. Su, W. E. Spicer: J. Vac. Sci. Tech. **15**, 1332 (1978)
8.94   M. G. Burt, V. Heine: J. Phys. C **11**, 961 (1978)
8.95   P. Williams, C. A. Evans, Jr.: Surf. Sci. **78**, 324 (1978)
8.96   R. Sahai, J. S. Harris, R. C. Eden, L. O. Bublac, J. C. Chu: CRC Crit. Rev. in Solid State Sciences **5**, 565 (1975)
8.97   J. S. Escher, R. D. Fairman, G. A. Antypas, R. Sankaran, L. W. James, R. L. Bell: CRC Crit. Rev. in Solid State Sciences **5**, 577 (1975)
8.98   J. S. Escher, R. Sankaran: Appl. Phys. Lett. **29**, 87 (1976)
8.99   J. S. Escher, P. E. Gregory, G. A. Antypas, R. Sankaran, Y. M. Houng: J. Appl. Phys. **49**, 447 (1978)
8.100 P. E. Gregory, J. S. Escher, S. B. Hyder, Y. M. Houng, G. A. Antypas: J. Vac. Sci. Tech. **15**, 1483 (1978)
8.101 J. S. Escher, P. E. Gregory, S. B. Hyder, R. Sankaran: J. Appl. Phys. **49**, 2591 (1978)
8.102 F. P. Landauer, J. R. Janesick, S. L. Knapp, M. M. Blouke, J. R. Hall: 1978 Government Microcircuit Application Conf. Digest of Papers, Vol. 7, 394 (1978) Joy Morrealle
8.103 M. M. Blouke, J. E. Hall, J. F. Breitzmann: Proc. of 1978 International Electron Device Meeting 412
8.104 R. H. Dyck: 1978 Government Microcircuit Application Conf. Digest of Papers, Vol. 7, 328 (1978) Joy Morrealle
8.105 W. F. Kosonocky, E. S. Kohso, B. R. Capone, S. A. Roosild: 1978 Conf. on the Applications of Charge Coupled Devices Proc., p. 2–27, (1978) Isaac Lagado
8.106 R. Balcerak, R. E. Flannery: "Staring Infrared Focal Planes for Smart Sensors", Proc. SPIE Tech. Symp. East '79, Washington, D.C., April, 1979
8.107 C. Carrison, B. Krzyzabiwsju, N. Foss: "Non-Uniformity Compensation Techniques for Staring IR Focal Planes", Proc. SPIE Tech. Symp. East '79, Washington, D.C., April, 1979
8.108 R. W. Helfrich: "Programmable Compensation Technique for Staring Arrays", Proc. SPIE Tech. Symp. East '79, Washington, D.C., April, 1979
8.109 R. A. Chapman, M. A. Kinch, A. Simmons, S. R. Borrello, H. B. Morris, J. W. Wrobel, D. D. Buss: "$Hg_{0.7}Cd_{0.3}Te$ Charge-Coupled Device Shift Registers", Appl. Phys. Lett. **32** (No. 7), 434–436 (1978)
8.110 R. A. Chapman, M. A. Kinch, S. R. Borrello, A. Simmons, D. D. Buss: "HgCdTe Charge Coupled Devices", CCD-78 Proc., pp. 2–1–2–17, San Diego, CA, Oct. 1978
8.111 R. D. Thom, F. J. Renda, W. J. Parrish, T. L. Koch: "A Monolithic InSb Charge Coupled Infrared Imaging Device", I.E.D.M. Technical Digest, pp. 501–504, Washington, D.C., Dec. 1978
8.112 E. E. Barrowcliff, L. O. Bublae, D. T. Cheung, A. M. Andrews, J. D. Blackwell, F. Cox, E. R. Gertner, W. E. Tennant, M. J. Lodowise, L. E. Wood: "Planar GaInSb CCDs", CCD-78 Proc., pp. 2–77, San Diego, CA., Oct. 1978
8.113 A. M. Andrews: "Hybrid Infrared Imaging Arrays", I.E.D.M. Tech. Digest, pp. 505–509, Washington, D.C., December 1978
8.114 J. C. Kim: "InSb Charge Injection Device Imaging Array", IEEE Trans. ED-**25**, 323–341 (1978)
8.115 M. D. Gibbons, W. E. Davern, J. Swab, R. W. Aldrich: "Status of InSb CID Arrays", Proc. IRIS Detector Specialty Meeting, Minneapolis, Minn., June 1979
8.116 R. A. Chapman, S. R. Borrello, A. Simmons, J. D. Beck, A. L. Lewis, M. A. Kinch, Jollynecek, C. G. Roberts: "Monolithic HgCdTe Charge Transfer Device Infrared Imaging Arrays", to be published IEEE Trans. Elect. Dev.
8.117 D. L. Weinberg, C. J. Gridly, A. F. Milton: "Evaluation of InSb Detector Array Performance", Proc. IRIS Detector Specialty Meeting, Minneapolis, Minn., June 1979

8.118 J. Farre, J. Buxo, D. Esteve: "Importance Relative des composantes de Courants de Charge dans les Capacites MIS destinees a Detecter le Rayonnement Infraronge". C. R. Acad. Sc. Paris, Serie B T. **283**, 103–106 (1976)
8.119 W. W. Anderson: "Tunnel Current Limitations of Narrow Bandgap Infrared Charge Coupled Devices". Infrared Phys. **17**, 147–164 (1977)
8.120 D. L. Spears, C. D. Hoyt: Solid State Research Report, Lincoln Laboratory, M.I.T. (1978 : 1) p. 1, 1978
8.121 D. L. Spears: Solid State Research Report, Lincoln Laboratory, M.I.T. (1978 : 3) p. 1, 1978
8.122 D. L. Spears, R. H. Kingston: Appl. Phys. Lett. **34** (9) 584, 1 May (1979)
8.123 J. F. Shanley, L. C. Perry: Proc. Intern. 1978 Electron Device Meeting, 424
8.124 W. L. Wolfe, G. J. Zissis: *The Infrared Handbook*, Environmental Research Institute of Michigan, pp. 3–82

Recent Publications

Review Articles

T. G. Blaney: J. Phys. E **11**, 856–881 (1978)
K. L. Chopra, D. K. Pandya: Thin Solid Films **50**, 81–98 (1978)
C. Huang: Opt. Spectra **12**, 47–50 (1978)
R. H. Kingston: *Detection of Optical and Infrared Radiation*, Springer Series in Optical Sciences, Vol. 10 (Springer, Berlin, Heidelberg, New York 1978) Chapt. 7
K. Shivanandan: SPIE **105**, 37–39 (1977)
W. E. Spicer: Appl. Phys. **12**, 115 (1977)

Subject Index

Absorption
 coefficient 131, 134, 137, 139
 constant 156
 cross section 134
 edge image converter 33, 96
Acceptor level 129, 139, 141
Acceptors 139, 141
Accumulation mode 219
AgMgO 185
Analog communication system 266
Anti-reflective coatings 111, 120, 129
Astronomical radiation 294
Atmospheric channel 282
Auger recombination 116, 120, 137, 138, 142, 307
Avalanche gain 149, 308, 309

Background fluctuation limit (BFL) 50, 56, 59
Background generation 204
Background radiation 104, 109, 120, 121, 125, 131, 133, 134, 140, 142
 diffuse emission 2
 Band-bending 159, 166
Bandgap 154, 175
Binary communications 270, 282
Binary compounds 174
BLIP 47, 133
Bolometer 82
 cryogenic 28
 metal 27
 superconducting 28
Bolometer effect 26, 302, 303
Buffers
 AlGaAs 176, 177
 GaAsP 176, 177
 InGaP 178
Bulk generation 204
Burstein-Moss effect 138

Carrier lifetime 111, 123, 135, 142, 173
Carrier sweepout 105, 108, 123, 127, 129, 133, 135, 143
Channel stops 202

Charge coupled devices (CCD) 3, 199, 310—313
 buried channel 203
 direct injection 199, 312
 infrared sensitive 199, 213, 310—313
 InSb 215
 signal processing 199
 silicon 199, 318
 two-phase operation 202
Charge injection devices (CID) 197, 220, 223, 312
 arrays 223, 312
 infrared sensitive 199, 310—313
 series-paralled scan mode 223, 311
"Classical" photoemitters 150, 156
Cold cathode 181
"Cold" electrons 157, 156
Composite background fluctuation limits 56
Composite signal fluctuation limits 56
Conduction band 154
Conduction band discontinuity 164, 172
CO_2 laser radar 269
Copper doped germanium 231
Correlated double sampling 210, 225
Cs^+ ions 167
CsSb 185
CuBeO 186
cw radar 249, 263

Dark current 182
Dark current charge-up time 203
dc-coupled staring sensor 198
Debye length 120
Dember effect 9, 23
Depletion region 204
Detection of remote species 288
Detectivity 46
 background limited 107, 109
 spectral 106
 vs wavelength 55, 58, 59, 60
Detectivity vs wavelength values of 0.1 µm to 1.0 µm photodetectors 59
Detector arrays 108, 304
Detector field of view 3, 56

322   Subject Index

Detector noise (see noise mechanisms)
Detector operating temperature 54, 314
Detector types
  elemental 7
  extrinsic 8, 9, 11, 133, 135, 143, 155, 308
  imaging 7, 310—313, 314
  intrinsic 8, 9, 129, 133, 138, 307
  optical 5, 314
  photoconductive 2, 56, 106, 108, 120, 129, 133, 307
  photoemissive 2, 149, 309, 310
  photovoltaic 2, 54, 104, 106, 108, 120, 134, 307
  thermal 2, 26—28, 32, 71, 72, 79, 82, 83, 301—306
Dielectric constant 137
Dielectric relaxation time 133
Diffusion current 110, 135
Diffusion length 115, 157
Direct injection: hybrid 199, 216
Direct injection (silicon) 216, 217
Direct radiation recombination 136, 141, 142
Distributed floating gate amplifier (DFGA) 210
Donor impurities 138, 141
Donor level 129
Donors 130, 144
Doped silicon 127
Double-quantum photoemission 238
Dynode 185
Dynode structures
  box-and-grid 186
  Rajchman 186
  venetian-blind 186
D^* ("dee-star") 44
D^{**} ("dee-double-star") 45
D^*f^* product 47
$D_\lambda^*(\lambda, f, 1)$ 44, 52, 54, 55, 109
$D^*(T, f, 1)$ 44
$D^*(T_S)$ 52, 54

Edge effect 206
Effective mass 137, 138, 143
Efficiency 180
Einstein relationship 115
Electron affinity 155, 163, 167
Electron-beam-interrogation 149
Electron emitters
  back illumination 151
  front illumination 151
  reflection mode 151
  semitransparent 151
  transmission mode 151
Electron escape depth 157
Electron lifetime 114

Electron mobility 127, 138
Electron multipliers 181, 184
Electron photoexcitation 103
Electronic integration 109
Energy barrier 166
Energy gap 104, 120, 137, 142, 154
Energy levels 143
Envelope detection 270
Error probability 270, 277, 278, 281, 284
Error probability modulation 280
Escape probability 158
Evaporagraph 33, 72
External photoeffect 8
Extrinsic long wavelength limit 11
Extrinsic photoconductor 218, 308

Figure of merit 42
Floating gate preamplifier 210
Focal plane filling factor 197
Free charge transfer 205
Frequency response 46
Frequency shift keying 279
Fringe fields 206
F^* 46, 312

GaAs 167, 183
Gain (noiseless) 2
GaP substrate 176, 309
Gas avalanche 9
Gas-filled condenser microphone 31
Gate electrodes 200
Germanium impurities 12
Golay cell 31, 89
Graded junction 114
g—r current 111, 114, 116, 136, 314

Hall coefficient 139, 140
Heterodyne detection 1, 3, 313, 314
Heterodyne signal processing 1, 3
Heterojunction 111, 309
Heterojunction photodiode 19
HgCdTe 118, 119, 125, 127, 129, 132, 133, 138, 139, 141, 142, 307, 308
Hole lifetime 116
Hole mobility 125, 138, 141
Hot electron bolometer 9, 23
"Hot" electrons 157, 161

"Ideal" infrared detector 1
Imaging 149, 151, 191, 310
Impurity ionization energy 104, 129
Impurity scattering 144
Impurity scattering mobility 144
InAs 114, 138
Index of refraction 137

Subject Index

Indirect bandgap 157
Infrared detectors 1, 5
Infrared heterodyne detection 1, 229
Infrared quantum counter 9, 25
Infrared response 150
Infrared sensitive phosphors 9, 25
InGaAs 178, 189
InGaAsP 177, 178, 183, 189, 308
Interface state generation 204
Interface states 174
Interfacial barrier 178, 181
Interstellar molecules 295
Ion bombardment 172
Ion implantation 141
Ionization energy 12, 133, 152
Ionized impurity scattering mobility 144
InSb 114, 138
Intrinsic
  carrier concentration 141
  infrared photoconductor 39
  photoconductor long wavelength limit 10, 11, 120, 216

Johnson noise 133
Josephson junction photoeffect 36

KCsSb 186

Laser cw radar 249, 263
Laser pulsed radar 270, 279, 282
Lattice constant 174, 175
Lattice parameter (a_0) 174
Lead-tin selenide photovoltaic detector 231
Linear expansion 95
Liquid crystals 32, 97
Lithium tantalate 92
Low-light level detection 149

Mercury-doped germanium 132
Metal-insulator-semiconductor structure (MIS) 200
Metal-metal oxide-metal photodiode 36
Metal-oxide-semiconductor structure 200
Minimum detectable power 49, 50, 57
Minimum resolved temperature difference 197
Minority carrier 110
Minority carrier lifetime 125
Minority carrier trapping 128, 129
MIS 200
MnO 161
Mobility ratio 134
Modulation bandwidth 49
Modulation transfer function (MTF) 204
Monolithic structures 132, 134

MOS 200
Mott transistor 143
Multifrequency selective heterodyne radiometry 288
Multiple photon photomixing 234
Multiple quantum
  direct detection 232
  heterodyne detection 242
  photoemission 232
  statistical effects 234
Multipliers 182
  channel 9, 120, 187
  crossed-field 186

Nd:YAG 179
Near intrinsic semiconductor 39
Negative electron affinity (NEA) 150, 159, 163, 165, 181, 185, 187, 189, 190
  GaP 185, 187
  silicon 181
Nernst effect 32
Noise equivalent power (NEP) 45
Noise mechanisms in photon detectors 36, 37, 72
  background 133
  electronic 106
  excess 37
  1/f 4, 212, 213
  generation-recombination (g−r) 37, 38, 105
  interface state trapping 215
  Johnson 37, 38, 133, 293
  KTC 210, 225
  quantum 293
  Nyquist 37
  photon 47, 53, 55
  shot 40, 41, 135, 208
  thermal 37
  transfer 209
Noise spectrum 47
n-type semiconductor 154

Observation time 49
Operating temperature effects 54, 314
Optical heterodyne detection 33, 229, 313, 314
Oxide capacitance 202

Parallel scan 197
Parametric effects 34
PbGeTe 116, 137
PbSnSe 137, 307
PbSnTe 115, 116, 119, 126, 134, 137, 218, 307
PbTe 116
Performance of detectors operating in the 1—1000 μm region 60, 64

Subject Index

Photocathodes
　conventional 9, 20
　negative electron affinity 9, 21
Photoconductive gain 103, 105, 123, 124, 129
Photoconductors (see detector types)
Photocurrent 103, 121, 122, 130
Photocurrent spectrum 238
Photoelectromagnetic effect 9, 23
Photoemissive detectors 2, 54, 56, 149, 309
Photoemissive effect 19
Photoemitter compounds containing cesium:
　AgCsO 156, 162
　Cs_2O 159, 168, 172
　Cs_3Sb 156, 180
　GaAs/Cs 165, 309
　GaAs/Cs:CsO 169
　metallic Cs 153
　NaKCsSb 156
　Si/CsO 159
　S-1 156
　S-20 156
Photoemitter fabrication 159, 162
Photoemitter substrate 174
Photoemitter thickness 174
Photoemitter types (also see electron emitters)
　opaque 151
　semitransparent 151, 173
　transparent 151
Photographic film 9, 25
Photomixing 231
Photomultipliers 9, 20, 182
Photon detectors 102, 104
Photon drag 9, 24
Photon effects 7, 8
　photoconductive 102, 122
　photovoltaic 102, 103
Photon noise limit 2, 47, 53, 55
Phototransistor 9, 24
Photovoltaic detectors (see detector types)
Photovoltaic effect 9, 14
p-i-n photodiode 18
Planck distribution function 50
Pneumatic infrared detector 32
p-n junction 15, 103, 109, 115, 116, 119, 135
Polarization modulation 280
Pollutants 295
Preamplifier (current mode) 209
p-type semiconductor 54, 173
Pulse-code modulation 276
Pulse height counter 187
Putley detector 23
PVF 92
PVF_2 92
Pyroelectric detectors 90, 96, 304, 305
Pyroelectric vidicon 72, 304

Pyromagnetic effect 28, 32
PZT 92, 305

Quaternary compounds 76, 309

Radiative recombination 114, 116, 136—138, 142
RA product 109, 112, 115, 134, 138
RC time constant 113, 119
Reflection mode materials 173
Response time 46, 108, 113, 134
Responsivity 43, 103
Richardson constant 113
Rise time 185

Saturation current 104, 110
Scattering (electron-electron) 157
Scattering length 173
Schottky barrier 103, 109, 111, 309, 311
Schottky diode 16, 18
Self-induced drift 200
Semiconductor 104, 154
Series scan 197
Series-parallel scan 199, 311, 312
Shockley-Read recombination 114, 116, 118, 120, 125, 135, 142
Signal fluctuation limit (SFL) 57
Signal fluctuations 47, 48, 50, 58, 59
Signal formats (nonorthogonal) 276, 279
Signal handling capability 203
Signal-to-noise ratio 237, 247, 255, 261, 265, 293
Silicon 104, 132, 143, 144, 308
Silicon impurities 12
Single- and two-electron events 187
Sodium 239
Solid solubilities 144
Space-charge layer 115
Spectral detectivity 59, 106, 109
Speed of response 14
Staring sensor, dc-coupled 204
State-of-the-art of infrared detectors responding in the 0.1—1000 μm region 59—63
Strontium barium niobate 92
Superconducting bolometer 28
Superinductor 28
Surface coating 180
Surface-dipole model 169
Surface states 167
S-1 156, 162, 183
S-20 (NaKCsSb) 156, 189

Temperature of operation 104, 108, 120—122, 126, 130, 132
Ternary compounds 176

TGS   92, 305
Thermal diffusion   206
Thermal effects   7, 26
   classification of   27
Thermal imaging system   197, 306, 311, 312
Thermal ionization   157
Thermal noise   2
Thermistor   83
Thermoelectric effect   31
Thermopile   79, 302
Three-frequency heterodyne detection   243, 288
Time constant   46
Time delay and integration (TDI)   197, 199, 213
Transfer inefficiency   204
Transmission mode   177
Trapping states (interface)   205
Traps   120

Ultraviolet response   15

Vacuum-level energy   153
Valence band   138, 154

Wave interaction   7
Wave interaction effects   34
Wavelength   104
Wavelength "cutoff"   103, 104, 108, 132, 133, 137, 139, 143, 144
White light sensitivity   169
Work function   154, 167

X-rays   150

Yield   173

Zinc blende crystal structure   125

R. H. Kingston

Detection of Optical and Infrared Radiation

1978. 39 figures, 2 tables. VIII, 140 pages
(Springer Series in Optical Sciences,
Volume 10)
ISBN 3-540-08617-X

Contents: Thermal Radiation and Electromagnetic Modes. – The Ideal Photon Detector. – Coherent or Heterodyne Detection. – Amplifier Noise and Its Effect on Detector Performance. – Vacuum Photodedectors. – Noise and Efficiency of Semiconductor Devices. – Thermal Detection. – Laser Preamplification. – The Effects of Atmospheric Turbulence. – Detection Statistics. – Selected Applications.

B. Saleh

Photoelectron Statistics

With Applications to Spectroscopy and Optical Communication

1978. 85 figures, 8 tables. XV, 441 pages
(Springer Series in Optical Sciences, Volume 6)
ISBN 3-540-08295-6

Contents: Tools from Mathematical Statistics: Statistical Description of Random Variables and Stochastic Processes. Point Processes. – Theory: The Optical Field: A Stochastic Vector Field or Classical Theory of Optical Coherence. Photoelectron Events: A Doubly Stochastic Poisson Process or Theory of Photoelectron Statistics. – Applications: Applications to Optical Communication. Applications to Spectroscopy.

Springer-Verlag
Berlin
Heidelberg
New York

Charge-Coupled Devices

Editor: D. F. Barbe

1980. 133 figures, 7 tables. XI, 180 pages
(Topics in Applied Physics, Volume 38)
ISBN 3-540-09832-1

Contents:
D. F. Barbe: Introduction. – *G. J. Michen, H. K. Burke:* CID Image Sensing. – *W. D. Baker:* Intrinsic Focal Plane Arrays. – *D. K. Schroder:* Extrinsic Silicon Focal Plane Arrays. – *D. F. Barbe, W. D. Baker, K. L. Davis:* Signal Processing with Charge-Coupled Devices. – *J. M. Killiany:* Radiation Effects in Silicon Charge – Coupled Devices.

Integrated Optics

Editor: T. Tamir

2nd corrected and updated edition. 1979. 99 figures, 11 tables. XV, 333 pages
(Topics in Applied Physics, Volume 7)
ISBN 3-540-09673-6

Contents:
T. Tamir: Introduction. – *H. Kogelnik:* Theory of Dielectric Waveguides. – *T. Tamir:* Beam and Waveguide Couplers. – *J. M. Hammer:* Modulation and Switching of Light in Dielectric Waveguides. – *F. Zernike:* Fabrication and Measurement of Passive Components. – *E. Garmire:* Semiconductor Components for Monolithic Application. – *T. Tamir:* Recent Advances in Integrated Optics. – Additional References with Titles. – Subject Index.

Laser Beam Propagation in the Atmosphere

Editor: J. W. Strohbehn

1978. 78 figures, 1 table. XII, 325 pages
(Topics in Applied Physics, Volume 25)
ISBN 3-540-08812-1

Contents:
J. W. Strohbehn: Introduction. – Laser Beam Propagation in the Atmosphere. – *S. F. Clifford:* The Classical Theory of Wave Propagation in a Turbulent Medium. – *J. W. Strohbehn:* Modern Theories in the Propagation of Optical Waves in a Turbulent Medium. – *M. E. Gracheva, A. S. Gurvich, S. S. Kashkarov, V. V. Pokasov:* Similarity Relations and Their Experimental Verification for Strong Intensity Fluctuations of Laser Radiation. – *A. Ishimaru:* The Beam Wave Case and Remote Sensing. – *J. H. Shapiro:* Imaging and Optical Communication Through Atmospheric Turbulence. – *J. L. Walsh, P. B. Ulrich:* Thermal Blooming in the Atmosphere. – Subject Index.

Semiconductor Devices for Optical Communication

Editor: H. Kressel

1980. 186 figures, 6 tables. XIV, 289 pages
(Topics in Applied Physics, Volume 39)
ISBN 3-540-09636-1

Contents:
H. Kressel: Introduction. – *H. Kressel, M. Ettenberg, J. P. Wittke, I. Ladany:* Laser Diodes and LEDs for Fiber Optical Communication. – *D. P. Schinke, R. G. Smith, A. R. Hartman:* Photodetectors. – *R. G. Smith, S. D. Personick, jr.:* Receiver Design for Optical Fiber Communication Systems. – *P. W. Shumate, jr., M. DiDomenico jr.:* Lightwave Transmitters. – *M. K. Barnoski:* Fiber Couplers. – *G. Arnold, P. Russer, K. Petermann:* Modulation of Laser Diodes. – *J. K. Butler:* The Effect of Junction Heating on Laser Linearity and Harmonic Distortion. – *J. H. Mullins:* An Illustrative Optical Communication System.

Springer-Verlag Berlin Heidelberg New York